Fluidization Technology

SERIES IN THERMAL AND FLUIDS ENGINEERING

EDITORS

JAMES P. HARTNETT and **THOMAS F. IRVINE, JR.**

Chang
- **Control of Flow Separation: Energy Conservation, Operational Efficiency, and Safety**

Keairns
- **Fluidization Technology**

IN PREPARATION

Begell
- **Glossary of Terms in Heat Transfer and Related Topics: English, Russian, German, French, and Japanese**

Brzustowski and Wojcicki
- **Aerosols in Combustion and Explosions**

Chi
- **Heat Pipe Theory and Practice: A Sourcebook**

Denton
- **Future Energy Production Systems: Heat and Mass Transfer Processes***

Eckert and Goldstein
- **Measurements in Heat Transfer, 2nd edition**

Ginoux
- **Two-Phase Flows with Application to Nuclear Reactor Design Problems†**

Goulard
- **Combustion Measurements: Modern Techniques and Instrumentation**

Hahne
- **Heat Transfer in Boiling**

Hartnett
- **Alternative Energy Sources***

Hsu and Graham
- **Transport Processes in Boiling and Two-Phase Systems, Including Near-Critical Fluids**

Moore and Sieverding
- **Two-Phase Steam Flow in Turbines and Separators: Theory, Instrumentation, Engineering†**

Pfender
- **High-Temperature Phenomena in Electric Arcs**

Richards
- **Measurement of Unsteady Fluid Dynamic Phenomena†**

van Stralen and Cole
- **Boiling Phenomena**

Yovanovich
- **Advanced Heat Conduction**

*A publication of the International Centre for Heat and Mass Transfer, Belgrade.
†A von Karman Institute Book, Brussels.

Fluidization Technology

VOLUME II

Fluidized Bed Performance with Internals
Solids Mixing and Transport
Fossil Fuel Processing
Application

EDITOR

Dale L. Keairns

IN COOPERATION WITH

M. A. Bergougnou
J. F. Davidson
J. M. Matsen
C. Y. Wen

HEMISPHERE PUBLISHING CORPORATION

Washington

IN ASSOCIATION WITH

McGRAW-HILL INTERNATIONAL BOOK COMPANY

New York   St. Louis   San Francisco   Auckland   Düsseldorf   Johannesburg
London   Mexico   Montreal   New Delhi   Panama
Paris   São Paulo   Singapore   Sydney   Tokyo   Toronto

"These proceedings are based upon the International Fluidization Conference which was sponsored by the Engineering Foundation at Asilomar Conference Grounds, Pacific Grove, California on June 15-20, 1975. The views presented here are not necessarily those of the Engineering Foundation, 345 East 47th Street, New York, New York 10017."

FLUIDIZATION TECHNOLOGY, Volume II

1 2 3 4 5 6 7 8 9 0   D O D O   7 8 4 3 2 1 0 9 8 7 6

Library of Congress Cataloging in Publication Data

International Fluidization Conference, Pacific Grove,
     Calif., 1975.
     Fluidization technology.

     An Engineering Foundation conference, co-sponsored by
American Institute of Chemical Engineers and other
organizations.
     (Series in thermal and fluids engineering)
     1. Fluidization—Congresses.   I. Keairns, D. L.,
1940–   II. Bergougnou, Maurice Amédée.   III. Engineering Foundation, New York.   IV. American Institute
of Chemicals Engineers.   V. Title.
TP156.F65I48 1975   660.2'8429   75-40106
ISBN 0–89116–006–X (v. II)

ENGINEERING FOUNDATION CONFERENCE

Proceedings of the International Fluidization Conference
held at
Asilomar Conference Grounds
Pacific Grove, California
June 15–20, 1975

Co-Sponsored by

American Institute of Chemical Engineers
Canadian Society of Chemical Engineers
Institution of Chemical Engineers, United Kingdom
Société de Chimie Industrielle, France
Society of Chemical Engineers, Japan

Financial Assistance was Provided by

Engineering Foundation
National Science Foundation

CONTENTS

PREFACE

The proceedings of the International Fluidization Conference held at Asilomar Conference Grounds, Pacific Grove, California, June 15–20, 1975 are presented in these two volumes. Volume I contains reports of investigations into fluidization fundamentals–bubble phenomena, gas exchange and fluid bed modeling, liquid phase fluidization, and three phase fluidization. The primary emphasis in Volume II is on fluidization processing–fluidized bed performance with internals, solids mixing and transport, fossil fuel processing, and applications.

The International Fluidization Conference was organized to disseminate present knowledge and to further the development of fluidized bed technology and applications of the technology. Investigators and practitioners from industry, universities, governments and research institutions throughout the world participated in the Conference.

The need for the international conference on fluidization arose out of a number of concerns. Fluidization technology is being applied in increasing numbers of new applications. However, the understanding of fluidization phenomena is far from adequate and the design of fluidization processes is primarily based on past experience. The extensive research work being carried out is illustrated by the thousands of papers and patents which have been issued. Previous international conferences, in particular the Toulouse meeting in 1973 and the Eindhoven meeting in 1967, provided effective means to exchange information. While there continue to be numerous symposia on fluidization fundamentals and applications, a major conference dealing exclusively with fluidization was considered timely. Discussions held with research investigators, personnel involved with large scale development programs, government contract monitors, and individuals responsible for fluidized bed production units resulted in this conference. The conference consisted of eight technical sessions, corresponding to the eight divisions in these two volumes. Parallel workshops were held on the day following the seven sessions on specific fluidization areas. The applications session was held on the final day followed by reports from the workshop discussion leaders. The conference site was selected to provide an atmosphere where informal discussion and exchange of ideas could be most readily fostered.

In order to achieve the conference objectives, the organizing committee adopted a rapporteur format. This was adopted to permit more information to be available for discussion and provided an opportunity to assimilate information through the use of reporters. Summary papers containing the scope, objectives, results and significance for each piece of work were provided to all conference participants. Extended manuscripts were available to the session organizers and reporters. The extended manuscripts form the basis for these proceedings.

It is my pleasure to record the important role of my colleagues on the organizing committee: M. A. Bergougnou, J. F. Davidson, J. M. Matsen and C. Y. Wen. The success of the conference was made possible by their work on the program organization, cooperative efforts, ideas, suggestions, and knowledge of investigators in the field. I personally express my appreciation to the following session chairpersons, rapporteurs, and workshop discussion leaders who provided the environment which resulted in a stimulating meeting with valuable technical insights.

Bubble Phenomena
    Chairperson   I. O. Molerus
    Rapporteur   P. N. Rowe
    Workshop discussion leader   P. H. Calderbank

Fluidized Bed Performance With Internals
    Chairperson   H. Angelino
    Rapporteur   J. S. Halow
    Workshop discussion leader   D. F. Wells

Gas Exchange and Fluid Bed Modeling
    Chairperson   D. Kunii
    Rapporteur   D. L. Pyle
    Workshop discussion leader   O. Levenspiel

Solids Mixing and Transport
    Chairperson   L. Massimilla
    Rapporteur   L. S. Leung
    Workshop discussion leader   J. S. M. Botterill

Liquid Phase Fluidization
    Chairperson   P. Le Goff
    Rapporteur   N. Epstein
    Workshop discussion leader   H. Littman

Three Phase Fluidization
    Chairperson   D. Harrison
    Rapporteur   C. G. J. Baker
    Workshop discussion leader   K. Ostergaard

Fossil Fuel Processing
    Chairperson   S. Freedman
    Rapporteur   D. H. Archer
    Workshop discussion leader   B. S. Lee

Applications
    Chairperson   F. A. Zenz
    Rapporteur   L. Reh

The session chairpersons were effective at establishing a positive format and were sensitive to potentially difficult situations. The technically thorough and comprehensive analyses presented by the rapporteurs provided an excellent basis for discussion. Their comments were effective at assimilating a diversity of information. The discussion leaders for the workshops enabled the prior discussion of specific investigations in each area to be extended and integrated with other experience and information.

On behalf of the organizing committee, I express our appreciation to the Engineering Foundation for enabling this conference to be held and to S. S. Cole, M. Keenberg, and W. E. Reaser for their efforts to implement the planning and arrangements. Financial support from the Engineering Foundation and the National Science Foundation which enabled many of the participants to attend is gratefully acknowledged. Also, I acknowledge the cooperation of N. Del Gobbo of the National Science Foundation. The technical societies and institutions who cosponsored this conference are acknowledged for their support and their assistance with publicity and providing contact with members.

The experience at Asilomar demonstrated to many participants the need for future exchange of information among those working on fluidization technology. As a result of this interest, a second Engineering Foundation conference will be held the first week of April in 1978 at the University of Cambridge, Cambridge, England.

It is our hope that the International Fluidization Conference and these proceedings effectively contribute to the concerns which led to the conference and provide for the exchange of information and ideas on fluidization technology.

Dale L. Keairns
Conference Chairperson

PART I

Fluidized Bed Performance with Internals

FLOW PATTERNS NEAR HORIZONTAL TUBES
IN A GAS-FLUIDIZED BED

N. M. ROONEY and D. HARRISON

INTRODUCTION

Experimental work has been carried out to obtain further insight into the flow patterns near horizontal tubes and tube arrays in a gas-fluidized bed, and to relate these to the mechanism of tube/bed heat transfer. Earlier work has been reported on two-dimensional systems (1) (2), and at low velocities in three-dimensional beds (3), but little published information is available at fluidizing velocities approaching those often used in practice, for example, in the fluidized combustion of coal.

EXPERIMENTAL

Two experimental techniques have been used:

(a) Cine-photography, using a number of tube diameters and superficial velocities up to 2.5 U_{mf}. Fig. 1 shows the experimental arrangement whereby qualitative information was obtained on particle and gas movement at the tube surface using a viewing mirror within the tube itself.

(b) Capacitance measurements of voidage fraction at the tube surface, using one tube diameter (37 mm) and two sand particle size ranges (150–180μm and 425–500μm).

The fluidized bed for all experiments was of square cross-section (278 x 278 mm). Experiments were carried out with a single tube and with an array of seven (37 mm) tubes in a centered hexagonal configuration, with an inter-tube spacing of 37 mm.

The most satisfactory capacitance probes were a needle probe (Fig. 2) for instantaneous voidage and an eight-pin probe (Fig. 3) for time-averaged voidage. These probes extended little, or not at all, into the bed so we believe their influence on flow patterns was negligible.

RESULTS

(a) Time-Average Voidage

For a single tube, Fig. 4 shows the time-average voidage measured by the eight-pin probe for fluidizing air flow-rates up to 4 U_{mf}. The time-average voidage generally increased with increasing air flow-rate, with the most marked change at the top of the tube as the defluidized cap of particles became mobile. Ciné-film of the downstream side clearly showed the mobility of the particles in contact with the tube. At just above minimum fluidization, the particles are not completely static, and particle residence times at the tube wall were estimated to be of the order of a few minutes. At 2 U_{mf}, bubbles occasionally swept across the top of the tube, completely displacing the defluidized cap; at 2.5 U_{mf} the cap was frequently dislodged. From measurements on the downstream side of the tube, where voidages were about 0.7, we inferred that the gas cushion,

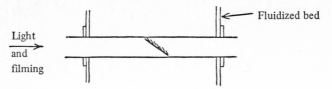

Fig. 1 Tube (with internal mirror) in fluidized bed.

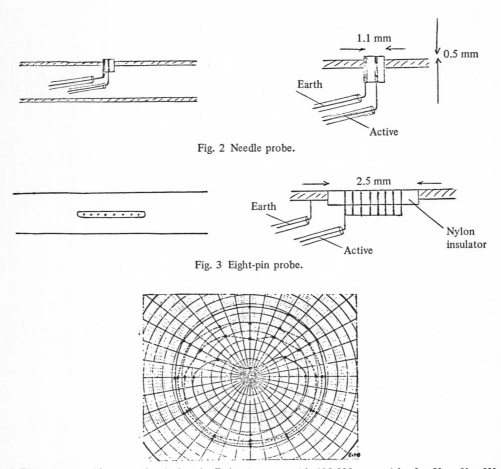

Fig. 2 Needle probe.

Fig. 3 Eight-pin probe.

Fig. 4 Time-average voidage round a single tube Eight-pin probe with 425-500μm particles $0 = U_{mf}$; $X = 2U_{mf}$; $\Delta = 3U_{mf}$; $+ = 4U_{mf}$.

often observed in two-dimensional equipment, was not so permanent or extensive in a three-dimensional bed.

Fig. 5 shows the time-average voidage for a tube at the center of an array, and comparison with Fig. 4 shows that the main differences between the single tube and the tube in an array occur at low fluidizing velocities. The tube array appeared to decrease the stability of the defluidized cap and, at an inter-tube spacing of a tube diameter, there was evidence of decreased particle contact at the sides of the tubes. At higher velocities the bubbles generated at one tube could come into direct contact with adjacent tubes.

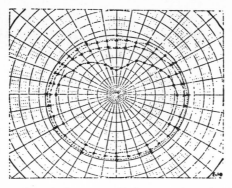

Fig. 5 Time-average voidage round a tube in an array Eight-pin probe with 425-500μm particles $0 = U_{mf}$; $X = 2U_{mf}$; $\Delta = 3U_{mf}$; $+ = 4U_{mf}$.

(b) Instantaneous Voidage

Instantaneous voidage measurements, on an arbitrary scale, have been used to count only changes in voidage which appeared to be sufficiently fast to be associated with particle exchange at the tube surface. Fig. 6 shows counts near a single tube as a function of fluidizing flowrate in the range 2–10 U_{mf}. The major disturbance with increase in air flowrate occurs at the top of the tube.

DISCUSSION

(a) Flow Patterns.

Our work suggests the defluidized cap in two-dimensional equipment is stabilized by the walls, and that the bubbling at low flowrates is more regular. Ciné-photography of the tube walls in our experiments at 2.5 U_{mf} shows that there is then little distinction between the bubble and the particulate phases. Instantaneous voidage measurements suggest values intermediate between $\epsilon = \epsilon_{mf}$ and $\epsilon = 1.0$; this voidage increases with fluidizing flowrate.

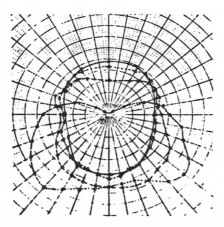

Fig. 6 Count frequency as function of air flowrate 150-180μm particles $0 = 2U_{mf}$; $+ = 5U_{mf}$; $\Delta = 7.5U_{mf}$; $X = 10U_{mf}$.

(b) Heat Transfer.

The heat transfer model developed by Mickley *et al* (4) considered unsteady state heating of "packets" of solid particles which were brought to, and removed from, the heat transfer surface by the bubbles. This model may be examined in terms of our voidage experiments, although differing experimental conditions prevent quantitative comparison. We have assumed that the fraction of time the tube contacted by bubbles may be calculated from the time-average voidage, and that the mean contact time may be taken as the reciprocal of the count frequency. The predicted trends of the Mickley model may be compared with heat transfer data (5), and these provide good support for the model. In particular, the high local heat transfer coefficient at the tube sides at low fluidizing velocities is predicted. Differences between the model and experiment appear to come from an assumption that the voidage (and hence thermal conductivity) of the "packet" is constant which, from our measurements of instantaneous voidage, seems doubtful.

REFERENCES

(1) Glass, D. H. & Harrison, D. (1964) *Chem. Eng. Sci. 19*, 1001.
(2) Fakhimi, S. (1969) *Ph.D. dissertation*, University of Cambridge.
(3) Rowe, P. N. & Everett, D. J. (1972) *Trans. Inst. Chem. Engrs. 50*, 42.
(4) Mickley, H. S., Fairbanks, D. F. & Hawthorn, R. D. (1961) *Chem. Eng. Prog. Symp. Ser. 57*, 51.
(5) Rooney, N. M. (1974) *Ph.D. dissertation*, University of Cambridge.

THE FLOW PROPERTIES OF FLUIDIZED SOLIDS

J. S. M. BOTTERILL and D. J. BESSANT

Experimental measurements of the non-Newtonian flow properties of a 200 μm diameter sand fluidized at 1.5 to 3 U_{mf} in open channel flow are reported.

Asymmetry in the velocity profile was found to change with change in the experimental conditions but it is thought primarily to be a consequence of air maldistribution which is partly generated by the bed flow in the experimental rig.

There was evidence of varying degrees of drag across the distributor from conditions of slip giving rise to a semi-plug flow profile at higher fluidizing velocities and bed flow rates to conditions where the drag exacted across the distributor at lower solids flow rates exceeds that at the vertical channel walls.

Bed flow properties deteriorate as the bed depth increases. This is a consequence of the deleterious effect of bubble growth within the bed.

Reasonable agreement was obtained between velocity profiles predicted following Wheeler and Wissler and those measured under conditions where the drag across the distributor and the vertical channel walls was of similar magnitude. Some estimate was made of the relative effects of power law index, channel aspect ratio and slip on the solids flow.

INTRODUCTION

It was thought that it should be possible to achieve
very short particle residence times, and hence high bed
to surface heat transfer coefficients (Botterill, 1975)
by constraining a fluidized bed to flow in a horizontal
plane past immersed tubular heat transfer elements.
However, although the flowing bed has very apparent
liquid-like properties, solids become defluidized at
upstream facing surfaces as they impact there and so the
bed looses its good flow properties and it takes somewhat
longer for the flow lines to recover after an obstruction
than with a true liquid (Botterill et al., 1970). The
result of this behaviour is that, although heat transfer
will be enhanced from those sections of the tube exposed
to the cross flow of the bed, at the upstream facing
section the solids residence time is increased with consequent
decrease in the local heat transfer coefficient there so
that the net advantage is small. Nevertheless, for some
heat transfer operations heat transport in the bulk of the
bed offers apparent advantage. Thus, for example, with
fluid bed steam raising plant, turn-down is something of a
problem and there could be advantage in separating the
combustor from the heat exchange section. Then, by
controlling the flow of fluidized solids between the combustor
and the heat exchanger, the rate of heat extraction from
the combustor can be regulated. In an earlier note
(Botterill and Elliott, 1964) we also suggested the possibility
of heat storage in large beds of particles and that this

could be used in meeting peak energy requirements more economically. Additionally, the study of the open-channel flow behaviour of fluidized solids can probably shed light on the gross circulation currents which become established close to the distributor in large-scale, deep beds (Whitehead et al., 1970).

This paper principally reviews the more recent work undertaken at Birmingham University on the open channel flow properties of fluidized solids (Bessant, 1973) additional to that previously published (Botterill and Bessant, 1973 and Bessant and Botterill, 1974).

GENERAL REPORTED BED BEHAVIOR

Geldart (1973) has suggested four principal groups into which powders may be divided according to their fluidization behaviour. With small particles interparticle forces are significant and the bed will not fluidize freely. Beds of particles of low density ($<$1400kgm/m^3) and of mean particle size in the range between 20 and 100 µm are characterized by considerable stable bed expansion beyond the point of minimum fluidization before the onset of bubbling. These materials generally flow freely in the fluidized condition. With larger (40 to 500 µm) and denser particles (1400 to 4000 kgm/m^3) the bed begins to bubble freely at or a little above the minimum fluidizing velocity (the 200 µm sand used in the experiments reported in this paper fall within this class). And fourthly, large and/or dense particles give rise to beds which behave unstably in the manner classified by Squires (1962) as "Teeter" beds.

De Groot (1967) measured particle diffusivities with catalyst material of two size distributions. His tests in beds of diameter up to 1.5m showed a marked difference between them. The beds of wider size distribution catalyst displayed better mixing and less sensitivity to the scale of the equipment. Geldart (1972) has suggested that the primary reason for this is because the sample of wider size distribution would have a lower mean size and be included within that class of material which is characterized by marked stable bed expansion and freer flow.

There is the general report that the apparent viscosity of a fluidized bed, as estimated from the drag exerted on a rotating viscometer element, decreases as the average size of the bed material decreases. With regard to materials of a single general class, McGuigan's experiments (McGuigan and Elliott, 1972), however, would suggest that there is little effect of particle size on the minimum apparent viscosity. Thus, earlier reported trends may be a consequence of comparing experiments under different fluidization conditions in that the minimum fluidizing velocity for the bed will change as the mean particle diameter of the material is altered.

EXPERIMENTAL EQUIPMENT

The rig used in our experiments has been described elsewhere (e.g. Botterill et al., 1970; Bessant, 1973). Essentially it consists of a closed-circuit, open, horizontal channel 300mm wide with a porous tile distributor base. The overall dimensions of the rig are length 3m and

width 1.3m. Solids flow is induced by the passage of a series of
paddles immersed in the bed along one of the straight sections of
the rig.

These paddles are carried on a belt and the belt drive
rate, and hence the bed flow rate, is regulated by a variable
speed drive. Average bed flow rates range up to 300 mms^{-1}
and tests are carried out from 1.5 to 3 U_{mf}. Experimental
measurements are made within a 2.5m long test section
mounted along the other long side of the circuit. This
section is made from two walls held accurately vertical and
parallel by spacers (so that the channel width can be
varied as required for the experiments) and with the
distributor as base.

The bed material used for most of the experiments
described in this paper was a narrow fraction dune sand
of 200 μm mean particle size. The bed flow properties
were estimated from the measured pressure drop developed
as the solids were constrained to flow along the channel.
(The assumption was made in interpreting these measurements
that the drag across the distributor is the same as that
developed across the vertical surfaces and this was obviously
not always so). Pressure drops were measured between
probe points using a Hilger-Watts micromanometer ($\pm$ 25.4mm WG).
Local flow velocity was measured using a small turbine
element whose rotor was made from an aluminium foil blank
14.29 X 6.35 mm and 0.254 mm thick and a direct measurement
was made of drag at the vertical channel wall surface using
a freely supported plate inserted between two sections of

the channel wall whose movement was balanced by stress in
a strain gauge transducer (Bessant, 1973). Additionally,
some measurements of shallow bed flow properties were made
with a simple inclined air-slide.

EQUIPMENT INFLUENCE

Longitudinal pressure drop traverses were carried out
to estimate the change in shape of the free surface of the
bed as it entered and flowed along the experimental channel
in the large circulating rig. As expected, there was a
larger pressure gradient in the entry section. Here there
are comparatively large frictional forces associated with
the acceleration of the bed into the channel. After the
first 0.3 - 0.4m from the channel entry, the hydraulic
grade line was very shallow, detected pressure drops being
less than 125 Nm^{-2} over a channel length of 0.83m when the
bed was fluidized at 2 U_{mf}. All the experimental measurements
were therefore made between 0.9m and 1.5m downstream from
the entry to the experimental section in order to be free
from its influence.

There was evidence that the maximum velocity occurred
beneath the free surface. In early measurements of flow
profiles a high degree of asymmetry was found. This was
most pronounced close to the bed surface and the flow became
more symmetrical deeper within the bed. Some tests were
carried out to determine the influence of distribution vanes
or meshes positioned at the channel entry on the downstream
velocity profiles but these   had negligible effect.

Indeed, when part of the entry section was deliberately blocked, the flow conditions in the experimental section very quickly stabilized. Movement of lateral position of the jigged experimental section showed that the profile was to some extent dependent on the relative position of the vertical walls about the longitudinal channel axis. Later experiments in the airslide using false walls to create a narrower channel also produced similar profile distortion if the false walls were not placed symmetrically about the channel axis. Thus it is concluded that distortion of the profile is, at least in part, a consequence of air maldistribution. By mounting the experimental section symmetrically within the channel more symmetrical profiles were obtained. Nevertheless, even having taken care over this, some degree of asymmetry persisted but now the most marked distortion was observed close to the distributor (see below, figure 2-5) and particularly at low solids flow rates. A consequence of air maldistribution could be that the fluidized solids in the main channel were fluidized at an air flow rate lower than supposed since the fluidizing gas flow rate was based on the measured total flow to the rig. Repeat measurements and measurements taken under conditions of different humidity were consistent within $\pm$ 5% at the worst.

BED FLOW PROPERTIES

In earlier tests (Botterill and van der Kolk, 1971) a bed of sand of mean particle size 185 μm was found to

display Bingham plastic properties and its plastic viscosity
passed through a minimum at 3 U_{mf}. Sand of mean particle
size 138 µm and Bauxilite of mean size 102 µm displayed a
range of behaviour from dilatant to pseudoplastic according
to the aspect ratio and fluidization conditions of the bed.
With this sand (mean size 200 µm), the apparent bed viscosity
went through a minimum at about 2 U_{mf} (0.5 Nm s^{-2} for a
shear rate of 10 s^{-1} compared with a value of 0.001 Nm s^{-2}
for water at 25°C (Bessant & Botterill, 1974). Above
2 U_{mf} there was a slight increase in apparent viscosity
and this seemed to be associated with the effect of greater
bubbling activity at the higher gas flow rates (Table 1).

A most important effect is the influence of bed height
on apparent viscosity. Thus, in earlier work with a bed
of Bauxilite (Botterill et al., 1972), a reduction in bed
height from 210 mm to 90 mm reduced the apparent viscosity
by a factor of 5. A similar effect was also observed in
viscometer experiments when a slight increase in depth
was found to have a disproportionate effect on the shear
stress for any particular rotor speed. Typical results
for the 200 µm sand are given in Table 1.
Such strong depth effect is a consequence of bubble growth
as they rise through the bed. It is thought that they can
also drain the fluidizing gas from the adjacent continuous
phase to some extent. The shallower beds of the air slide
(generally less than 20 mm deep) are not deep enough for the
occurence of significant bubble growth. In those experiments
there was an increasing tendency towards Newtonian behaviour

Table 1 Effect of Bed Depth on Apparent Viscosity

$\dfrac{U}{\bar{U}_{mf}}$	μ_{app} at 10 s^{-1} shear rate for 180 mm channel width, Nsm^{-2}	
	H = 118 mm	H = 77 mm
1.75	0.57	0.37
2	0.55	0.25
3	0.60	0.32

as the fluidizing air flow rate was raised from 1.25 to 2.5 U_{mf}.
Associated with this was a tenfold reduction in the system's
apparent viscosity from 0.4 Nsm^{-2} at 1.25U_{mf} to 0.045 at
2.5 U_{mf} at a shear rate of 20 s^{-1}. Incidentally, for similar
experimental conditions and a bed of spherical aluminium
oxide of 350 μm mean size, the apparent viscosity was more
than four times greater. Matheson et al. (1949) earlier
reported their observation of more viscous behaviour with
beds of spherical rather than angular material.

It is thought that the minimum fluidizing velocity
has to be exceeded before good air distribution is obtained
throughout the base of the bed, close to the distributor,
and that continuing bed expansion is some indication of more
uniform air distribution. Counter to this is the detrimental
effect of growth in size and heterogeneity of the bubble
population as the gas flow rate is further increased.

With the 200 μm sand, the form of the shear curves
was similar to that for a pseudoplastic system. In the
flowing bed, the shear thinning behaviour appears to be a

consequence of air redistribution which largely results
from some degree of suppression or retardation of bubble
growth as the solids flow rate is increased. The shear
curves with different channel widths indicated that there
was a varying degree of distributor influence which is
attributed to slip across the distributor base. This also
may result from some redistribution of air and a complex
particle separation mechanism from the fluid in the dense
phase in this region.

ESTIMATED AND MEASURED VELOCITY PROFILES

Wheeler and Wissler (1965) solved the equations of
motion for the flow of a power law fluid through a rectangular
duct using finite difference approximations. It is to
be expected that this simple model is likely to be directly
applicable under those circumstances where the drag across
the distributor and along the vertical walls are of similar
nature. A limit to this calculation is the extreme case
of very high aspect ratio which approximates to the
situation where there is negligible drag across the
distributor. Profiles for representative aspect ratios
and power law fluids were calculated (Bessant, 1973).
Additionally, because the measured flow profiles displayed
a range consistent both with the effects of a varying
power law index and degree of drag across the distributor,
the calculation was modified to investigate the effect
of varying degree of slip across the distributor in terms
of a velocity there, a given fraction of the free surface

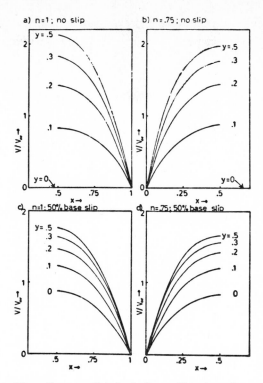

Fig. 1 Effect of base slip on predicted velocity profiles; channel aspect ratio 0.5.

value.   Figure 1 reproduces the predicted profiles for power
law fluids of index n = 1 (i.e. Newtonian) and n = 0.75
with a channel aspect ratio of 0.5.   Comparison is made
between the situation with no slip at the distributor and
with some allowance for distributor slip by imposing the
restriction that the velocity at the base shall be 50%
of that at the free surface.   Table 2 summarizes the
estimated effect of aspect ratio and slip on the total flow
rate.   It can be seen that the degree of slip across the
distributor is more important in controlling the bulk flow
properties of the bed than is the power law index until one
reaches the deeper beds of high aspect ratio.   Calculation

Table 2 The Effect of Aspect Ratio and Slip on the Total Flow Rate

Channel aspect ratio.	$V_{max}V_{av}^{-1}$, dimensionless			
	$n' = 1.0$		$n' = 0.65$	
	no slip	50% base slip	no slip	50% base slip
0.430	2.13	1.76	1.91	1.61
0.655	2.11	1.76	1.90	1.62
0.840	2.07	1.75	1.88	1.62
25.0	1.57	1.57	1.45	1.45
Parallel plates	1.51	1.51	1.48*	1.48*

* For an n value of 0.9

only predicts small increase in the velocity gradient near
to the boundary as a consequence of the marked pseudoplastic
nature of a fluid with an index as low as 0.55.

As reported in the Toulouse paper (Bessant and Botterill,
1974), the measured velocity profile was found to vary according
to the fluidization and bed flow conditions. Thus, there was
a range from observations of fully developed gradients in both
the x, horizontal direction perpendicular to the line of bulk
flow, and y, in the vertical plane, to those of semi-plug
flow with little gradient in the vertical direction, for the
widest channel tests (180 mm) at the higher fluidizing velocities
and solids flow rates studied. This variation is thought to be
a consequence of the differing degrees of distributor influence
because of slip close to the distributor.
More direct measurements of slip could only be made in the

air slide experiments. Estimated velocity values then
varied between 0 and 218 mm s^{-1} being greatest for the
sand fluidized at 2 U_{mf}. Generally for both the sand and
the alumina, base slip was lowest at the lowest fluidizing
air flow rate. Above 2 U_{mf} for the sand and 2.5 U_{mf} for the
alumina, the slip velocity appeared to reduce at any
particular angle of inclination.

Where the shape of the profiles was consistent with
little or no slip at the distributor, comparison was made
between the measured profiles and the predicted behaviour
for a power law fluid describable by

$$\tau = k' \left| \frac{dV}{dy} \right|^{n'}$$

where the values of the constants k' and n' were estimated
from a linear regression analysis of the logarithmic shear
curve data for the individual series of flowing bed experiments.
Four profiles for given bed aspect ratio and fluidization
conditions are reproduced in figure 2 to **5,** each at two shear
rates. Corrsponding values of the estimated power law constants
and the correlation coefficient, r, are given in Table 3. It
can be seen (Figures 2 to 5) that all the measured velocity
profiles were comparatively flat and similar to those expected
from a pseuodoplastic system with V_{max}/V_{av} in the range 1.5-1.8
for V_{av} computed on the basis of zero slip at both the
vertical and base surface. Gel'perin et al., (1968) have
observed values in the same range whilst statistically
examining the flow distribution of a similar sand across

a weir. It can be seen between the pairs of curves that
the prediction is generally closer to the measured profile
at the lower shear rate. This is to be expected if there
is varying slip at the distributor for this would increase
with increasing bed shear rate.

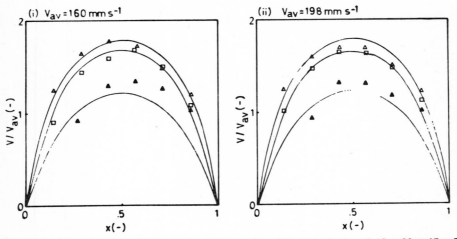

Fig. 2 $2U_{mf}$; H = 118 and W = 140 mm. Comparison between measured velocities (points: △ 10, ○ 30, ● 45, □ 50, ■ 60, ▽ 70, ▲ 90, ◇ 105 mm below free surface) and predicted profiles (continuous lines) for various conditions.

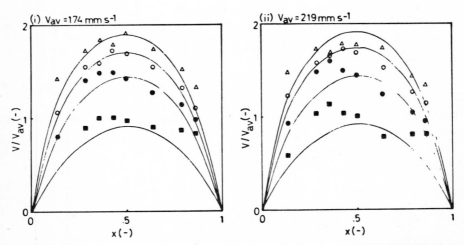

Fig. 3 $2U_{mf}$; H = 77 and W = 140 mm. Comparisons between measured velocities. (See Fig. 2).

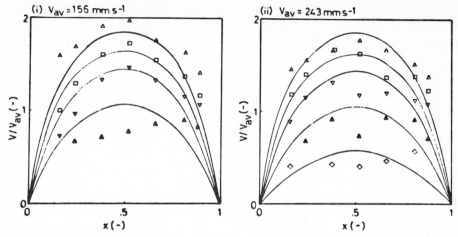

Fig. 4 $1.75U_{mf}$; H = 118 and W = 140 mm. Comparisons between measured velocities. (See Fig. 2).

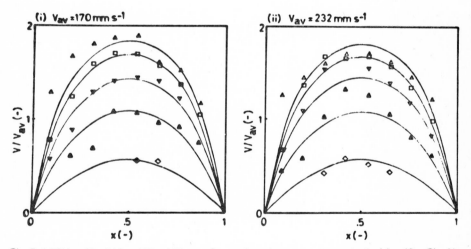

Fig. 5 $1.75U_{mf}$; H = 118 and W = 180 mm. Comparisons between measured velocities. (See Fig. 2).

WALL DRAG MEASUREMENTS

For the plug flow conditions, a more realistic assumption would seem to be that drag across the distributor is negligible. The wall drag estimated on this basis from the pressure loss measurements $\tau_{\Delta P \text{ corrected}}$, has been compared with the directly measured value, τ_g, obtained

Table 3 Shear Curve Data

Fluidization condition $(\frac{u}{U_{mf}})$	Bed Height H mm	Bed Width W mm	n'	k'	r
2	118	140	0.53	1.38	0.99
2	77	140	0.66	0.52	0.99
1.75	118	140	0.61	0.40	1.00
1.75	118	180	0.56	1.60	1.00

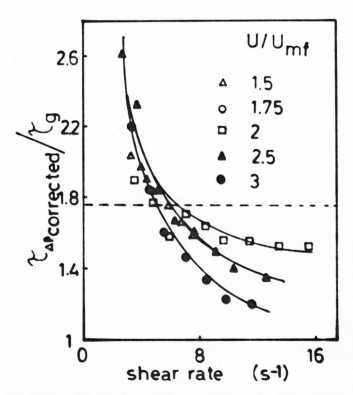

Fig. 6 Effect of fluidization conditions on distributor drag. H = 118 mm.

from the force exerted on the movable wall section as
measured by the strain gauge (Figure 6). It can be seen
that the ratio tends towards unity for the higher fluidizing
gas flow rate (3 U_{mf}) and higher shear rates as expected.
However, what is more significant is the fact that at low
shear rates ($\sim 4s^{-1}$) the ratio reaches higher values than
would be expected even for the case of the drag force across
the distributor being equal in magnitude to that at the
vertical surfaces (a ratio of 1.75 would then be expected
for the 180mm wide and 118mm deep bed used in the experiment).
The results therefore indicate that there are conditions
under which the drag force across the distributor can exceed
that at the vertical surface i.e. at lower fluidizing
velocities and shear rates. This is likely to be a
consequence of some degree of defluidization over zones
of the distributor and could result from local segregation
of larger particles within the bed.

It is interesting to note that reasonable correspondence
has been found between the direct measurements of wall
drag and rotating viscometer measurements carried out at
the University of Aston on a sample of the bed material
(McGuigan, 1973). The problem remains, however, that it is
generally impractical to apply the equivalent diameter concept
as a link between open channel flow and rotating viscometer
measurements because of the variable nature of the drag
obtaining across the distributor.

CONCLUSIONS

The flowing fluidized bed of sand possesses a non-

Newtonian character which changes with the fluidization
conditions.

   The experimentally measured velocity profiles exhibited
some degree of asymmetry which was affected by the experimental
arrangements. This is thought to be partly a consequence
of air maldistribution which, in turn, is
partly generated by the action of the bed flow.

   Under tests at the higher fluidizing velocities and
solids flow rates with the widest experimental channel,
semi-plug flow was observed i.e. with negligible drag being
exerted across the distributor. Particularly at lower
bed flow rates and closer the the minimum fluidizing velocity,
there was evidence of greater drag across the distributor
than that exacted at the vertical channel walls. Thus the
degree of drag exacted across the distributor is dependent
on the fluidization conditions and the cross-flow rate of
solids there. This prevents simple use of the equivalent
diameter concept in applying rotating viscometer measurements
in the prediction of open-channel flow losses.

   Under conditions where the velocity profiles were
consistent with drag at the distributor and the vertical
channel walls being of similar magnitude, reasonable agreement
was obtained between profiles predicted from a power law
model following Wheeler and Wissler (1965) and those measured;
the flow properties being estimated from the pressure drop
measurements made as the solids circulated through the
experimental channel. Calculations show that slip at the
distributor is more important in controlling the bulk flow

properties of the bed than is the power law index until one reaches beds of high aspect ratio.

Bed height is a very important variable. In deeper beds bubbles can grow to significant size and they then have a deleterious effect on bed flow properties.

ACKNOWLEDGEMENTS

The authors gratefully acknowledge the support given by the S.R.C. for this work. We are also indebted to I.C.I. Corporate Laboratory for their help in the design and construction of the turbine element.

REFERENCES

Bessant, D.J. (1973). Ph.D. Thesis, University of Birmingham.

Bessant, D.J., and Botterill, J.S.M. (1974), in Angelino "La Fluidization et ses Applications" Toulouse, p 81, Societe Chimie Industrielle.

Botterill, J.S.M. (1975), "Fluid-Bed Heat Transfer", Academic Press, London.

Botterill, J.S.M. and Bessant, D.J. (1973), Powder Technol 8, 213.

Botterill, J.S.M., Chandrasekhar, R. and van der Kolk, M. (1970). Chem.Eng.Prog.Symp.Sci. 66, (101) 61.

Botterill, J.S.M. and Elliott, D.E. (31st July, 1964) Engineering 198, 146.

Botterill, J.S.M., Elliott, D.E., van der Kolk M. and McGuigan, S.J. (1972). Powder Technol. 6, 343.

Botterill, J.S.M. and van der Kolk, M. (1971). Chem.Eng.Prog.

Symp. Series 67 (116), 70.

Geldart, D. (1972). Powder Technol. 6, 201.

Geldart, D. (1973). Powder Technol, 7, 285.

Gel'perin, N.I., Einstein, V.G., Lapshenkov, G.I. and Koslovski, A.I. (1968). Khim Prom 6, 415.

de Groot, J.H. (1967) in Drinkenburg, "International Symposium on Fluidization", p 348. Netherlands Univ.Press.

Squires, A.M. (April 1962). Chem.Eng.Prog. 58, No.4 66.

Matheson, G.L., Herbst, W.A. and Holt, P.H. (1949), Ind. Eng.Chem. 41, 1099.

McGuigan, S.J. (1973). private communication.

McGuigan, S.J. and Elliott, D.E. (1972). "The viscosity of shallow fluidized beds" 4th International Congress CHISA, Prague, Sept. 11-15.

Wheeler, J.A. and Wissler, E.H. (1965), Am.Inst.Chem.Eng.J. 11, 207.

Whitehead, A.B., Gartside, G. and Dent, D.C. (1970). Chem.Eng.J. 1, 175.

NOMENCLATURE

H height of flowing bed over distributor, mm

k' constant in the power law equation, $Nm^{-2} s^{n'}$

n' index in the power law equation, dimensionless

U fluidizing velocity, ms^{-1}

U_{mf} minimum fluidizing velocity, ms^{-1}

V_{av} average velocity of flowing fluidized solids, ms^{-1}

V_{max} maximum velocity of flowing fluidized solids, ms^{-1}

W   channel width, mm

X   dimensionless co-ordinate, x'/W

x' distance from inner channel wall, mm

τ shear stress, Nm^{-2}

$\tau_{\Delta P \text{ corrected}}$ shear stress at vertical wall estimated from
          pressure drop measurement on the assumption that there
          is no drag across the distributor, Nm^{-2}

τ_g directly measured shear stress at vertical wall, Nm^{-2}

μ_{app} apparent bed viscosity, Nsm^{-2}

ENTRAINMENT REDUCTION BY LOUVERS IN THE DILUTE PHASE OF A LARGE GAS FLUIDIZED BED

Y. MARTINI, M. A. BERGOUGNOU and C. G. J. BAKER

ABSTRACT

The effect of louvers on entrainment above a gas fluidized bed has been studied. Experiments were carried out in a column 0.61 m in diameter and 10 m high containing 142 μm silica sand.

The presence of louvers located close to the bed surface reduced the flux of entrained particles reaching the cyclones by 33%. When the louvers were positioned in the middle or upper part of the column, entrainment was either increased or unaffected. The mean size of the entrained particles decreased more rapidly with height in the presence of louvers, but remained within the same limits.

INTRODUCTION

Most large commercial fluidized beds operate at superficial velocities many times greater than the minimum fluidization velocity. Large bubbles bursting on the surface of the bed entrain a considerable number of particles of all sizes into the dilute phase above the bed. The larger particles, having a terminal velocity greater than the superficial velocity, eventually fall back into the bed; the finer particles follow the gas stream to the cyclones. The flux of entrained particles initially shows a sharp decline on moving away from the bed surface. It then becomes approximately constant above a characteristic height above the bed surface, the transport disengaging height, TDH (1).

The TDH is generally large in beds of large diameter. The cyclones have to be positioned above the TDH to prevent them flooding, and thus a lot of space is wasted inside the reactor. This has the following drawbacks. First, the cost of the reactor is higher than necessary because of the longer vessel shell. Secondly, if the reaction is strongly exothermic, a large hold-up of solids in the dilute phase might result in high temperatures above the bed. There is, thus, a strong incentive to reduce the hold-up of solids in the dilute phase by means of de-entrainment devices.

The objective of the present stage of this study was to evaluate de-entrainment equipment already used in industry. Louvers were chosen because they are simple and also because they have been found to be reasonably effective, at least under certain conditions.

ENTRAINMENT REDUCTION DEVICES

A considerable amount of work has been carried out on entrainment but very little on its reduction. Large bubbles are mainly responsible for entrainment when they burst at the bed surface and eject solids into the dilute phase. It was therefore thought that breaking large bubbles into smaller ones would reduce particle carry-over. Horizontal baffles have thus been used to divide the bed in a number of compartments (2). They were sometimes found to be effective in decreasing entrainment above the highest compartment. Unfortunately, because of the dilute phases present under each baffle, the total column length was larger than for the unbaffled bed and no saving in reactor height was achieved.

Wire obstructions in the dense phase resulted in an increase in entrainment instead of the expected decrease (3). The use of a stirrer in the bed on the other hand led to a lower entrainment (3). The reduction in the carry-over was more pronounced with a more sophisticated stirrer and higher stirring speeds. Screen packing also reduced entrainment when located in the dense phase and increased it when placed in the dilute phase (4). The above devices have several disadvantages. First, they are mechanically complex and since they obstruct the bed, might lead to bed plugging if the solids are in the least tacky or cohesive. Also, the devices will probably be eroded quickly and give rise to maintenance problems. Other entrainment suppressors located in the bed or floating on its surface have been found to be either ineffective or troublesome from a mechanical point of view (2).

The above devices were based on the principle that entrainment could be reduced by breaking the bubbles in the bed. Another approach is to de-entrain the particles in the dilute phase above the bed. For instance, perforated grids and tubular arrays may reduce entrainment, particularly when they are located close to the bed surface (5). The use of a vertical array of slanted baffles also gave rise to a reduction in entrainment (6,7). The results are difficult to apply to commercial situations because either the fines were not returned to the bed or closely sized particles were used. Such systems are convenient in the laboratory but are not representative of commercial operations. Furthermore, these results were obtained in small bench-scale units and it is virtually impossible to scale them up to give reliable results for commercial beds.

EQUIPMENT AND PROCEDURE

The fluidization column shown in Figure 1 was 0.61 m in diameter and about 10 m high. Two rows of sampling ports, 0.05 m in diameter and 0.305 m apart, were drilled diametrically opposite each other up the column wall. Static pressure taps were positioned

A :   High pressure air
B :   Air from metering system
C :   Conical baffle
D :   Windbox
E :   Grid plate
F :   Solids discharge port
G :   Trickle valve
H :   Solids loading port
I :   Sampling ports
J :   Pressure taps
K :   Metallic sections
L :   Primary cyclone
M :   Secondary cyclone
N :   Air exit
O :   Cyclone dipleg
P :   Plexiglass sections
Q :   Pneumatic lift line
R :   Ejector

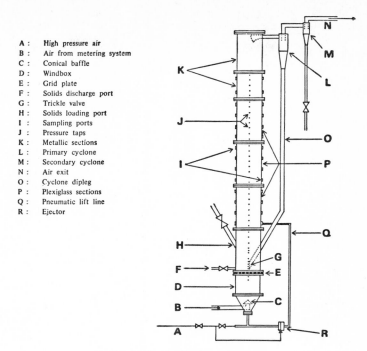

Fig. 1 The fluidization column.

along the column 0.076 m apart in the dense phase region of the bed and 0.152 m above it. They were connected to a series of water manometers. This enabled pressure profile measurements to be made from which bed height and bed density could be determined.

The grid consisted of a 3.18 mm thick steel plate with 102 holes each 3.18 mm in diameter, on a regular pitch. A fine metallic screen was laid on top of the grid to prevent particle leakage through grid holes into the windbox when the column was not operated. A pneumatic lift was provided to recycle any solids in the windbox back into the bed.

The column was equipped with a two-stage cyclone system. Essentially all the particles were collected in the primary cyclone and returned to the bed through a dipleg equipped with a trickle valve which was completely immersed in the bed. Therefore the column could be operated at steady state. All the measurements were made at least half an hour after the initial start up.

The de-entrainment device shown in Figure 2 was basically a horizontal array of louvers. It consisted of a row of 12 panels, 68 mm wide and 5 mm thick, located 48 mm apart with a 45° inclination to the horizontal. It was fitted inside a 0.58 m diameter, 2 mm thick and 0.178 m high aluminum cylinder. The apparatus was positioned by means of three cables passing through the top of the column. The cylinder was then completely

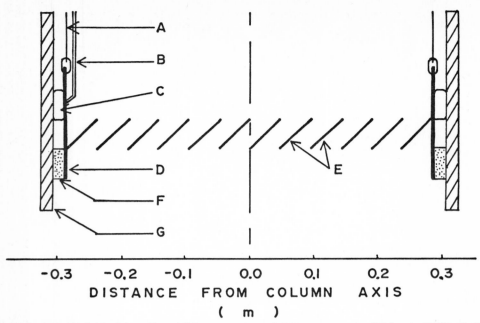

Fig. 2 The louvers. (A) Cable; (B) Air Pipe; (C) Pneumatic joint: (D) Metallic cylinder; (E) Panels; (F) Plastic foam; (G) Column wall.

immobilized by pressuring a bicycle inner tube squeezed between the column and the cylinder. A ring of plastic foam was located under the inner tube to protect it from abrasion by the particles. The height of the louvers above the bed was defined as the distance between the top of the bed and the bottom of the louver panels.

Oil-free air at room temperature passed through a rotameter and was fed to the column windbox. Its superficial velocity was 0.31 m/s under dilute phase conditions. The solids employed in the bed were 391 kg of silica sand having a mean particle size of 142 microns on a weight basis. The particle size distribution of this sand is shown in Figure 3. At the constant operating velocity, the expanded bed height was 1.03 m.

Local upward entrainment fluxes in the dilute phase region were measured by means of the isokinetic suction probe shown in Figure 4. The probe was mounted on a bracket and fitted into a sampling port with the aid of a continuously backflushed air bearing. The probe mouth was 0.040 m in diameter. The collected solids were accelerated in a reducing section to prevent saltation in the short transfer line. The particle-loaded air entered a paper filter shaped like a thimble, where even the finest particles were trapped (Figure 4). The difference in weight of the filter before and after an experiment gave the amount of solids collected. After passing through the paper thimble filter, which was enclosed in a larger glass jar, the air flow through the sampling probe was measured by means of a rotameter and was then exhausted through a vacuum pump. The pressure in the system was read on a water manometer. To ensure isokinetic sampling, the air velocity at

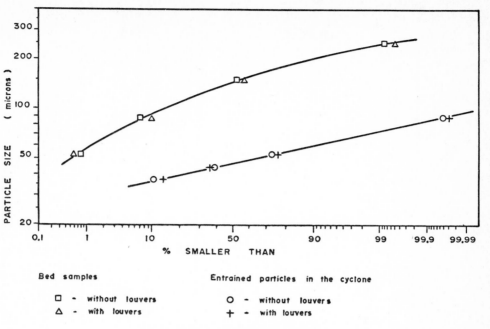

Bed samples   Entrained particles in the cyclone

☐ - without louvers   ○ - without louvers
△ - with louvers   + - with louvers

Fig. 3 Typical particles size distributions.

A :   Column wall
B :   Probe mouth
C :   Transfer line
D :   Backflushed air bearing
E :   Probe arm
F :   Positioning scale
G :   Sweeping air
H :   Sampling valve
I :   Paper thimble filter
J :   Water manometer
K :   Rotameter
L :   Vacuum pump

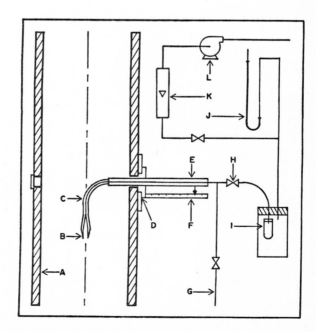

Fig. 4 The local sampling system.

the tip of the probe mouth was maintained at the same value as the superficial velocity in the column.

The probe tip was accurately positioned by means of a graduated scale at various distances from the column axis. These ranged from 0.05 m from the axis on one side to 0.25 m on the other side. Samples were taken at increments of 0.05 m. A complete radial profile of the flux could be obtained by fitting the probe into the opposite port in the column. In a similar manner, axial entrainment profiles were measured by moving the probe from port to port along the column.

The total flux of entrained solids reaching the cyclone was also determined by collecting those particles which fell down the dipleg into a container during a given length of time and then weighing them. A special gate valve could be closed upstream to prevent the escape of solid particles and air from the dipleg while changing containers. Due to the large mass of solids in the bed, the amount elutriated during each experiment was negligible. After weighing and sieve analysis, the collected samples were returned to the bed through the unused lower section of the dipleg. In all experiments, the entrainment was determined by weighing the solids collected and the sampling time. Particle size distributions were obtained by sieving.

RESULTS AND DISCUSSION

Radial Profiles

Although radial velocity profiles in the dilute phase were probably not flat, especially when the louvers were used, the suction velocity in the probe mouth was maintained constant at the average superficial gas velocity while sampling along a column radius. This can be justified by the fact that the particles were relatively coarse and heavy. Under these conditions the collected samples were representative of the true entrainment flux provided the suction velocity did not vary too significantly from the isokinetic point (8).

Typical radial entrainment flux profiles at different heights above the bed in the absence of louvers are shown in Figure 5. Close to the bed they were flat because of the random bursting of the bubbles at the surface. Further up the column they had a tendency to become convex in shape. Figure 6 presents similar data when louvers were located 0.15m above the bed surface. The radial profiles show that the particles and gas were deflected towards one side of the column and hit the wall about 1 m above the louvers. Beyond that point, the gas stream expanded again to eventually fill the column cross-section at a distance of about 3 m from the louvers. These complex patterns could be observed visually through the plexiglass wall of the column. They explain the radical variation in shape of the radial flux profiles as one moves up the column.

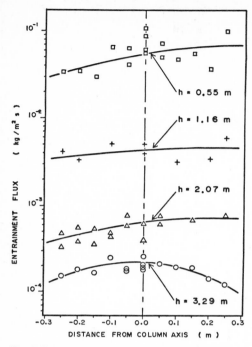

Fig. 5 Radial entrainment profiles, without louvers, for different sampling heights h above the bed.

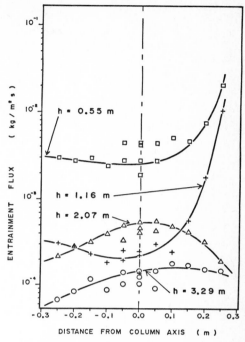

Fig. 6 Radial entrainment profiles, with louvers, for different sampling heights h above the bed. L = 0.15 m.

Axial Profiles

In Figure 7, the axial entrainment flux is plotted against the height above the bed. The flux of particles reaching the cyclone, as measured by collecting the solids passing down the dipleg, is also shown for the sake of comparison. In the absence of louvers, the upward flow of solids leaving the bed was large because of the intense bubble bursting action. It decreased to about 0.1 kg/m^{2}s at a distance of 0.55 m above the bed surface. Above the TDH, which was about 3 m, the entrainment became relatively small, about 0.0002 kg/m^{2}s. This is because only about 2% of the bed solids was entrainable above the TDH at a superficial velocity of 0.31 m/s.

When the louvers were located 0.15 m above the bed surface, the entrainment rate was drastically reduced, about 50 times at a height of 0.55 m. As may be seen in Figure 7, on moving away from the surface of the bed entrainment first decreased to a minimum, corresponding to the region in which both the solids and gas were deflected towards the wall of the column. It then increased to a maximum at the point at which the gas and solids returned to the center of the column. Entrainment finally decreased asymptotically towards a constant value which was approximately one third smaller than in the absence of louvers.

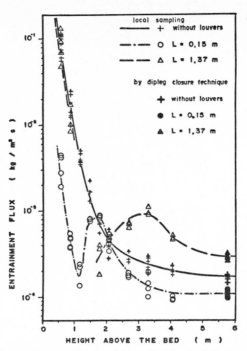

Fig. 7 Axial entrainment flux profiles for different heights L of the louvers above the bed.

When the louvers were located 1.37 m above the bed surface, isokinetic sampling could be carried out below them. The entrainment, in this case, was exactly the same as in the absence of louvers. Above the louvers, the situation was similar to that described above and illustrated in Figure 7. It should be noted, however, that in this case, the final entrainment was greater than that in the absence of louvers.

Integrated Flux Profiles

Because the local axial entrainment rate was not generally representative of the whole cross sectional area, the local radial entrainment profiles were integrated. For this purpose, the column was divided into eleven, basically rectangular, sections (8). In Figure 8, this average entrainment is plotted against the height above the bed. The inverse Froude number is also indicated on the abscissa to enable comparison to be made with data previously published in the literature.

Both with and without louvers, entrainment decreased drastically with height above the bed and ther. remained essentially constant above the TDH which was about 3.3 m. These experimental results could not be tested against theoretical models such as the one proposed by Zenz and Weil (9) since the parameters involved are almost impossible to measure, particularly in the large column used here. However, Zenz and Weil (9) also proposed an empirical correlation to estimate the TDH based on data obtained in columns up to 5 meters in diameter. This method gave a TDH of 2.1 m, the closest value to the present experimental results of all methods tried. Amilin *et al.* (10) have suggested the following semi-empirical equation for predicting the transport disengaging height :

$$TDH = 0.85 \; V^{1.2} \quad (7.33 - 1.2 \log V)$$

where V is the superficial gas velocity. This formula gives only 1.7 m in our case. Based on experimental studies, Fournol *et al.* (11) defined the TDH as the height above the bed surface at which the inverse Froude number becomes equal to 1000. In contrast to the two preceding methods, this one gives a TDH of 9.8 m which is higher than our experimental result. Above the TDH the experimental entrainment is 0.18×10^{-3} kg/m^2s which is reasonably close to the value of 0.26×10^{-3} predicted by the correlation of Zenz and Weil (9). When the louvers were located 0.15 m above the bed, the entrainment was reduced by 90 percent at a height of 0.55 m above the bed. It is interesting to note that the entrainment was also reduced by about one third above the TDH.

Effect of Louver Position

Figure 9 shows the flux of entrained solids which reached the cyclone as a function of the height of the louvers above the bed. Data were obtained by taking samples from the

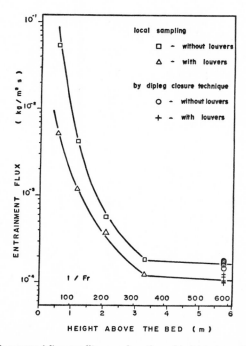

Fig. 8 Integrated flux profiles as a function of height above the bed.

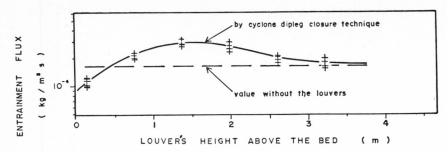

Fig. 9 Entrainment at the cyclone mouth as a function of the location of the louvers above the bed.

primary cyclone dipleg as explained above. When the louvers were located less than 0.4 m above the surface of the bed, the entrainment reaching the cyclone was smaller than in the absence of louvers. It should be noted that entrainment at the cyclone actually increased when the louvers were positioned higher up. This finding agrees with the results of Blyakher and Pavlov (5) which were obtained in the region close to the bed surface and can be explained by the intermittent gas jet theory of Zenz and Weil (9). When the louvers were close to the bed, they not only de-entrained particles but also destroyed the high velocity jets which are responsible for the large entrainment rate. When the louvers were located relatively far from the bed surface, the entrainment actually increased because of the deflection of the gas-solid stream. Also, the particles which initially passed through the louvers may have been re-entrained as they attempted to fall back through them. Similar situations leading to increases in entrainment have also been observed by several investigators (2-4). Finally, when the louvers were located in the upper part of the dilute phase, the entrainment decreased to the value in the absence of louvers. In the present experiments, this occurred at a height above the bed approximately equal to that of the cyclone inlet.

Entrained Particle Size Distribution

In Figure 10, the mean diameter of the entrained particles calculated on a weight basis is plotted against both the height above the bed surface and the inverse Froude number. In general the data reported in this figure were obtained from point samples taken on the axis of the column. However, where the louvers deflected both gas and particle streams towards one side of the column, the samples were collected at this point since it was felt these would be more representative of the column cross-section. Because of bubbles bursting at the surface of the bed, solids were ejected into the dilute phase and thus the mean diameter was about the same in the bed and just above it. Without the louvers it decreased from 142 microns in the bed to 47 microns in the cyclone. When the louvers were located close to the bed surface, the mean particle size decreased more rapidly up the column. However, the final value of 46 microns was essentially unchanged. Comparison of the two cases shows that, as expected, the coarse particles were de-entrained more easily by the louvers. It should also be noted that the mean diameter attained its final value at a lower

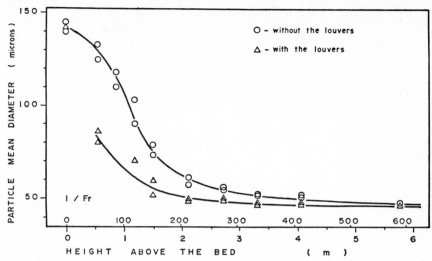

Fig. 10 Entrained particle size distribution as a function of height above the bed.

height than did the entrainment flux (Figures 8 and 10). Thus the mean particle diameter became constant below the TDH. This was also observed by Fournol *et al.* (11). A typical size distribution of particles collected in the cyclone is shown in Figure 3.

CONCLUSION

Within the limitations of this study, it can be said that louvers are moderately effective in decreasing entrainment flux provided that they are located relatively close to the bed surface.

ACKNOWLEDGEMENT

Grateful acknowledgement is given:

1)   to the Canada Council for a scholarship to Mr. Y. Martini

2)   to the National Research Council of Canada for a grant-in-aid of research to Dr. M.A. Bergougnou.

Their financial help made this work possible.

NOMENCLATURE

Fr   :   Froude number, V^2/gh

g   :   acceleration due to gravity, m/s^2

h   : sampling height above the bed, m

L   : height of the louvers above the bed, m

TDH: transport disengaging height, m

V   : superficial gas velocity, m/s

REFERENCES

1. Zenz, F.A. and D.F. Othmer, "Fluidization and Fluid-Particle Systems", Reinhold
    New York, (1960).

2. Overcashier, R.H., D.B. Todd and R.B. Olney, A.I.Ch.E. Journal, $\underline{5}$, 1, 54, (1959).

3. Lewis, W.K., E.R. Gilliland and P.M. Lang, Chem. Eng. Prog. Symp. Series, $\underline{58}$, 38
    65, (1962).

4. Tweddle, T.A. C.E. Capes and G.L. Osberg, Ind. Eng. Chem. Process Des. Develop.,
    $\underline{9}$, 1, 85, (1970).

5. Blyakher, I.G. and V.M. Pavlov, Int. Chem. Eng., $\underline{6}$, 1, 47, (1966).

6. Harrison, D., P.N. Aspinall and J. Elder, Trans. Instn. Chem. Engrs., $\underline{52}$, 213, (1974).

7. Harrison, D., Chem. Eng. Prog. Symp. Series, $\underline{69}$, 128, 14, (1973).

8. Martini, Y., M.E.Sc. Thesis, The University of Western Ontario, London, Canada,
    (1975).

9. Zenz, F.A. and N.A. Weil, A.I.Ch.E. Journal, $\underline{4}$, 4, 472, (1958).

10. Amitin, A.V., I.G. Matyushin and D.A. Gurevich, Khimiya i Tekhnologiya Topliv
    i Masel, 3, 20, (1968).

11. Fournol, A.B., M.A. Bergougnou, C.G.J. Baker, Can. J. Chem. Eng., $\underline{51}$, 4, 401
    (1973).

A MODIFICATION OF FLUIDIZING BEDS BY INSERTING PARTITION WALLS AND A MODIFIED DISTRIBUTOR

MASAHISA FUJIKAWA, MASAO KUGO and KOHNOSUKE SAIGA

It is well known that the contact between gas and solids in industrial fluidized beds is not quite adequate due to the presence and formation of bubbles throughout the bed. In order to eliminate this deficiency, we attempted using a method consisting of inserting partition walls into the bed. Another attempt was made to use a low-pressure-drop gas distributor in multistage fluidized beds.

THE EFFECT OF WIDTH BETWEEN PARTITION WALLS

If parallel plates were inserted in the bed to divide the bed into narrower beds, a gas velocity distribution between the plates, as shown in Fig. 1, was observed. This can be expected to increase the lower gas velocity zones which are contiguous to the plates and to promote the circulation of particles along the plates. In such a bed, with inserted parallel plates, a higher rate fluidization gas velocity would be applicable than for common fluidized beds, because of an elutriation effect of the plates resulting from the increase of downward motion of particles. In a cylindrical bed, this device can be used by inserting parallel smaller tubes or concentric cylindrical plates, and closely placed parallel plates in a box bed.

It would be necessary to determine the appropriate distances between tubes or plates before the insertion. As a preliminary experiment, therefore, the effects of bed diameter or width on the contact efficiency were studied.

The Effect on Cylindrical Bed

In order to examine the effect of bed diameter, the efficiency of contact was designated by the mass transfer coefficient (k_G), which was obtained from batchwise drying of silica gel at the constant rate. As Fig. 2 shows, the coefficient was correlated with the ratio of cylinder diameters to particle diameters (D_T/d_p). The result shows that the mass transfer coefficient decreases with the ratio; and the relation can be expressed to show that the coefficient is inversely proportional to the square of the ratio. With cylinders of 20–65 mm in diameter, and particle diameter of 0.2–1.7 mm, the larger the particle diameter, the higher the mass transfer coefficient.

The Effect of Width in Box Bed

The three beds used each had a height of 550 mm and a flank of 500 mm; their widths were 10, 20 and 30 mm. The particles of silica gel and glass bead were fluidized by air. To illustrate the elutriation effect of the wall, the ratio of fluidizing height (L) to the initial one (L_0) was kept

41

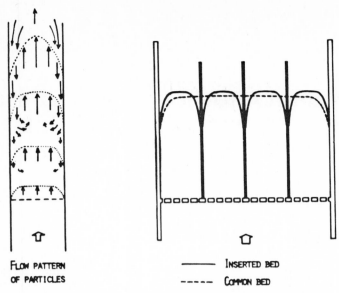

Fig. 1   Gas flow pattern in fluidized bed inserted with partition walls.

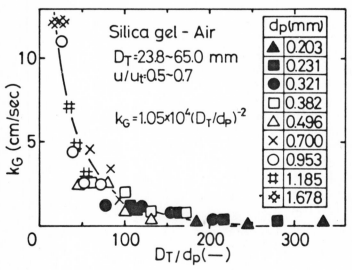

Fig. 2   Mass transfer coefficient in cylindrical bed.

constant. The results are shown in Fig. 3, where the narrowest bed of 10 mm has the lowest height.
The mass transfer coefficient (k_G) was obtained in the same manner as for the cylindrical bed and
is shown in Fig. 4, curve I. The coefficient decreases with an increase of the ratio; this relation
expresses the coefficient as being inversely proportional to the ratio. When the width was less than
10 mm, however, the uniformity of fluidizing seemed to disappear. It might depend on an excess
of the wall effect. This excess of wall effect is evident in the high value of the pressure drop of bed
(ΔP) at less than 30 of δ/d_p shown in Fig. 5, where ΔP_{th} is the theoretical, that is, the equivalent
value of the weight of fluidizing particles on the unit bed area. On the other hand, the beds with

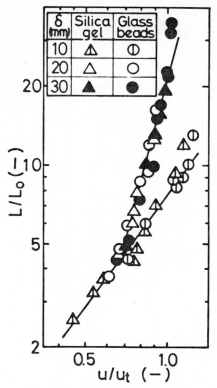

Fig. 3 Bed height in single box bed.

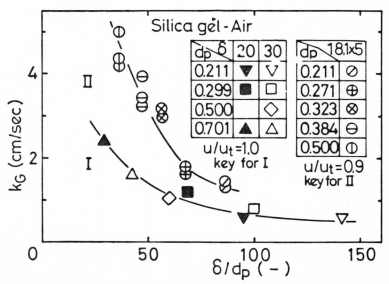

Fig. 4 Mass transfer coefficient in box bed.

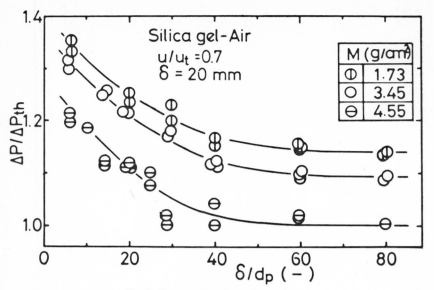

Fig. 5 Pressure drop in single box bed.

wide widths of more than 80 of the ratio, gave low values of the coefficient. These results suggest an optimum value of the ratio which may be 30 to 50, but the width has to be more than 15 mm.

Effect of Plates Insertion

The mass transfer coefficient (k_G) was obtained with various diameters of particles in a bed composed of 18.1 mm as δ in the same way as mentioned before. In Fig. 4, curve II shows the values expected from the previous results. To examine the uniformity of fluidizing, the ratio L/L_0 of this bed was also determined with various particles. As shown in Fig. 6, the value of the ratio is high for rather small δ, but becomes much lower for smaller values of δ, such as 10 mm. Therefore, as in single beds, the optimum value of the ratio δ/d_p can be suggested to be between 20 and 50.

The effect of the partition walls on the contact efficiency was also confirmed by heat transfer. The experimental apparatus is shown in Fig. 7. These beds were of the flank (B) of 400 mm and their widths were 20, 40 and 80 mm. Parallel plates, inserted into the two beds, were 40 and 80 mm wide, to keep the space δ 20 as well as 40 mm. The diameters, density and specific heat of sand used as the fluidizing material as well as the feed rate and the fluidizing mass per unit bed area are given in Table 1. The velocities of air as fluidizing gas were 0.4–0.6 of the terminal velocities.

The results are summarized by a relation between the Nusselt number and some dimensionless groups. The variables considered to affect the heat transfer are shown in Table 2 (p. 46). Equation (1) in the table is the ratio of the heat carried away with the feed per unit bed area (F_s) to a kind of fiducial heat to be accepted by a single particle. Equation (2) is the ratio of the heat capacity of fluidizing particles per unit bed area to that of a single one. In addition, the ratio of δ/d_p, the flank (B) to δ, Re number and Pr number are considered in the equations. Generally, heat transfer data have been correlated by a dimensionless equation, such as Eq. (3). From the experimental results, we obtained Eq. (4), where 0.1 in the ratio of 0.1 to δ means that the effect of width would be positive up to 100 mm of width and that at over 0.1 might become constant. The relation of calculated values to the observed ones is shown in Fig. 8.

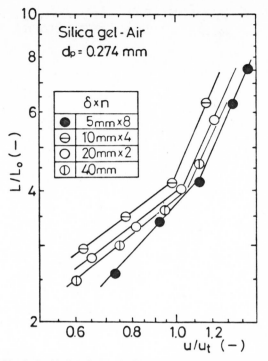

Fig. 6   Bed height in box bed inserted with partition walls.

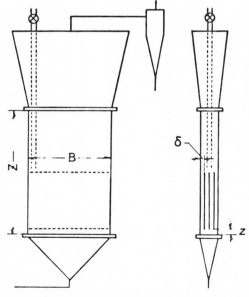

Fig. 7   Box bed inserted with partition walls.

Table 1. Experimental Conditions of Heat Transfer.

Particles: Sand		
Diameter	(d_p)	0.320, 0.505, 0.599, 0.815 and 1.33 mm
Density	(ρ_s)	3,040 kg/m^3
Specific heat	(c_{ps})	0.216 kcal/kg °C
Feed rates	(F_s)	5 to 50 kg/hr
Mass of fluidizing particles		
per unit area	(M)	20 to 159 kg/m^2
Gas: air		
Heat conductivity	(k_f)	0.0246 kcal/m hr °C
Velocities	(u/u_t)	0.4 to 0.6
Inlet temp.	(t_{in})	100 to 110 °C

Table 2. Variables of Heat Transfer.

Coefficient of heat transfer	h_p	Mass of fluidizing particles		
Flank of bed	B	per unit area	M	
Effective flank of bed	B*	Density of particle	ρ_s	
Width of parallel plates	δ	Specific heat of particle	c_{ps}	
Diameter of particle	d_p	Heat conductivity of particle	k_s	
Gas velocity	u	Density of gas	ρ_f	
Terminal velocity of particle	u_t	Specific heat of gas	c_{pf}	
Feed rate of particle	F_s	Heat conductivity of gas	k_f	
Acceleration of gravity	g	Viscosity of gas	μ	
		Section	n	

Some dimensionless groups

Ratio of the heat carried away with F_s to a kind of fiducial heat to be
accepted by a single particle per unit surface area

$$(F_s\ c_{ps}\ \Delta t)\ /\ [\Delta t\,(k_f/dp)] = (const)\ (F_s\ c_{ps}\ d_p/k_f) \tag{1}$$

Ratio of the heat capacity of fluidizing particles per unit area to
that of a single particle

$$(Mn\ \delta\ B^*c_{ps}\Delta t)\ /\ (\pi\ d_p^3\ \rho_s\ c_{ps}\ \Delta t/3) = (const)\ (M/d_p\rho_s)\ (\delta/d_p)\ (B^*/d_p) \tag{2}$$

Nusselt number $Nu_p = h_p\ d_p/k_f$
Reynolds number $Re_\delta = \delta\ u\ \rho_f/\mu$
Prandtl number $Pr\ = c_{pf}\ \mu/kf$

Result of dimensional analysis

$$Nu_p = (const)\ Re\ \delta^{\,a}(F_s\ c_{ps}\ d_p/k_f)^b\ (M/d_p\ \rho_s)^c\ (\delta/d_p)^d\ (B^*/\delta)^e$$
$$X\ (u/u_t)^f\ (B^*/d_p)^g\ (k_f/k_s)^h\ (c_{pf}/c_{ps})^i\ (\rho_f/\rho_s)^j$$
$$X\ (u^2/d_p\ g)^k\ Pr^m \tag{3}$$

Experimental Equation

$$Nu_p = 8.65\ X\ 10^{-10}\ Re\ \delta^{\,1.44}\ (F_s\ c_{ps}\ d_p/k_f)\ (d_p\ \rho_s/M)\ (\delta/d_p)$$
$$X\ (B^*/\delta)\ (0.1/\delta)^{1.4} \tag{4}$$

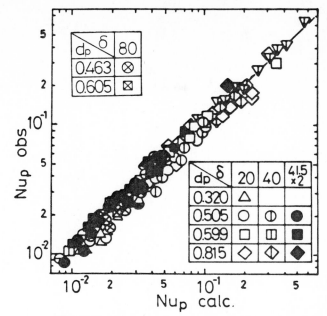

Fig. 8   Nusselt number _____ obs vs calc.

In this experiment, the inlet and outlet of charging particles were situated at the opposite sides and the inserted plates were parallel to this flank. Therefore, the higher the feed rate, the more active is the horizontal movement of the particles for a certain bed width. When feed rate, holding mass and properties of particles and gas properties are all the same, this horizontal movement of particles would promote the heat transfer as shown by Eq. (4) as in proportion to the flank (B); the bed with the longer flank could give higher efficiency for equivalent bed areas.

A MODIFIED DISTRIBUTOR

The behavior of fluidizing particles is affected considerably by the construction of the gas distributor; nowadays perforated plates and bubble cap type distributors are most commonly used in industrial fluidized beds. Such distributors, however, have a serious drawback of high pressure drop across the distributor to prevent the blockage of opening caused by the inclination of loading particles. This may increase the power consumption of blowers in operations of fluidized beds and especially in those with more than two stages; sometimes it may be beyond the performance of the blower, if the bed load is so high. This is the reason for suggesting a new distributor.

The proposed distributor gives a lower pressure drop through it without the blockage. A schematic drawing of the distributor is shown in Fig. 9. The distributor is composed of a vertical gas opening from the base level of the bed and a skirt with a throat to blow the falling particles away from the openings. The low pressure drop, characteristic of the distributor is due to the rather large opening area, which is larger than 8% of the bed area. Uniform gas flow through the opening is achieved in such a manner that when the blockage occurs, the particles in the blockage immediately flow down into the skirt, breaking the blockage, since the angle α (Fig. 10) is larger than the repose angle of the particles. Besides, the downflow of the particles may be expected to tend to wash out the adhesion of fine particles entrained with the gas form the lower stage.

Fluidizing tests were carried out by a rectangular bed of 500 mm flank, 400 mm width and 1000 mm height with sand as fluidizing particles as well as air at room temperature as the gas. The

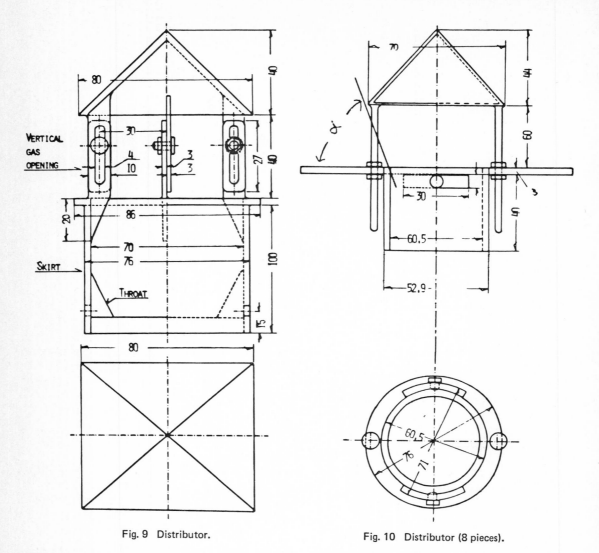

Fig. 9 Distributor.   Fig. 10 Distributor (8 pieces).

disposition of the distributor is shown in Fig. 11. Plate A has a single rectangular distributor in the center. Plate B has 4 rectangular distributors, and plate C has 8 round ones. The pressure drop across the tested plate is designated as ΔP_0. The result was 10–200 mm H_2O, when the gas velocity at the vertical opening was up to 16 M/ sec as about 8 times of u_t and the opening area was 3–20% of the bed area.

    Fig. 12 shows the pressure drop ratios of the stable fluidization, where the ratio $(\Delta P/\Delta P_0)$ was defined as the ratio of the entire pressure drop through a fluidizing bed to that through the plate alone. The result was a ratio of 10 to 30 and in the plate C of 8 distributors, the highest value was obtained in spite of the same opening area ratio; because of greater uniformity of gas distribution, this is considered to be 5 to 10 times that of common distributors. It is also confirmed that such higher loads of the solids can be held in the bed by use of the proposed distributor. As mentioned before, the special feature of this distributor is downflow of the particles through it. It is

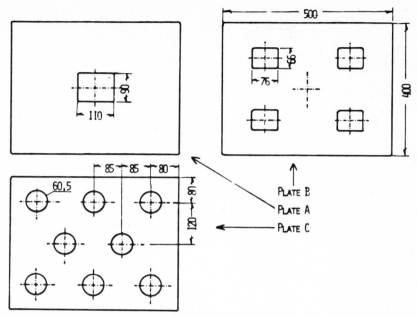

Fig. 11   Disposition of distributor

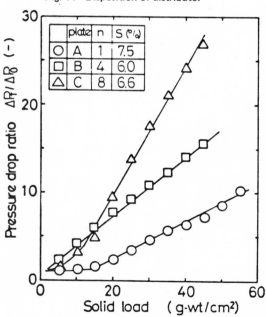

Fig. 12   Pressure drop ratio.

very important, therefore, to find the gas velocity at onset of a discontinuous downflow of solids as well as a continuous flow through the skirt. The gas velocities of the discontinuous and continuous downflow, based on the vertical opening area, are shown in Fig. 13 as u_d and u_c, respectively, where u_d increases from 1 to 4 times of u_t with the solid load. On the other hand, u_c decreases

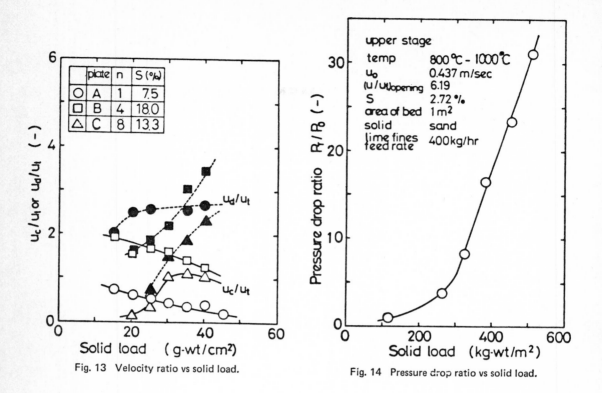

Fig. 13 Velocity ratio vs solid load.   Fig. 14 Pressure drop ratio vs solid load.

from 2 to 0.5 of u_t with the load, but for the plate A with an opening area ratio of less than 2%, it was observed that the particle downflow did not occur at velocities less than 0.3 u_t, in spite of high particle loading of more than 40 g/cm². This suggests that enlargement of the opening area is necessary for the performance of this distributor, and that the merit of this proposed distributor is its low pressure drop.

From these results, it could be concluded that an adequate gas velocity through the vertical opening is 1 to 3 times of u_t, but that an optimum opening area must be determined to keep appropriate rates of the solids downflow as well as the pressure drop, according to the circumstances of the operation of equipment.

This proposed distributor was applied to a 2-stage fluidized bed as a pilot-scale calciner, for the recovery of the waste lime cake in a beet sugar manufacture. The cake contained 7–10% of organic compounds and became pasty while heating up to 500°C. In addition, the diameter of particles in the cake was as fine as 2 to 100 μm; with 10% being less than 10 μm. From the two points above, the fluidization of the particles alone was almost impossible. Therefore, a 2-stage bed, fluidized with the limestone of 40–60 mesh, was used to feed the dry cake powder into the lower bed and then passed through the upper bed with the fluidizing gas heated up to 1000°C. The experimental conditions and the pressure drop are shown in Fig. 14. Where the ratio of about 30 indicates a rather large capacity of this fludized bed. The calcined product was of good quality giving more than an 85% conversion. This would suggest that the proposed distributor can be successfully applied to large-scale fluidized beds.

COMPARISON OF RECIRCULATING FLUIDIZED BED PERFORMANCE IN TWO-DIMENSIONAL AND THREE-DIMENSIONAL BEDS

W. C. YANG and D. L. KEAIRNS

ABSTRACT

A recirculating fluidized bed coal devolatilizer/desulfurizer is being developed as part of an advanced coal gasification process. Design information was previously reported based on data obtained from a two dimensional model. A semi-circular unit has been constructed and operated to investigate the scale-up of data from the 2-D unit. Data obtained from the two units are compared. The results show that a 2-D bed can provide reliable design and operating data for larger scale facilities if similarity can be identified. A generalized mathematical model is proposed to calculate both solid circulation and gas by-passing in recirculating fluidized beds.

INTRODUCTION

Westinghouse is developing a recirculating fluidized bed coal devolatilizer as part of a multi-stage fluidized bed coal gasification process to produce low Btu gas for combined cycle electric power generation.[1,2] A 1200 lb/hr fluidized bed gasification process development unit (PDU) has been constructed to investigate and develop the integrated process.[3] The recirculating bed devolatilizer in the PDU is 20 inches I.D. and 30 ft. height.

Crushed coal is introduced into the devolatilizer unit. A schematic representation is shown in Figure 1. There the devolatilization and partial hydrogasification are combined in a single recirculating fluidized-bed reactor operating at 1300 to 1700°F. Desulfurization can also be achieved in the devolatilizer. Dry coal is introduced through

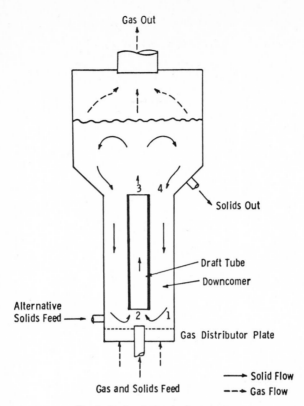

Fig. 1 Recirculation bed concept.

a central draft tube in this reactor. Inside this tube, the raw coal
and large quantities of recycled solids -- char or lime sorbent -- are
carried upward by gases flowing at velocities greater than 15 ft/sec.
The recycle solids that continually dilute the feed coal and also temper
the hot inlet gases descend in an annular downcomer -- a fluidizied bed
surrounding the draft tube. The recirculating solids effectively prevent
or control agglomeration of the coal feed as it devolatilizes and passes
through a phase in which it becomes sticky. Volatile products are driven
off the coal in an atmosphere containing hydrogen, which reacts with the
coal and char to form methane and higher hydrocarbons and release heat.
The dry char that results from devolatilization has a particle size and
density that cause it to concentrate in the top section of the fluidized
bed where it can be withdrawn.

Lime sorbent can be added near the top of the bed and mixes
with the recirculating solids, removing hydrogen sulfide from the fuel

gases. Spent (sulfided) sorbent is withdrawn from the reactor after
stripping out the char, either in the reactor, in the char transfer line
or in a separator of special design. Although some heat is provided by
the hydrogasification and devolatilization reactions, much of the heat
required is supplied by the high-temperature fuel gas fed from a gasifier/
combustor unit. Supplementary heat can be supplied by burning some char
in the downcomer around the draft tube.

        Two-dimensional laboratory scale cold-model tests were used to
identify critical design parameters and to obtain operating data for
the PDU.[4] The recirculating fluidized bed concept and data obtained
in the two-dimensional (2-D) unit (8.5 in. x 1.5 in. x 6 ft.) have been
reported.[4] Critical variables were identified using the 2-D unit and
a mathematical model was developed to predict performance.[4] The results
indicated that solids circulation rates in the bed can be controlled at
weight rates up to 100 times the coal feed rate. However, extension of
the model to a large scale three-dimensional unit was uncertain due to
the scale-up of 2-D bed data and the limited range of data obtained on
the 2-D unit. The two dimensional model was limited in that the dimensions
were small compared to the PDU and the model did not permit tests with
draft tube heights greater than 3 feet.

        A semi-circular cold model, Figures 2 and 3 was conceived to
investigate scale-up of data from the 2-D unit and to study important
design variables such as draft tube height, relative area of draft tube
and downcomer, and draft tube inlet design. The semi-circular unit
(11.25 inch I.D., 20 ft. height) was sized to approximate the dimensions
of the recirculating bed reactor in the PDU.

EXPERIMENTAL RESULTS

        Data were collected to compare solid circulation rates obtained
in the 2-D and 3-D units. Ottawa sand, 606μ average particle size, was
the solid and air was the fluidizing gas. The following variables were
investigated: gas flow rates to the downcomer and draft tube, bed height,
draft tube height and distance between the distribution plate and draft
tube inlet. The relationship between the solid circulation rate and the
gas bypassing rate is shown in Figure 4. Gas bypassing is the amount
of gas fed to the downcomer which bypasses into the draft tube.

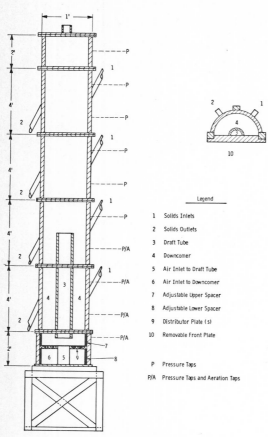

Legend

1   Solids Inlets
2   Solids Outlets
3   Draft Tube
4   Downcomer
5   Air Inlet to Draft Tube
6   Air Inlet to Downcomer
7   Adjustable Upper Spacer
8   Adjustable Lower Spacer
9   Distributor Plate (s)
10 Removable Front Plate

P   Pressure Taps
P/A Pressure Taps and Aeration Taps

Fig. 2 Schematic of the semi-circular unit (not to scale).

Fig. 3 Assembled semi-circular plexiglas column.

The gas bypassing rate was calculated by applying the modified Ergun equation to the downcomer side substituting gas-solid slip velocities for gas velocities as shown in Equation (A-2) (see Appendix). The shape factor, ϕ_s, of the Ottawa sand was taken to be 0.86 determined in a separate 3 in. bed using the method described by Leva (1959). This value is consistent with that reported in the literature. The voidage, ε_d, was calculated to be 0.467 from Equation (A-2) by operating the downcomer of the recirculating bed at the minimum fluidizing condition and assuming that $U_{fd} = U_{mf}/\varepsilon_d$. These values, ε_d = 0.467 and ϕ_s = 0.86, were used throughout for both 2-D and 3-D beds. Once the downcomer

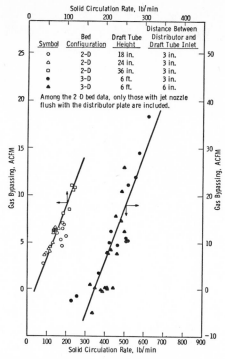

Fig. 4 Comparison of the dependance of solid circulation rate on gas bypassing rate for the 2-D and 3-D beds.

pressure drop was known by actual measurements, U_{fd} could be calculated from Equation (A-2) since U_{pd} was known by following tracer particles with a stop watch. The gas bypassing rate was the difference between the total gas supplied to the downcomer and the amount of gas actually passed through the downcomer.

The solid circulation rate was shown to be dependent on gas bypassing rate in both 2-D and 3-D beds and could be approximated by straight lines (Fig. 4). Both straight lines had similar slopes but did not pass through the origin. This indicated that the contribution to the total solid circulation in the bed came from two sources. One contribution was from the intrinsic solid circulation rates represented by the intersections of the straight-lines with the abscissa at zero gas bypassing rate. The other source was from the solids associated with the bypassing gas. High speed movies (1000-1500 frames per sec.) taken at the inlet and the mid-section of the draft tube in the 3-D bed revealed that solid transport inside the draft tube was not in a

conventional pneumatic transport, where uniform solid suspension prevailed,
but in a slugging type of transport. The high speed movies taken at
the inlet showed that the air jet issued from the jet nozzle supplying
air to the draft tube took the form of bubbles rather than a steady jet.
The bubble grew from the mouth of the nozzle until its roof reached the
draft tube, then the sudden suction from the draft tube punctured the
roof. A continuous stream of dilute solid suspension passed through the
roof into the draft tube. Simultaneously, another bubble was initiated.
When the bubble grew, it pushed a slug of solid into the draft tube.
The high speed movies taken at the mid-section of the draft tube exhibited
alternate sections of dilute solid suspension and solid slug occupying
the total cross-section of the draft tube. Observed bubble frequency
ranged from 7 to 12 bubbles per sec. The bubble frequency and shape
seem to depend on the nozzle velocity and the distance between the nozzle
and the draft tube inlet; however, there are not yet enough data available
for establishing a reliable correlation.

　　　　From what was revealed in the high speed movies, it is reasonable
to state that the intrinsic solid circulation rate probably relates to
the frequency and size of solid slugs being pushed into the draft tube
by the air bubbles at zero gas bypassing and the solid circulation rate
associated with bypassing gas may be contributing to the dilute solid
suspension. However, this distinction is by no means complete. The
bypassing gas can also help the bubble growth and thus effect the frequency
of bubbles, but this and other secondary interactions are too complicated
to analyze at this moment. The physical phenomena described here are
supported by the data. The data are replotted based on per unit area of
the jet nozzle supplying air to the draft tube in Figure 5. All data fall
on a common straight line with a single intersection with the abscissa.
The intersection represented the intrinsic solid circulation rate at zero
gas bypassing for the recirculating bed. This representation indicates
that the intrinsic solid circulation rate depends only on the size of
the jet nozzle. In the 2-D bed, a copper tube of 3/8 in. I.D. (area=0.11
in^2) was used as the jet nozzle supplying air to the draft tube and in
the 3-D bed, a semi-circular section of a 2 in. Sch. 40 pipe (area=1.68
in^2). The bed heights studied were not an important factor.
The similar slopes for both 2-D and 3-D data implied a constant voidage
of ∿ 0.95 in the gas bypassing stream (determine from their slopes in

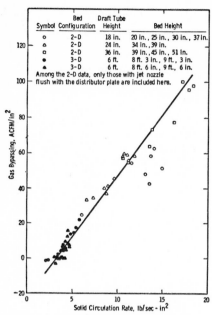

Fig. 5 Comparison of the 2-D and the 3-D
beds data based on unit jet nozzle area.

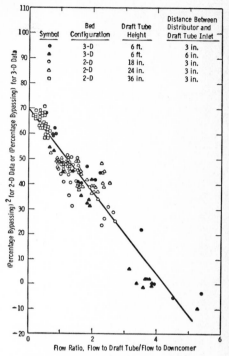

Fig. 6 Comparison of gas bypassing rate
in two-dimensional and three-dimensional
beds.

Fig. 4). Negative gas bypassing where the gas bypassed from the draft
tube gas supply to the downcomer was also shown to be possible. If this
occurred the solid circulation rate was reduced to below the intrinsic
rate due to countercurrent flow of solid and gas. The distance between
the distributor plate and the draft tube inlet (3 in. and 6 in.) in the
3-D bed did not seem to play an important role.

Gas bypassing data are correlated with the gas flow ratio in
Figure 6. The gas flow ratio was defined as the ratio of gas supplied
to the draft tube to that supplied to the downcomer assuming no gas
bypassing. The percentage gas bypassing was the percentage of downcomer
flow which bypassed into the draft tube. The percentage gas bypassing
was used as the ordinates for the 3-D bed data and the square of the
percentage gas bypassing was used for the 2-D bed data. Experimental
evidence indicated that flow supplied to the downcomer could be assumed
to be uniformly distributed across the bed cross-section. The fraction
of gas bypassing is $(b \cdot \alpha)/(a \cdot \alpha)$ for a two-dimensional bed of cross-

section a·α (α << a) and with the grid area b·α beneath the draft tube
being the area supplying the gas which flows into the draft tube. If this
2-D bed was to become a 3-D bed of diameter a, then the percentage gas
bypassing would be $(b/a)^2$. To place the data on a common geometrical
basis, the 2-D data had to be squared.

The 2-D bed used for the present experiments had an area ratio
of downcomer and draft tube of 5 and the 3-D bed had a ratio of 7.7. The
effect of downcomer/draft tube area ratio does not appear to be important
for this range of area ratios.

MATHEMATICAL MODEL AND PERFORMANCE PROJECTS

The mathematical model developed from the 2-D bed data identified
the pressure drop at the bottom of draft tube and downcomer as the primary
driving force for the solid circulation in the bed and utilized it as the
primary correlation parameter.[4] Experiments with higher draft tube
heights (> 18 inches) in both 2-D and 3-D beds indicated that this
pressure drop was increasingly difficult to obtain due to pressure drop
fluctuation. This pressure drop would also be difficult to measure in
commercial practice involving high pressure and high temperature. The
earlier model also had a limitation in that it is only applicable when
the downcomer is minimally fluidized. A new generalized model is proposed
which requires only the knowledge of the pressure drop in the downcomer
and is not restricted to minimum fluidization velocity in the downcomer.
Details of the model are described in the Appendix.

The new model balances the pressure drops through the downcomer
and the draft tube. The pressure drop between the bottom of the draft
tube and downcomer is negligible in comparison to these pressure drops.
The draft tube pressure drop correlation is extended to include particle
acceleration. The solids circulation rate is predicted from a trial and
error solution of the two pressure drop equations and the material
balance given the experimental downcomer pressure drop and the empirical
correlation for gas bypassing. This permits calculation of the solids
circulation rate utilizing available experimental data and operating with
downcomer velocities above and below the minimum fluidizing velocity.

The model predicts solid circulation rates 0 to 100% higher
than the observed rates. This may be the result of particles close to

the wall, which were used to measure the circulation rate, moving at a lower rate and/or the equations used for the draft tube pressure drop might be underestimating the actual pressure drop due to a slugging type transport observed in the draft tube. Recent tests with a radio pill indicate the circulation rate is ∿ 50% higher than the rate based on observing particles at the wall.

CONCLUSION

The data show that a 2-D bed is useful to obtain quick and reasonably accurate results for design and scale-up if the similarities between the 2-D and 3-D beds are identified. The comparison presented shows that the 2-D bed data can be extrapolated to predict performance in a 3-D bed. A generalized mathematical model is proposed to calculate both solid circulation rate and gas bypassing rate in recirculating fluidized beds.

APPENDIX

The Generalized Mathematical Model

The mathematical model proposed earlier[4] recognized the pressure drop at the bottom of the draft tube and downcomer (points 1 and 2 in Figure 1) as the primary driving force for the solid circulation in the bed and utilized it as the primary correlation parameter. Experiments with higher draft tube heights in both 2-D and 3-D beds indicated that this pressure drop was increasingly difficult to obtain due to fluctuation. This pressure drop would also be difficult to measure in commercial practice operating at high pressure and high temperature. The earlier model also had a limitation since it is only applicable when the downcomer is minimally fluidized. A new generalized mathematical model is proposed here which requires only the knowledge of pressure drop in the downcomer and is also applicable for different amounts of downcomer aeration. The new model is described below.

At steady state, the downcomer pressure drop (refer to Figure 1), ΔP_{1-4}, can be expressed as

$$\Delta P_{1-4} = \Delta P_{1-2} + \Delta P_{2-3} + \Delta P_{3-4} \tag{A-1}$$

All experiments conducted so far indicated that ΔP_{3-4} fluctuated around
zero with different amplitudes and could be safely taken as zero without
introducing considerable error. ΔP_{1-2} is difficult to measure in actual
practice but can be assumed to be in the same order of magnitude as the
downcomer pressure drop per unit length. The pressure drop in the downcomer, ΔP_{1-4}, was correlated with the modified Ergun equation substituting
gas-solid slip velocities for gas velocities as shown in Equation (A-2):

$$\Delta P_{1-4} = \frac{L}{g_c} \left[150 \ \frac{\mu (U_{fd} + U_{pd})(1 - \epsilon_d)^2}{d_p^2 \phi_s^2 \epsilon_d^2} \right. $$

$$\left. + \ 1.75 \ \frac{\rho_f (U_{fd} + U_{dp})^2 (1 - \epsilon_d)}{d_p \phi_s \epsilon_d} \right] \qquad (A-2)$$

The pressure drop inside the draft tube, ΔP_{2-3}, is more complicated because
it involves acceleration of solid particles from essentially zero vertical
velocity. The acceleration term is especially significant when the draft
tube is short. The acceleration length is calculated through numerical
integration of the following equation:

$$\Delta L = \int_{U_{p1}}^{U_{p2}} \frac{U_p \, dU_p}{\frac{3}{4} C_{DS} \epsilon_r^{-4.7} \ \frac{\rho_f (U_f - U_p)^2}{(\rho_p - \rho_f) dp} - (g_c + \frac{f_p U_p^2}{2D})} \qquad (A-3)$$

Equation (A-3) is obtained by force balance similar to the one presented
earlier.[5] The solid friction factor, f_p, can be evaluated with the
equation proposed by Yang[6,7]

$$f_p = 0.0206 \ \frac{(1 - \epsilon_r)}{\phi_s^2 \epsilon_r^3} \left[(1 - \epsilon_r) \ \frac{(Re)_t}{(Re)_p} \right]^{-0.869} \qquad (A-4)$$

The lower limit of integration, U_{p1}, is derived from

$$W_s = U_p \rho_s (1 - \epsilon_r) \tag{A-5}$$

with $\epsilon_r = 0.5$ and the upper limit, U_{p2}, by Equation (A-6)[5]

$$U_{p2} = U_f - U_t \cdot \sqrt{(1 + \frac{f_p U_p^2}{2g_c D}) \cdot x \, \epsilon^{4.7}} \tag{A-6}$$

The total pressure drop in the draft tube can be expressed as

$$\Delta P_{2-3} = \int_o^\ell \rho_s (1 - \epsilon_r) dL + \int_o^\ell \frac{2f_g \rho_g U_o^2}{g_c D} dL$$

$$+ \int_o^\ell \frac{f_p \rho_s (1 - \epsilon_r) U_p^2}{2g_c D} dL + \left[\frac{\rho_s (1 - \epsilon_r) U_p^2}{g_c} \right]_{at \, \ell} \tag{A-7}$$

If the draft tube height is less than the acceleration length, the solid
particle velocity at the exit of draft tube and the total solid circulation
rate can be readily calculated from Equations (A-1), (A-2) and (A-7) with
known downcomer pressure drop, ΔP_{1-4}. If the draft tube height is larger
than the acceleration length, the integration of Equation (A-7) is carried
out for the total acceleration length and the extra pressure drop for
the rest of the draft tube can then be included to give the total pressure
drop in the draft tube. By trial and error between Equations (A-1)
through (A-7), total solid circulation rate can be obtained. The amount
of gas bypassing can also be calculated by the same method.

        The aforementioned mathematical model was applied to the data
obtained in both 2-D and 3-D beds. The model predicted solid circulation
rates from 0 to 100% larger than the observed rates for 80% of the data.
The maximum deviation was +170%. The consistent prediction of larger
solid circulation rates by using the mathematical model was due to two
reasons: (a) the particles close to the wall were moving at velocities
20% to 80% lower than the particles farther from the wall due to wall
effect based on results from radio pill measurements and, (b), the
equations used for calculating the draft tube pressure drop might be

underestimating the actual pressure drop due to a slugging type of solid transport frequently observed inside the draft tube. The deviation of calculated gas bypassing rates was less than $\pm$ 10% of that observed in all cases.

NOTATION

C_{DS}	drag coefficient on a single particle
D	draft tube diameter, m
d_p	mean particle size, m
f_g	Fanning friction factor
f_p	solid friction factor
g_c	gravitational acceleration, m/sec^2
L	height of the draft tube, m
ΔL	distance required to accelerate particle, m
ΔP_{1-2}	pressure drop between 1 and 2 (see Figure 1) kg/m^2
ΔP_{1-4}	pressure drop between 1 and 4 (see Figure 1) kg/m^2
ΔP_{2-3}	pressure drop between 2 and 3 (see Figure 1) kg/m^2
ΔP_{3-4}	pressure drop between 3 and 4 (see Figure 1) kg/m^2
$(Re)_p$	Reynolds number based on the slip velocity and defined as $d_p(U_f - U_p)\,\rho_f/\mu$
$(Re)_t$	Reynolds number based on the terminal velocity of the solid particles
U_o	superficial fluid velocity, m/sec.
U_{fd}	actual fluid velocity in the downcomers, m/sec.
U_f	actual fluid velocity in the draft tube, m/sec.
U_{pd}	solid particle downward velocity in the downcomers, m/sec.
U_p	solid particle velocity in the draft tube, m/sec.

U_{sl} slip velocity between fluid and solid particle in the draft tube, m/sec.

U_t terminal velocity of the solid particle, m/sec.

W_s solid flow rate per unit area, kg/sec-m^2

Greek Letters

ϵ_d voidage in the downcomers

ϵ_r voidage in the draft tube

μ viscosity of the fluid, kg/sec-m

ρ_f density of the fluid, kg/m^3

ρ_s solid particle density, kg/m^3

ϕ_s sphericity of the solid particle

REFERENCES

1. Archer, D. H., E. J. Vidt, D. L. Keairns, J. P. Morris, J.L.P. Chen, "Coal Gasification for Clean Power Production", Proceedings of the Third International Conference on Fluidized Bed Combustion, Hueston Woods, Ohio, 1972, EPA 650 12-73-053 (NTIS PB 231977).

2. "Advanced Coal Gasification System for Electric Power Generation", Annual Technical Reports 1973, 1974 to Office of Coal Research, U.S. Department of the Interior by Westinghouse Electric Corporation, Contract No. 14-32-0001-1514.

3. Archer, D. H., D. L. Keairns, E. J. Vidt, "Development of a Fluidized Bed Coal Gasification Process for Electric Power Generation", Energy Communications, 1(2), 115-134 (1975).

4. Yang, W. C. and D. L. Keairns, "Recirculating Fluidized Bed Reactor Data Utilizing a Two-Dimensional Cold Model", AIChE Symposium Series, Vol. 70, No. 141, 27-40 (1974).

5. Yang, W. C., "Estimating the Solid Particle Velocity in Vertical Pneumatic Conveying Lines", Ind. Eng. Chem. Fundam., 12 (3), 349, (1973).

6.   Yang, W. C., "Correlations for Solid Friction Factors in Vertical
     and Horizontal Pneumatic Conveyings", AIChE Journal 20 (3),
     605 (1974a).

7.   Yang, W. C., "Additional Notes on Solid Friction Factors in
     Pneumatic Conveyings", Paper presented at 77th AIChE National
     Meeting at Pittsburgh, Pennsylvania, June 1974.

8.   Leva, M., Fluidization, McGraw-Hill Book Co., New York, 1959.

CIRCULATION OF LARGE BODIES
IN AN AGGREGATIVELY FLUIDIZED BED

B. B. PRUDEN, D. CROSBIE and B. J. P. WHALLEY

INTRODUCTION

Heated beds of inert fludized solids are of considerable interest
as heat-transfer media for physical and chemical reactions and for
industrial processes such as immersed heat-treating of metals and
immersed combustion of coal and other solid materials.   They have
also been used for heat-treating coal briquettes to make substitutes
for conventional slot-oven coke as solid reductant in iron-making blast
furnaces.   In the latter application, briquettes, typically of coal (or
partially carbonized coal) are heated to temperatures about 450°C
for several hours to increase their mechanical strength sufficiently
to withstand breaking and attrition in transferring them into the blast
furnace.

At the Canadian Centre for Metals and Mineral Technology
(CAN MET) a program of investigation of Canadian coals for formed
coking has been underway for several years and recently a sand fluidzed
bed reactor has been constructed and used to heat-treat briquettes of
coal and char in the 450-700°C range and is intended also to carbonize
coal briquettes at a temperature of about 1000°C[1].   This unit is a
batch unit, with the briquettes lowered into the sand fluidized bed inside
a wire cage for the required time.   To make it compatable with other
continuous units, and to improve it's efficiency, modifications have been
studied to make it continuous.   One such modification is that in which
briquettes circulate through a fluidized bed in the required residence
time, the bed being divided into downflow and upflow sections by a
partition.   Another possibility is to use a medium in which the difference

in density of untreated and treated briquettes can be used for separation. To these ends an investigation was initiated into the circulation behavior of relatively large bodies simulating coal briquettes. This paper deals with the preliminary results of this investigation.

Two important design variables for fluidized beds are minimum fluidization velocity and pressure drop. Minimum fluidization velocity not only sets a lower limit on gas flow rate to the fluidized bed, but also is useful for prediction of bed expansion, for calculating heat transfer rates and in calculating pressure drop[2]. Pressure drop across a fluidized bed determines the sizing of the blower or compressor supplying the gas to the fluid bed and is useful as an index of fluidized solids inventory.

Since the behaviour of large bodies in a fluidized bed is obviously dependent upon its turbulence, use of the minimum fluidization velocity, V_{mf}, for comparing conditions of fluidization was mandatory. In the course of investigation of some of the properties of fluidized beds of sand, an apparently new method of defining V_{mf} was developed.

Initially, work was done to determine V_{mf}, and to compare V_{mf} measured using several methods. Following this, a detailed study was made using a nominal 35 mesh sand (see Table 1) and hollow cylinders with diameter and height of one inch made from aluminum. Based on this study, other work was done with 2" cylinders and 1/2" cylinders in sand and 1" cylinders in glass beads. As described earlier cylinders were chosen because they resemble the shape from laboratory presses and approximate the pillow shape from commercial briquetting presses.

The third part of this work deals with forced circulation of cylinders with diameter and height of one inch in a fluidized bed.

THEORY

Development of V_{mf} Determination

Minimum fluidization velocity may be determined by computation[3, 4, 5, 6, 7, 8] or by experimental methods. One of the most commonly used experimental techniques is based on the method of Parent et al.[9] in which V_{mf} is the gas velocity at which the bed pressure drop is just equal to the net weight of solids per unit area. Another non-graphical technique consists in observing the point at which bubbles first appear.

These non-graphical methods appear to be theoretically plausable, but actually the bed may bubble at lower gas velocities, because of bypassing around stagnant areas, or may even remain stationary at rates above the theoretical minimum, owing to the extra force required to overcome electrostatic forces.

This most commonly used graphical method is based on measuring the pressure drop through the bed of solids. Some authors[2] have even observed the change of heat transfer coefficient with gas velocity as a criteria for the minimum fluidization velocity.

Conventional graphical methods are quite useful if the behavior of the system under study is of the form shown in Fig. 1 (a) that is, with a definite inflection. However if the behaviour of the system is of the form shown in Fig. 1 (b) i.e. a smooth curve, the location of a distinct definite point of inflection becomes less precise and the method of intersecting lines is often used. When the curve is of the form in Fig. 2 (a) the inflection may be in considerable doubt especially if it is not well defined.

The principle of the method developed during this work is illustrated by considering a bed packed with particles to a height h with two pressure taps, one near the top of the bed (at 1) and one at the bottom. In an increasing flow rate of fluidizing gas, in the fixed bed region (assuming the bulk density, ρ_b to be constant throughout the bed) the ratio of the pressure drops to the corresponding bed heights will be equal or

$$\frac{-\Delta P_1}{x} = \frac{-\Delta P_o}{L} \qquad (1)$$

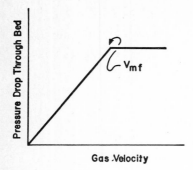

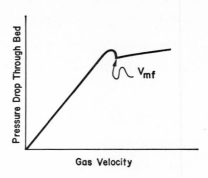

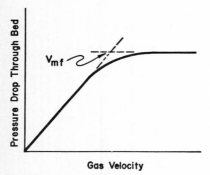

Fig. 1 Typical pressure drop-flow
diagram for (a) ideally fluidized
solids, (b) slugging solids.

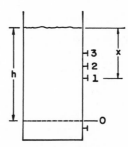

Fig. 2 (a) Typical pressure drop-flow
diagram, moderately channeling solids
(b) arrangement of pressure taps.

and

$$\frac{-\Delta P_1}{P_{w_1}} = \frac{-\Delta P_0}{P_{w_0}} \qquad (2)$$

where P_{w_0} is the total weight of solids divided by the bed cross sectional
area, and P_{w_1} is the weight of solids above pressure tap number 1 (after
the bed has been fluidized and then collapsed slowly) divided by the bed
cross sectional area. Equation 2 is equal to equation 1 with both sides
multiplied by $1/\rho_b$. Beyond incipient fluidization, equation 2 no longer
holds because the changes in x and L as the bed expands are dispropor-

tionate. It follows that if $-\Delta P_o / P_{w_o}$ and $-\Delta P_l / P_{w_l}$ are plotted versus gas fluidization velocity the curves coincide up to V_{mf} and diverge thereafter, suggesting a way to measure V_{mf}.

APPARATUS

A simple apparatus was used to obtain experimental data. A schematic diagram is shown in Fig. 3. At the outset, the system consisted of a fluidization column, a suitable rotameter, two oil-filled manometers and a pressure gauge.

The fluidization column, 9.72 in. ID and 53 in. high was made of Plexiglas to permit visual observation of the fluidized solids. The bed was supported on a porous 0.35-in. thick stainless steel plate (nominally 200 mesh pores) held between bolted Plexiglas flanges. Seven pressure taps with isolation valves were located in a vertical line on the column, one just below the porous plate, the others at intervals of 6 in., the lowest being 3 in. above the plate. There were two air inlets at the base of the column 180° apart to ensure even distribution of air.

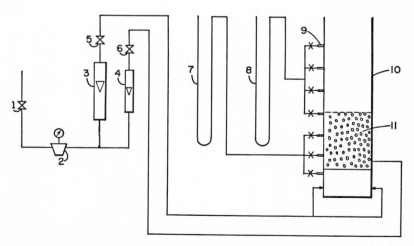

Fig. 3 Apparatus: 1. valve; 2. pressure regulator; 3,4. rotameters; 5,6. needle valves; 7,8. manometers; 10. column; 11. bed of particles.

Table 1 Properties of Sand and Glass Beads. Sand — "Ottawa Flint Silica Coarse". Sieve Analysis

Canadian Standard* Sieve	Standard Sieve	Percent in Range
- 25 x 35	500 - 707 μ	15%
- 35 x 50	297 - 500 μ	82%
- 50 x 70	2 10 - 297 μ	3%
- 70 x 0	Less than 2 10 μ	0. 05%

Particle Density 165. 4 lb/ft^3

GLASS BEADS

SIZE; - 60 x 80 Canadian Standard Sieve or 177 to 250 micron standard sieve

Particle Density 152 lb/ft^3

* same as U.S. standard sieve

Silica sand (Ottawa flint silica coarse, Table 1) was used for the bed. Filtered dried air was used as the fluidizing gas at a pressure of 40 psig.

RESULTS AND DISCUSSIONS

Part 1 Fluidization Characteristics of Sand Bed and Evaluation of V_{mf}

In this part of the experimental work, beds of sand of different weights and initial bulk densities were fludized. High initial bulk densities were obtained by vibrating and tapping the bed or allowing the fluidized bed to collapse abruptly.

Pressure drops were measured across different incremental depths of bed stepwise with rising and with falling fluidizing gas velocities. For all cases in the fixed bed region the pressure drop ($-\Delta P$)

increased linearly with superficial gas velocity (V_s) in the usual way
and was less than the weight (mass) of the bed per unit area.

Transition region between fixed bed and fluidized bed

In general the ($-\Delta P$) vs V_s curves for the total bed depth of
the higher initial density beds (i.e. before fluidization and all from the
same sample of sand) exhibited the characteristic hump near V_{mf} of
such beds, whereas those of low density exhibited relatively smooth
curves. In Fig. 4 curve A represents an increasing flow in a high-bulk-
density (108 lb/ft^3) bed, whereas B represents increasing flow from 94.3
lb/ft^3. Curve B shows as well the ($-\Delta P$) vs V_s for a decreasing flow of
gas. It can be seen that the V_{mf} could not have been determined from
curve A alone and that its determination from curve B would have been
indefinite unless the flow was increased well beyond 1.0 ft/sec. Simul-

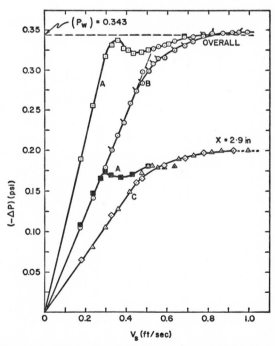

Fig. 4 Pressure drop — V_s diagram for 25.5 lb of sand (A) Forward flow, initial bulk density 108 lb/ft^3, □ — overall
pressure drop ■ — pressure drop at X = 2.90. (B) ⚬ — Backward flow from curve A and then ⊙ forward flow, overall
pressure drop. (C) ◊ Backward flow from curve A and then △ forward flow, pressure drop at x = 2.90.

taneous pressure-drop readings taken at x = 2.90 inches show similar
behaviour in Fig. 4 curves A and C. That is, that the backward-flow
curve of a high-density bed coincided with the forward flow curve of the
lowest density bed. When the depth-normalized pressure drops ($-\Delta P/P_w$) vs V_s were plotted for low-density beds, those portions corresponding
to the fixed bed region were essentially coincident. As the gas flow in-
creased there was a marked divergence of slope as shown in Fig. 5.
This point of divergence corresponded to the V_{mf} (0.56 ft/sec) determined
by conventional plotting (i.e. intersection of the extrapolated straight
line portions of the curve traced out by a diminishing gas flow).

Figs. 6, 7 & 8 show the pressure-drop-normalized curves for whole
beds and fractional portions of the beds. The curves have been separated
in each case for clarity. The inflection of the top curves in Figs. 6 and 8
reflect the pressure-drop changes in the top 20% bed depth; the top curve
in Fig. 7 by a 40% depth. The bottom curves in Figs. 6, 7 & 8 reflect the
normalized pressure drop changes in the total bed.

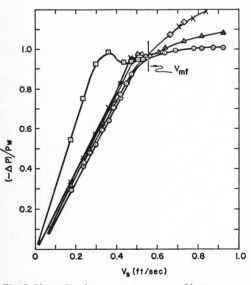

Fig. 5 Normalized pressure drop versus V_s. □ —
Forward flow from ρ_b = 108, 25.5 lb. sand, overall
ΔP. ○ Backward flow, 25.5 lb. sand, overall ΔP. ◊,
x, △, pressure drop at tap 2.9 inches below bed
surface, 75.6, 51 and 25.5 lb. of sand respectively.

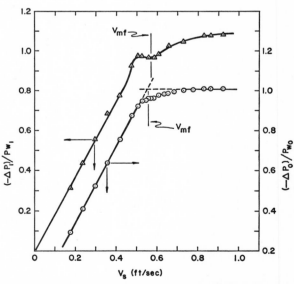

Fig. 6 Normalized pressure drop versus V_s for 25.5 lb of sand.
Top curve — pressure drop at x = 2.90 inches. Bottom curve —
overall pressure drop. Comparison of graphical methods.

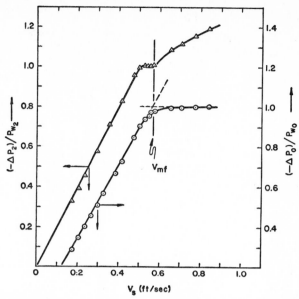

Fig. 7 Normalized pressure drop versus V_s for 51 lb of sand. Top curve — pressure drop at $x = 2.90$ inches. Bottom curve — overall pressure drop. Comparison of graphical methods.

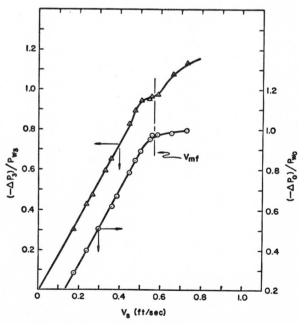

Fig. 8 Normalized pressure drop versus V_s for 75.6 lb of sand. Top curve — pressure drop at $x = 2.90$ inches. Bottom curve — overall pressure drop. Comparison of graphical methods.

It can be shown that the V_{mf} - indicating inflection for the curves up to V_{mf} can be made more definite by measuring the pressure drop in a shallower (say 10%) depth of bed.

Although the pressure drop at the top taps were depth-normalized for the purpose of showing coincidence of the curves, an arbitrary multiplying factor applied to the pressure drop value before plotting manually (and by inference by a gain adjustment in a recording instrument such as a strip-chart recorder) is evidently all that is required to accentuate the inflection of the shallow-tap pressure drop.

The graphical approach may be further developed as follows: If equation 2 is written and assuming fixed bed conditions, (i.e. no expansion), then it can be seen that plotting the ratio of the pressure drops should result in a straight line of slope zero in the fixed bed region and that beyond V_{mf} an inflection in the curve will occur, because of the increasingly greater pressure change in the top of the bed relative to the whole bed caused by bed expansion.

When the ratio was plotted against V_s for increasing flow rates, curves as shown in Fig. 9 resulted for beds of normal density. Fig. 9B shows the character of the curve corresponding to decreasing gas flow rates. Similar curves are shown in Figs. 10 & 11. It can be seen that these minima agree well with those done by conventional graphical and by the normalized pressure-drop methods. These graphs also show that the V_{mf} was independent (within experimental error) of bed height in accordance with theory.

Because of the sharp inflection at V_{mf}, it would appear possible to detect it instrumentally by, for example, using pressure transducers with a millivolt output for taps at shallow and at full depths. By adjusting the initial voltages to match approximately the differential signal could be used to trace out the fluidization curve for a linearly-increasing and decreasing gas-flow rate.

Figure 12 is a forward flow curve for an initially tightly packed

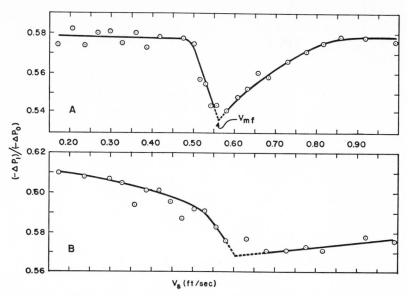

Fig. 9 Ratio of pressure drop at x = 2.90 inches to overall pressure drop for 25.5 lb. sand. (A) Forward flow from bed bulk density of 94.3 lb/ft³. (B) Backward flow.

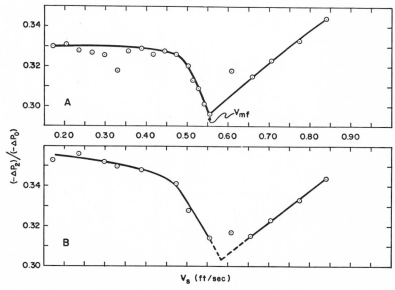

Fig. 10 Ratio of pressure drop at x = 2.90 inches to overall pressure drop for 51.0 lb. of sand. (A) Forward flow from bed bulk density of 95.6 lb/ft³. (B) Backward flow.

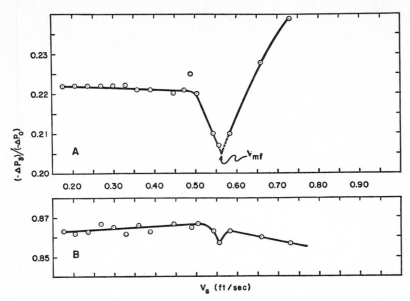

Fig. 11 Ratio of pressure drop at x = 2.90 inches to overall pressure drop for 75.6 lb. of sand. (A) Forward flow from bed bulk density 93.8 lb/ft³. (B) Backward flow.

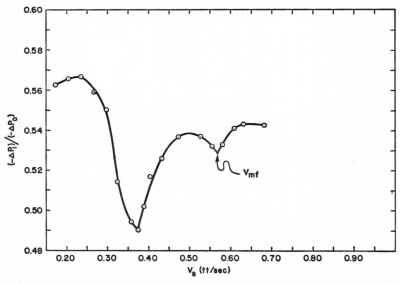

Fig. 12 Ratio of pressure drop in upper portion of bed to overall pressure drop versus V_s for high initial bulk density bed (108 lb/ft³) forward flow.

bed. Note that V_{mf} is predicted by the second minimum in the curve.

V_{mf} was also measured as the point at which bubbles first appeared and found to be 0.47, 0.49, and 0.50 ft/sec for 25.5, 51 and 76.5 lb of sand respectively. The graphical intersection method in a plot of $-\Delta P_o$ versus V_s yielded V_{mf} values of 0.54, 0.56 and 0.56 ft/sec for 25.5, 51 and 76.5 lb of sand (6.3, 12.6, and 18.9 inches bed height) respectively.

Of the three methods described in this work for using the pressure drop at the top of the bed, the method of Figures 9-12, plotting the ratio of pressure drops versus V_s is clearly superior.

Part II Behavior of Cylinders in an Air Sand Quiescent and Fluidized Bed

A known weight of sand (51 lb) was poured into the column and fluidized. The gas flowrate was then decreased slowly and the particles allowed to settle freely. The resulting bed height was 12.6 inches corresponding to a bulk density of 94.1 lb/ft^3. This was repeated to measure the bulk density of the bed when fluidized and, as shown earlier, this value was reached from a more densely packed state as well.

In all studies with cylinders the air flow was first adjusted to give V_s = 0.75 ft/sec and then reduced slowly to give the lowest value of V_s, used, 0.4 ft/sec. The cylinder was then placed face down on the bed surface. (Five different positions were chosen for measurement: the centre and four radial positions 90° apart near the wall). After a cylinder was placed it's motion was observed and it's vertical position measured after movement ceased. The air flow was then adjusted upward stepwise and the cylinder position was measured as a function of V_s as shown in Figure 13. The depth to which the cylinder sank on the surface was observed on a scale on its side. To determine its position below the surface of the bed, a fine thread was attached.

Twenty-five different one-inch cylinders varying in density were studied. All cylinders were made of hollow aluminum and loaded

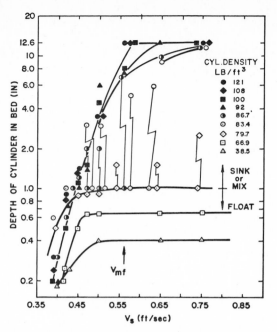

Fig. 13 Depth to which cylinder sinks in bed (measured from lowest side to bed surface) versus V_s. Cylinders one inch in diameter and height in 51 lb. of sand, 20°C, 1 atm. pressure.

symmetrically to produce densities of from 38 to 169 lb/ft^3.

It should be noted that when the bed was fluidized in the manner described above there was excellent gas distribution and, when bubbling, the gas bubbles appeared at random at the surface of the bed. If the bed was tilted approximately five degrees, or if the fluidizing gas was turned on suddenly, maldistribution of gas in the bed was apparent from the increased bubbling in one section of the bed surface. This maldistribution would continue, once started, even if the bed was returned to a vertical position, and had to be corrected by raising and then lowering the gas flowrate.

In initial experiments using 1-inch cylinders, it was observed that they would turn on their side and float, sink, or circulate at flow velocities less than V_{mf}. It is of interest that results for the fixed-bed region, where difficulty in repeating results would be expected, were found to be reproducable provided that experimental conditions were carefully repeated. For example it was found that the bed had to be completely mixed top to bottom from time to time to overcome segregation of the bed particles which affected the position of the cylinders.

It can be observed from Figure 13 that, for all superficial gas velocities, cylinders of density 38.5 lb/ft^3 and 66.9 lb/ft^3 floated on the surface of the bed. It is not shown on the graph, but these cylinders would return to the surface from any position and float provided V_s was greater than V_{mf}. Experimentally they were placed at various positions under the bed surface (near the support, etc.) and they would return to the surface in less than one minute.

Cylinders with density 79.7 lb/ft^3 would float most of the time at $V_s > V_{mf}$, but occasionally they would submerge and then reappear. If these cylinders were pushed about three inches below the surface at $V_s = 0.56$ ft/sec they would surface in from one to twenty minutes. At $V_s = 0.65$ they would surface in less than one minute if they were initially three inches below the surface. If placed at the bed support (12.6 ") these cylinders would not surface after 20 minutes at any V_s up to 0.75 ft/sec. Recall that the bed bulk density at rest was 94.1 lb/ft^3, it was 94.1 lb/ft^3 at V_{mf}, and decreased to 90.7 at $V_s = 0.65$ ft/sec. It follows that as V_s was increased from 0.56 to 0.65 ft/sec the density difference between the bed and the particles ($\Delta \rho$) decreased from 14.4 to 11 lb/ft^3. This could account for the greater tendency to mix as V_s was increased. It is also of interest that $\Delta \rho$ had to be over 10 lb/ft^3 before the particles would float, whereas in a liquid this difference is very small. Cylinders of density 83.4 lb/ft^3 floated and mixed in the top portion of the bed up to $V_s \sim 0.65$ ft/sec and then sank as V_s was further increased. The behavior of these cylinders was not simple. For example the cylinders would occasionally float to the surface if placed six inches below the surface at $V_s = 0.65$ ft/sec; however if placed on top of the bed at this V_s they would ultimately sink to the bottom. This could be explained by the "Gulf Stream" circulation in fluidized beds described by J. F. Davidson [10], and Talmar and Benenati [11], and by the density difference between the fluidized bed and the cylinders, which is the driving force for floating. Cylinders of this density were very sensitive to the quality of fluidization and their behavior is not adequately described in Figure 13. However the lower limit of $\Delta \rho$ of about 7 lb/ft^3 for sinking irreversibly in the bed should be noted.

Cylinders of density 86.7 lb/ft^3 floated and mixed in the upper portion of the bed until V_{mf} was reached, and then sank irreversibly at higher V_s. At the higher V_s values these cylinders would occasionally float for a few minutes then sink and not reappear. Once submerged at any depth with the gas superficial velocity above V_{mf} these particles would not reappear and float at the top of the bed.

Cylinders of density 92 lb/ft^3 and higher sank irreversibly to the bottom of the bed as V_s was increased. Similar results were obtained for 1/2-in. and 2-in. cylinders. The effect of diameter of the cylinders on their behaviour in the bed is shown in Figure 14, which shows that for approximately the same cylinder density, the small cylinders sank more readily in the sand bed than the larger cylinders. Clearly if the particle diameter were further decreased the behavior would be described as in the work of Chen and Keairns [12].

Studies were also made with 1-inch particles in glass beads. These glass beads had a particle density of 152 lb/ft^3 and bulk density 93.4 lb/ft^3 (measured as described above for the sand bed). Results are identical to the results for cylinders in the sand bed for the glass beads [13] if V_s in the abscissa is replaced by V_s/V_{mf}. V_{mf} for the glass beads was 0.07 ft/sec.

One general observation from the above discussion is that particles in fluidized beds will float only if their density is much less

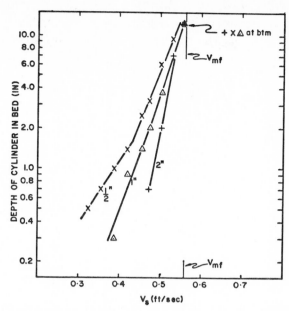

Fig. 14 Depth to which cylinder sinks in bed (measured from lowest side to bed surface) versus V_s for ½", and 1" and 2" cylinders with densities 164, 169 and 169 lb/ft³ respectively.

than the bed bulk density. Furthermore bed bulk density decreases as V_s is increased, so that the tendency to float is decreased as the superficial gas velocity is increased. Particles which have a density above a certain value (86.7 lb/ft³ in this study) simply sink irreversibly as V_s is increased, and particles within a certain range of densities will float, mix or follow "gulf stream" circulation.

Part III Forced Circulation of Cylinders in an Air-Sand Bed

Circulation of 1-inch diameter cylinders in the fluidized sand bed was promoted by an internal Plexiglas divider, one quarter of an inch thick, eight inches high and nine inches wide cemented to the inside walls of the cylinder with the lower edge two inches above the porous plate forming Columns A and B as shown in Figure 15. An auxiliary air flow two inches above the porous plate entered the smaller column (A) in a central position. The total amount of sand in the bed was 37.9 pounds, with sand level with the top of the divider.

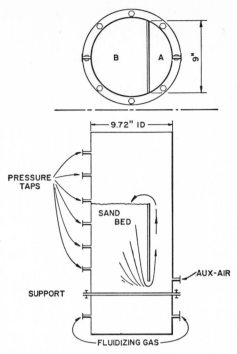

Fig. 15 Schematic diagram showing partition and circulation in sand bed.

As shown in Table 1 only a small auxiliary air flow was required
to induce circulation of sand at air velocities above V_{mf} (0. 56 ft/sec).
The higher the flow rate of the fluidizing gas the lower was the auxiliary
flow required to start effective circulation of sand. A total superficial
velocity of over 0. 82 ft/sec in column A was required to start circulation
The flowrate in column A was made up of its portion of the main
fluidizing gas plus the auxiliary flow. Once the circulation was started,
the sand flow downward in column B forced some fluidizing gas from
column B into column A under the baffle, thus increasing the total
flow of gas in column A. For the higher fluidizing gas velocity
(0. 655 ft/sec) there was violent slug flow in column A, sufficient to
circulate high density (169 lb/ft^3) cylinders.

The density range of cylinders which circulated depended on the
fluidizing and auxiliary flow rates. It can be seen by comparing
Table 2 with Fig. 13 that for the same V_s of 0. 66 ft/sec, a range

Table 2 Cylinder Forced Recirculation

Fluidizing Gas Flowrate Q	Superficial Velocity based on Q	Auxiliary Air Flowrate Qa	Auxiliary Air Superficial Velocity	Total Air Superficial Velocity	Range of cylinder Densities which circulated freely	Cylinder circulation Time Range (sec)				Bed Circulation Rate
						Side *		Centre +		
Q	V_s	Qa	V_{as}	Column A		range (sec)	mean (sec)	range (sec)	mean (sec)	
(ft^3/min)	(ft / sec)	(ft^3/min)	(ft / sec)	(ft/sec)	(lb/ft^3)					(lb/hr)
17.2	0.555	2.91	0.355	0.91	82 - 91	27 - 32	30	45 - 50	48	1100
17.2	0.555	4.31	0.525	1.08	66 - 104	12 - 18	15	17 - 23	20	2200
18.9	0.610	2.91	0.355	0.97	68 - 102	16 - 19	18	18 - 22	20	1700
18.9	0.610	4.31	0.525	1.14	65 - 112	-	-	13 - 15	14	3100
20.3	0.655	2.91	0.355	1.01	65 - 112	10 - 13	11	10 - 13	11	2100
20.3	0.655	4.31	0.525	1.18	35 - 169	9 - 11	10	8 - 11	10	3500

* Cylinder placed adjacent to partition and side of vessel with largest area
+ Cylinder placed at axis of cylinder

of 1-in. cylinders of densities from 68 - 112 as against 85-88 lb/ft^3 could be circulated for an additional air flow of only 15% for the dimensions of the forced-circulation fluidization unit. The practical significance of forced circulation of briquettes or other large bodies in a fluidized bed is the increased capacity of the bed to contain such bodies within the bed under controlled temperature and gas conditions in contrast to their floating on top of a bed and the ease of retreival in contrast to their remaining at the bottom of the bed.

The circulation time (or the time for one cycle around the divider) of the cylinders was short, reflecting the high sand circulation rates (which were measured independently). For example at Q = 20.3 , Q_a = 4.31 ft^3/min (V_s = 0.655, V_{as} = 0.525 ft/sec) the circulation of sand was 3500 lb/hr: the sand residence time based on the total weight of sand in the bed was [37.9 x 3600 / 3500 =] 39 sec, whereas the cylinder circulation time was 8-10 sec, indicating that not all of the sand was circulating, as was the case. Under most conditions the sand in column B was divided into a stagnant and a circulating portion with stagnant sand near the wall opposite the divider and increasing downward flow of sand near the divider. There was apparently no such division in column A which was visually observed to be well mixed.

CONCLUSIONS

Three variations of the method of determining V_{mf} by plotting pressure drop in a portion of the bed vs fluidizing gas V_s for a sand fluidized bed have indicated that determination of V_{mf} can be made with more precision than the conventional one of plotting pressure drop through the whole bed against V_s. These variations are: (1) graphical comparison of the depth-normalized pressure drop through a small fraction at the top of the bed against that of the whole bed to indicate V_{mf} by the divergence of these curves; (2) amplification of pressure-drop for the top of the bed plotted against V_s to give a well-defined inflection at V_{mf}; (3) plotting of the ratio of pressure drop through a small fraction at the top of the bed (say 10%) to that of the whole bed against V_s.

The third variation also promises to characterize the type of fluidized bed under a variety of conditions such as initial compaction, particle shape, size distribution and changes in the character of a bed during fluidization such as segregation.

The sensitivity of the circulation behaviour of the cylinders to their density in the sand bed relative to the prevailing bulk density of the bed suggests that this feature may be of considerable use in controlling the residence time for bodies of uniform size and shape such as briquettes or pellets in a fluidized bed of inert material. For example in the carbonization of coal briquettes, by choosing a bed of material of a density close to but denser than, the carbonized (less dense) briquettes, they could be made to rise to the surface as they decreased in density.

It has been shown that, for a small secondary flow of fluidizing gas a large sand circulation is obtained in a fluidized bed with an internal partition. This flow permits a controlled circulation of large bodies over a wide density range relative to that of the bed.

NOMENCLATURE

A	Cross sectional area of column used, 0.515 ft^2, 0.0479 m^2
h	Height of sand bed above support, ft (1 ft = 0.3048 m)
ID	Column internal diameter, 9.72 inches, 0.245 m
ΔP	Pressure drop, psi (1 psi = 6895 nt/m^2)
ΔP_o	Pressure drop measured from just above the support to the top of the bed, psi. (1 psi = 6895 nt/m^2)
$\Delta P_1, \Delta P_2, \Delta P_3$	Pressure drop measured from position 1, 2, or 3 respectively (Figure 2) to the top of the bed, psi.
P_w, P_{w_o}	Weight of sand in bed divided by 144A, psi.
$P_{w_1}, P_{w_2}, P_{w_3}$	Weight of sand above tap 1, 2, or 3 respectively divided by 144A, psi. Measured when bed is just fluidized.
V_s	Superficial gas velocity, ft/sec (1 ft/sec = 0.3048 m/sec)
V_{mf}	Minimum V_s at which bed of sand is just fluidized
x	Distance defined in Figure 2
ρ_b	Bed bulk density, lb/ft^3 (1 lb/ft^3 = 16 Kg/m^3)
$\Delta \rho$	Density difference between cylindrical particles and bulk density of bed of small particles lb/ft^3 (1 lb/ft^3 = 16 Kg/m^3)

REFERENCES

1. Canada Dept. of Energy, Mines and Resources, CANMET "Development of a Fluidized Sand Coker", Divisional reports MREC 74/4 and MREC 74/94, B.B. Pruden.

2.   Zabrodsky, S. S. , "Hydrodynamics and Heat Transfer in Fluidized Beds", MIT Press, Cambridge, Mass. (1966).

3.   Leva, M. , "Fluidization", McGraw Hill, New York (1959).

4.   Miller, O. C. , and Logwinuk, A. K. , Ind. Eng. Chem. , 43, No. 5, 1220 (1950).

5.   Van Heerden, C. , Nobel, A. P. P. , and Van Krevelen, D. W. , Chem. Eng. Sci. 1, No. 1, 37 (1951).

6.   Frantz, J. F. , Chem. Eng. Progr. Symp. Ser. 62, No. 62, 21 (1966).

7.   Narishiman, G. , A. I. Ch. E. J. , 11, No. 3, 550 (1965).

8.   Wen, C. Y. and Yu, H. Y. , Chem. Eng. Prog. Symp. Ser. 62, No. 62, 100 (1966).

9.   Parent, J. D. , Yagol, N. , and Steiner, G. S. , Chem. Eng. Prog. 43, 429 (1947).

10. Davidson, J. F. , AIChE Symp. Series, 69, No. 128, 16 (1973).

11. Talmar, E. , and Benenati, R. F. , A. I. Ch. E. J. , 9, No. 4, 536 (1963).

12. Chen, J. L. P. and Keairns, D. L. "Particle Segregation in a Fluidized Bed", Can. J. Chem. Engr. , 53, 395-402 (August 1975).

13. Crosbie, D. , M. A. Sc. , Thesis, Dept. of Chemical Engineering, University of Ottawa, Ottawa, Ontario, Canada.

EXPERIMENTAL STUDIES AND THE INDUSTRIAL PROSPECTS OF THE SCREEN PACKED BATCH AND COUNTERCURRENT GAS-SOLID FLUIDISED BED

G. CLAUS, F. VERGNES and P. LE GOFF

The conventional gas-solid fluidised contactor has serious draw-backs : gas by-passing through bubbles, difficulties in scaling-up, small height to diameter ratios and short contact times. Many authors (1-6) proposed a solution to these problems : the filling of the bed with a fixed wire screen Raschig ring packing of high voidage. In these conditions, a pseudo-particulate fluidisation occurs (4,8), allowing a good expansion very similar to that of a liquid fluidised bed. Moreover the most useful properties of the fluidised bed : its thermal properties are relatively well preserved (7). As a first step, towards forecasting the industrial applications of this reactor we have studied its hydrodynamical behavior in the two cases of the batch and the continuously fed system.

EXPERIMENTAL SYSTEM

Packing and particles : The only packing studied is a random stack of open ended screen cylinders with a mean voidage of 0.972. The diameter and the length of each ring is 2 cm, the mesh opening 1.04 mm and is made from stainless steel wire with diameter of 0.5 mm. The properties of the particles are listed in the table 1.
The batch and the continuous apparatus are described in appendix.

RESULTS AND DISCUSSION

In the batch system : we observe a very different behavior between the different materials : with our coarse particles, the fluidisation is very uniform with a lot of little bubbles uniformly distributed along the whole bed and a perfectly linear pressure profile. With the fines : glass beads and cracking catalyst, the fluidisation is very

Table 1

	dp (μ)	ρ_s g/cm^3	u_t cm/s	Re_t
Alumina 1	150	1.27	58	5.53
Alumina 2	370	1.27	173	40.6
Glass beads	58	2.40	23.8	0.88
Cracking catalyst	50	1.19	8.9	0.46
Sand	235	2.65	179	26.7
Polystyrene	424	1.05	177	49
Vegetable abrasive	362	1.38	185	53

bad with a large amount of solid agglomerating on the packing.

The relation between pressure drop and velocity is given in
Fig. 1 for the two aluminas. The ordinate 1 corresponds to the normal
pressure drop in a fluidised bed i.e. the ratio of the weight of the
bed on the section of the column : ΔP_o. It is obvious from Fig. 1
that this normal pressure drop is attained at the minimum fluidisation velocity, but on increasing the velocity, the pressure drop decreases down to 80 % of this normal value for a fluidising velocity
10 to 20 times greater than this which corresponds to the incipient
fluidisation. Very similar curves are obtained with the other well
fluidised materials. This odd behavior shows already that a fundamental role of the packing is in supporting the grains. Fig. 3 gives the
expansion of the bed as a function of the air rate. The porosity of
the bed progressively reaches very high values unexpected for a normal gas fluidised bed. The curves log ϵ vs log u become linear for
voidages greater than 0.6 and can be represented by a power law of

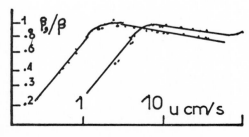

Figure 1.

the Richardson and Zaki type : $u_g/u_i = \epsilon^n$ 1)   where u_i is
the extrapolated value of u_g for $\epsilon = 1$. Capes and cw. (8) have proposed some relations giving the parameters u_i and n from the Reynolds
numbers of the particles under free fall conditions R_t. Our results
agree well with these relations for the largest particles, but not
for the finest. This is not surprising, the formula proposed in 8)
being established for $Re_t > 40$.

Countercurrent system : Generally in two-phase systems, the authors
choose as key variable the relative velocity defined as :

$$u_r = u_g/1{-}\beta + u_s/\beta \qquad \text{where } \beta = 1 - \epsilon$$

If the voidage is a function of this velocity only as in the liquid
fluidised systems (9,10) equation 1) becomes : $u_r/u_i = \epsilon^{n-1}$
The Fig. 2 related to the sand shows that it is not true : in logarithmic coordinates, the curve is neither linear nor independent of
the flowrate of the solid. On Fig. 3 we have plotted the volumic
hold-up β as a function of the superficial velocity of gas, for different flowrates of solids. On each curve and for each gas velocity
two working points occurs : one with low, the other with high solid
concentration : the dilute bed and the dense phase bed. This behavior
is very similar to many other countercurrent two-phase flows, such as
liquid-liquid or gas-liquid (11). All these facts lead us to assumme
that the bed is not supported by the gas alone but also by the packing. The balance of forces over unit volume can be written :

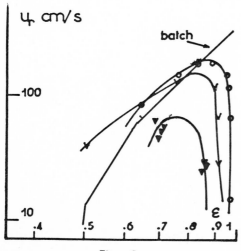

Figure 2.

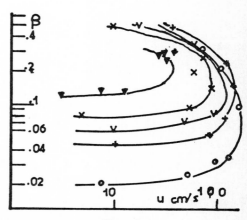

Figure 3.

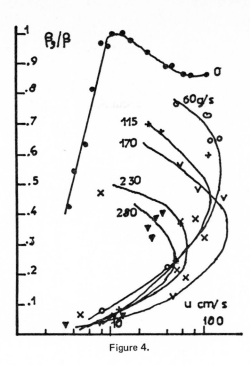

Figure 4.

$$\beta \ (\rho_s - \rho_g) \ g = - \ dp/dz - F_g + F_s \qquad 2) \quad \text{where the}$$

friction on the packing is called F_s for the solid and F_g for the gas this last being often negligible. Dividing by $(\rho_s - \rho_g)$ g the relation 2) is rewritten

$$\beta = \frac{- \ dp/dz - F_g}{(\rho_s - \rho_g) \ g} + \frac{F_s}{(\rho_s - \rho_g) \ g} = \beta_g + \beta_s$$

showing the contribution of the gas and of the packing in the supporting of the global volumic concentration. The fundamental design parameter β_g/β equal to the relative pressure drop $\Delta P/\Delta P_o$ is plotted vs the gas flowrate on the Fig. 4. All increase in the flow of the solid or in the dilution of the bed leads the packing to support the greater part of the grains. These results allow us to forecast which conditions and additional studies will be required for a correct use in industrial operations of the three kinds of screen packed fluidised beds : batch, dense countercurrent bed, dilute countercurrent bed. Firstly in all cases a good fluidisation requires larger particles than for a conventional fluidised bed. High H/D ratios will be possible and favourable to a sharp residence time distribution.

In batch beds, it is theorically possible to look upon all exothermic

catalytic reactions as partial oxidation of hydrocarbons. Sharp RTD
with a good flexibility in the contact time, high wall heat transfer
and thermal diffusivity can lead to a favourable compromise between
fixed and ordinary fluidised reactors. Such a plant should be simpler
than other possible competitor : the cocurrent transported bed. The
great advantage of the dilute phase bed is its very low pressure drop
in some cases smaller than 1 % of the mass of contacting solid. Such
a device can advantageously compete with the multistage fluidised
bed, for instance in adsorption of pollutants from flue gases. But
for a such countercurrent operation, the H.E.T.P. remains to be
determined.
The countercurrent dense phase bed should be very useful for the
continuous treatment of solids such as the reduction of oxides or
regeneration of used adsorbents.
But if the "chemical" behaviour of the screen packed fluidised bed
seems simpler than that of the conventional fluidised bed, its
hydrodynamical behavior is conditioned by many more factors such as
the geometry of the packing in relation to the size of the grains,
the forces between packing and particles which varying with the
flowrates. Much work is yet necessary for the optimal design of a
such vector.

APPENDIX

Experimental System

Batch apparatus. The batch apparatus is a very conventional Perspex
column 92 mm I.D. and 4.5 m height, with a disengaging zone and a
recovery system for the elutriated fines. Pressure taps distributed
at 0.5 m intervals and an orifice on the feeding line allow the
pressure profile and the air flow rate to be measured. The voidage
is deduced from the height of the bed.

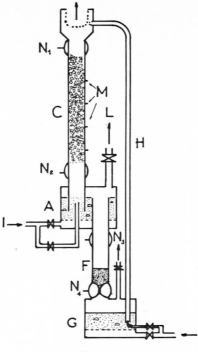

Figure 1a.

Continuous apparatus. Fig. 1a shows the principle of the system :
the packing is stacked in the column C (92 mm I.D. 3 m height) between two straight crossing valves N_1 & N_2 whose simultaneous closing
allows the hold-up of the column to be measured. The column is fed
continuously from the classical fluidised bed G by the air lift H
and drained off by the fluidised siphon A. The air-lock F between
valves N_3 & N_4 allows the flow rate of solid falling back in the bed
G to be measured without air loss. The superficial velocity in the
column is given by a volumetric balance between the inlet I and the
outlet L, and the pressure profile by six pressure taps M.
Some additional details are given in Ref. 12.

ACKNOWLEDGEMENT

        One of us (G.C.) is very grateful to ELF - ERAP Co for finan-
cial support during this work.

BIBLIOGRAPHY

1) T. ISHII & G.L. OSBERG, A.I.Ch.E.J., 11 (1965) 279

2) W.K. KANG & G.L. OSBERG, Can. J. Chem. Eng., 44 (1966) 162

3) J.P. SUTHERLAND, G. VASSILATOS, H. KUBOTA, G.L. OSBERG, A.I.Ch.E.
     J., 9 (1963) 437

4) B.H. CHEN & G.L. OSBERG, Can. J. Chem. Eng., 45 (1967) 90

5) W.K. KANG, J.P. SUTHERLAND, G.L. OSBERG, I. & E.C. Proc. Des.
     Dev., 6 (1967) 499

6) C.E. CAPES & A.E. Mc ILHINNEY, A.I.Ch.E.J., 14 (1968) 917

7) C.E. CAPES, J.P. SUTHERLAND, A.E. Mc ILHINNEY, Can. J. Chem. Eng.,
     46 (1968) 473

8) C.E. CAPES & A.E. Mc ILHINNEY, Trans. Instn. Chem. Engrs., 50
     (1972) 1

9) T.S. MERTES & H.B. RHODES, Chem. Eng. Progr., 51 (1955) 429, 517

10) B.G. PRICE, L. LAPIDUS, J.C. ELGIN, A.I.Ch.E.J., 5 (1959) 93

11) G.B. WALLIS, "One dimensional two phase flow" Mc Graw Hill edi-
     tor, 1969

12) G. CLAUS, Thesis Nancy 1974

A STUDY OF THE ACTION OF FLOATING BUBBLE BREAKERS IN FLUIDIZED BEDS BY INTERACTIVE COMPUTER GRAPHICS

S. A. KEILLOR and M. A. BERGOUGNOU

Many commercial processes in the petroleum, mining and chemical industries rely heavily on the use of the fluidized bed. In such reactors the presence of large gas bubbles may lead to a reduction in gas particle contact and thus to process inefficiency. In large reactors it is frequently necessary to enhance the contacting of the components by increasing the bubble population through the use of fixed baffles within the bed. However, no generally satisfactory design for such reactor internals has yet been developed.

C. E. Capes of the National Research Council of Canada studied the use of a fixed helical packing material as a means of improving bed efficiency through bubble breakup. It was noted that, under certain conditions, the packing material would begin to move within the bed. This observation prompted the concept of the mobile or floating bubble breaker. Such bubble suppressors freely moving within a bed could be easily added or removed as dictated by the behaviour of the bed.

In evaluating the feasibility of the mobile bubble breaker concept the following physical parameters were considered:

    1)   the ratio of bubble breaker to bed density

    2)   bubble breaker shape

    3)   bubble breaker size.

The initial study was conducted in a thin two-dimensional fluidized bed of sand and air. A 16 mm photographic record was made of each experiment. The filmed action was then analyzed, initially by hand, and later by means of a moderately sophisticated interactive computer graphic technique specifically developed for this purpose.

FLUIDIZED BED

To facilitate observation of the dynamics of the bubble population and the affect of the bubble breakers all experimental work was conducted in a thin two-dimensional plexiglas bed. The bed, illustrated in Figure 1, was constructed of two aluminum bars, one inch thick (2.54 cm) and of square cross-section sandwiched between two one half inch thick (1.27 cm), nine and one half feet long (290 cm) and three feet wide (91.44 cm) plexiglas plates. This configuration provides for a bed thickness of one inch (2.54 cm). An expansion section, four feet long (122.0 cm) and four feet high (122.0 cm) was mounted on top of the bed to de-entrain particles carried over. A windbox, below the grid, dissipated the momentum of the entering gas. The grid, a perforated stainless steel plate, was covered with a 400 mesh stainless steel screen

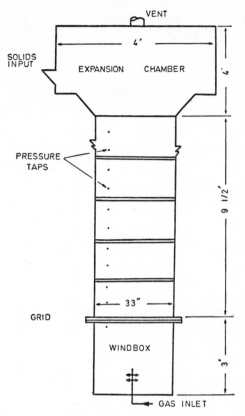

Fig. 1 Fluidized bed.

which prevented the bed particulate from falling into the
windbox during bed shut down. The number and the diameter
of the holes in the grid plate were chosen so as to have
a pressure drop across the grid sufficiently high to ensure a uniform distribution of air over the bed cross-
section.

Pressure taps located at six inch intervals were connected to manometers containing either Mericam fluid
(sg 1.0) or mercury.

A one inch square reference grid was drawn on the

front plate of the bed as an aid to the later frame by
frame analysis.

BED PARTICULATE

Pure white Ottawa sand, having a measured bulk density of ninty-nine lb/ft^3 (1587 kg/m^3) was sieved to obtain particles in the size range of 500 to 595 microns.
A static bed height of sixty-one inches (154.9 cm) applies
to all experiments.

CONTACTOR FABRICATION

Fabrication of the floating bubble breakers was initially restricted to those which would be easily and in-
expensively constructed in the laboratory. The two-
dimensional nature of the fluidized bed placed substantial
restrictions on the possible dimensions of the breakers.
A thickness slightly less than one inch (2.54 cm) but
slightly greater than one-half inch (1.27 cm), was required if the contactors were to fit in the bed and at the
same time avoid the problem of bridging or jamming of two
or more contactors within the bed.

Table 1

BUBBLE BREAKER SHAPE	DIMENSIONS
Cylindrical	1 1/2 inch O. D. (3.81 cm) 5/8 inch thick (1.59 cm)
Rectangular	4 inches long (10.16 cm) 5/8 inch thick (1.59 cm) 2 inches wide (5.08 cm)

Two basic contactor shapes were constructed:
cylindrical and rectangular.

Construction of Cylindrical Contactors

Construction of Cylindrical Contactors

Five eighth inch (1.59 cm) sections of one and a half
inch (3.51 cm) outside diameter acrylic tubing were filled
with measured amounts of reduced iron powder mixed with a
liquid synthetic resin. The variation in contactor density was achieved through the addition of the reduced
iron powder (Figure 2).

Construction of Rectangular Contactors

Aluminum dishes were used as molds for the large
rectangular contactors. The forms were filled with the

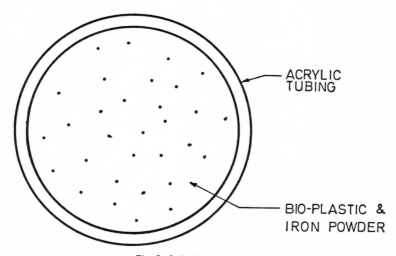

ACRYLIC
TUBING

BIO-PLASTIC &
IRON POWDER

Fig. 2 Cylindrical contactor.

casting plastic and measured amounts of the iron powder.
In general, control of the thickness was less reliable in
the rectangular molds and several problems with the bridging of contactors within the bed did occur.

FILM EQUIPMENT AND RECORDING TECHNIQUE

Sixteen millimeter motion pictures of the visible
bubbles in the bed were taken with a Bolex H16 reflex cine
camera at a speed of 16 frames per second. The camera,
equipped with a wide angle lens was positioned to face the
central part of the bed. Due to the difficult lighting
conditions a 16 mm black and white high speed negative
film was used.

Illumination was provided by several 750 watt flood-
lights. For each experiment two film records were made.
The first used weak front lighting and strong through the
bed back lighting. This provided sufficient illumination
 to enable the viewing of the surface grid and gave good
definition for larger bubbles. However, front illumination made the definition of the smaller bubbles difficult.
Keeping the camera positioned as for the front of the bed
illumination a second record was made using only the
strong through-the-bed rear illumination.

ANALYSIS TECHNIQUE

The quantity and density of experimental information
contained on several hundred feet of 16 mm film places

considerable strain on the data handling and analysis
task. The initial attempt to extract information manually
was later replaced by a computer assisted digitization of
the film information.

Manual Data Analysis

The analysis of the collected film information was
initially conducted using a LW Photo Optical Data Analyzer
having a variable frame rate. Initially the photographic
image was projected onto large sheets of graph paper. The
projected bubbles were then traced out by hand and various
parameters (area, velocity) determined.

This manual technique is a tedious time consuming
exercise especially if information on consecutive frames
is to be correlated.

Interactive Computer Graphics

Experimental information held by visual storage media
such as 16 mm film generally gives rise to considerable
data handling problems. To facilitate the digesting and
digitization of the data an interactive computer graphic
data collection and analysis system has been and continues
to be developed.

The system is configured around a Digital Equipment
PDP-12 computer having 16 k of core, two floating head
disc packs, an IBM compatible magnetic tape unit and a
floating point processor. This is interactively coupled
to a Computek 450 Graphics Display console and digitizing

Fig. 3 Interactive Computer Graphic Facility.

tablet (Figure 3).

Prior to digitizing the projected images a user is
requested to enter via the tracking pen, tablet and console keyboard the appropriate scale factors (Figure 4).
Once this information is available to the computer the
user can begin the task of digitizing via the tracking pen
the images projected on the tablet.

As each bubble is traced out the X-Y co-ordinate positions of the tracking pen are continuously sampled. As
each set of co-ordinates (X, Y position on the tablet) is
received it is stored twice; once in the memory field
using the Y co-ordinate as the address and the X co-
ordinate as the data and once in field 2 using X as the
address and Y as the data. In field 3 a series of

counters is set up to record the number of times an individual Y co-ordinate is used as an address. Also, the
number of times that co-ordinate has already been used as
an address is noted. If this is odd, the entering X co-
ordinate is subtracted from the contents of the memory
location; as specified by Y; if it is even, the new X is
added. A similar process is carried on in field 2 with
the co-ordinates reversed.

On completion of the tracing out process field 3
will contain the incremental areas $\pm$ X dy, each stored at
location Y, the sign of the areas being dependent upon a
clockwise or counterclockwise direction of drawing.

The total area or bubble volume is then determined
by summing the absolute value of the incremental areas.
The centre of gravity is calculated by summing the products of the incremental areas times their locations in
memory and dividing by the total area.

A similar process is carried out in field 2 to determine an area and centre of gravity with respect to the
X axis.

These results are then scaled as initially indicated
by the user.

The results are printed as shown in Figure 4.

Additional software to determine bubble velocity,
direction and coagulation and breakup is presently being
developed.

All information relating to the experiment is stored
on magnetic tape for future processing.

TOUCH THE PEN TO TWO POINTS ON THE TABLET

ENTER X AND Y VALUES

AREA[1]	ERROR[2]	DELTA X[3]	DELTA Y[4]
.377	.001	3.463	3.498
.843	.009	.202	.779

[1]average of the two areas calculated

[2]difference between the two areas calculated

[3]location of centre of gravity relative to X axis

[4]location of centre of gravity relative to Y axis

Fig. 4 Computer output.

EXPERIMENTAL WORK

The action of the bed was observed over a superficial velocity range of 1.0 ft/sec (32.0 cm/sec) to 2.3 ft/sec (71.0 cm/sec). Based on the observation of the bubble patterns as to

1) bubble size

2) length of bubble track (3 ft. [91. cm])

3) bubble definition

4) non-excessive coagulation of bubbles within one
    and a half feet of the bed surface

all experimental work was conducted at a superficial velocity of 1.7 ft/sec (51.5 cm/sec). For this particular gas velocity the fluidized bed was observed to have a density of 87.9 lb/ft^3.

Contactor Density

The effect of the ratio of contactor to bed density was the first of the physical parameters to be considered. The action of three contactor density ratios, .83, .95 and 1.3, were initially observed.

$$\text{density ratio} = \frac{\text{contactor density}}{\text{bed density}}$$

Contactors of the ratio 1.3 fell slowly through the fluidized bed and finally came to rest on the surface of the supporting grid. However, during their descent they were observed to exhibit a positive bubble breaking potential.

Contactors having a density ratio of .83 were observed to float or ride within the fluidized media and at no time were they noted within the bottom two feet of the bed. However, the lighter contactors did not demonstrate as strong a bubble breaking effect. The contactor appeared at best to slice off the side of the bubble and at worst, for larger bubbles, to ride around the outer surface of the bubble.

Bubble breakers having a ratio of .94 appeared in general to ride slightly lower in the bed and exhibited an increased breaking potential over the lighter contactors. On occasion the edge slicing effect was noted, especially for some of the larger bubbles, but in general the surface riding effect was less prominent and a moderately satisfactory breaking potential was noted.

Contactor Shape and Size

A study of the effect of contactor shape was restricted due to the limitations of contactor fabrication.
For this reason, as well as the time and cost factors,
two fairly simple shapes were selected: cylindrical and
rectangular (See page 5 for details).

The cylindrical contactors, having approximately the
same density as the bed, functioned fairly well. However, the 'riding' phenomenon, which did appear on occasion, suggested that a rougher contactor surface, or perhaps contactors having protruding side arms which could
penetrate the bubble surface might be more effective.
This latter suggestion however was not pursued·for two
reasons. Firstly, the difficulties associated with the
casting of such shapes. Secondly, contactors suspended
in the bed are subjected to vigorous interactions and observation of damage to some of the cylindrical contactors
suggested that plastic breakers with extended arms would
not be sufficiently strong to withstand the interactions
within the bed.

The very large rectangular contactors did not
exhibit the 'riding' or 'side slicing' effects of the
smaller cylindrical breakers. They tended to fall through
the bubbles on impact thus splitting them in two. However, due to the configuration of the experimental equipment it was very difficult to add or remove these larger
contactors. A careful control of the contactor's thick-

ness was essential to:

a)   permit removal of the contactors from the bed

b)   prevent the wedging or bridging of contactors

Unfortunately, as the size of these contactors was at the upper limit for use within the bed, problems associated with contactor jamming in the bed were frequently encountered.

Bubble breaking potential

Using the cylindrical contactors, successive experiments were run in which the ratio of the volume of the contactors to the volume of the bed was increased at the rate of approximately 1%. At the upper limit, a contactor volume equivalent to approximately 20% of the volume of the fluidized bed was observed. At this limit 156 cylindrical bubble breakers of approximately the same density as the fluidized bed had been added. The effect of these mobile breakers on the bubble population is presented in Figure 5.

Film data suggests that the presence of the mobile breakers results in a noticeable increase in the raining of particulate matter within the bubbles. Methods of evaluating this parameter are currently being investigated.

SUMMARY

The results of the preliminary investigation has supported the basic concept of a mobile bubble breaker or

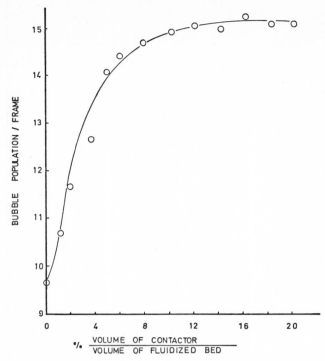

Fig. 5 Effect of contactors on bubble population.

contractor and suggests several areas for future research. It is essential that the subsequent experimental work be conducted in a three-dimensional system in order that the problems associated with the wall affects of the two-dimensional system be overcome.

The presence of the floating contactor leads to a substantial increase in bubble population up to approximately 16% volume of contactors to volume of fluidized bed where it appears that a leveling out of the effect begins to take place.

The studies investigating the effect of variations in the density ratio suggest that there exists a narrow

range of desirable densities and that the optimum is approximately equivalent to that of the fluidized bed.

Visual observation of the respective actions of the
cylindrical and rectangular contactors should prompt further investigation into the possible increase in bubble
breaking efficiency which may be associated with the
contactor shape.

Work is continuing on the development of the interactive computer graphic facilities. The software for
bubble analysis is being extended to provide for the determination of bubble velocity and direction. Consideration is presently being given to the incorporation of a
computer compatible scanning microdensitometer to determine the extent of raining of the particulate matter
within the bubbles.

PART II

Solids Mixing and Transport

THE FLOW OF FLUIDIZED SOLIDS

R. D. LaNAUZE and J. F. DAVIDSON

A large number of fluid-bed processes require the bulk transfer of fluidized solids. In these systems it is important to be able to predict the circulation rate of the bed material. This paper outlines the use of a draft tube apparatus for studying induced circulation and presents experimental results which support a simple "two-phase" model for predicting the circulation rate in which the energy dissipation occurs as a particle shear loss at the walls of the bed.

THE DRAFT TUBE

The draft tube assembly is shown in Fig.1. The solids are transported up the tube by bubbles or slugs, and because of the density difference between the tube and the annulus, the solids flow. The flow of solids down the annulus is largely bubble-free. Circulation is aided by an independent air supply, U_a, to the base of the annulus to avoid choking ($U_o < U_a < 1.5 U_o$). Keairns et al.[1] give considerable emphasis to such a system for fluidized combustion of coal in which the solids are transported up the tube as a dilute phase. We operate the bed at lower gas velocities and expect the behaviour in the riser to be that of a bubbling or slugging bed and in the downcomer that of a bed close to the incipient fluidizing velocity, U_o. We argue that the circulation rate is a balance between the density driving force between the annulus and draft tube and the particle shear loss at the wall. Calculations based on the flow of solids in the captive wake, following arguments by Rowe[2], did not produce satisfactory results for the flow rate of particles down the annulus, since wake shedding and circulation of particles may occur within the draft tube.

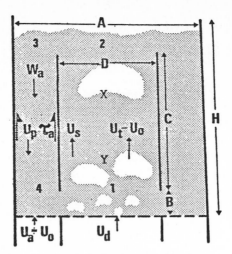

Fig. 1 Draft tube apparatus.

DEVELOPMENT OF THE MODEL

Consider the system in Fig.1: around the circulation path 1,2, 3,4,1 the pressure drop is zero. Experiments show that $P_2 = P_3$ and it is also assumed that the pressure is not recovered when the particles lose kinetic energy, i.e. $P_4 = P_1$. Assuming a wall shear loss we can sum the pressures around the circulation path to give

$$\rho_o g H (\varepsilon_d - \varepsilon_a) = \frac{\tau_a S_a}{A_a} + \frac{\tau_d S_d}{A_d}$$

where ρ_o is the bed density at U_o, ε_a the <u>bubble</u> voidage in the annulus, cross-section A_a and total wall area S_a over which the particle-wall shear stress is τ_a. ε_d is the <u>bubble</u> voidage in the draft tube of cross-section A_d and wall area S_d. For the entrance section into the draft tube we have assumed that the pressure loss due to particle acceleration can be neglected. Assuming that the solids flow down the annulus close to incipient fluidization, then $\varepsilon_a = 0$. The bubble voidage ε_d may be predicted by re-arranging an equation due to Matsen[3] based on the velocity of a slug rising relative to a moving particulate system,

$$\varepsilon_d = \frac{(U_t - U_o) - U_s \frac{\varepsilon_o}{1 - \varepsilon_o}}{(U_t - U_o) + U_s + u_b}$$

in which U_s is the mass flux of particles in the draft tube/particle

density, U_t the total superficial gas velocity in the tube
including the incoming gas entrained in the emulsion phase, and u_b is
the absolute rising velocity of an isolated slug in a non-flowing
system, $0.35\sqrt{gD}$. U_s is related to the particle flowrate in the
annulus by a mass balance. The mass flux, W_a, in the annulus at any
given air velocity supplied to the base of the draft tube, U_d, may be
obtained by solving the above equations provided the wall shear is known
as a function of particle velocity. Several relationships for shear
stress have been proposed[4] but these are not yet general and the shear
stress has to be measured and is expressed here as a function of
particle velocity.

EXPERIMENTAL

The dimensions of the bed of circular cross-section are given in
Table 1 and correspond to letters in Fig.1. Particle properties are
also listed. Particle velocities were measured by a radio pill[5] and
the shear stress by suspending a plate from a calibrated spring in the
downflowing, bubble-free solids.

RESULTS AND DISCUSSION

(i)   The particle-wall shear stress is plotted against particle
velocity on Fig.2. The results are of a similar magnitude, though
consistently higher than those reported[4] for horizontal flow of a
similar sand in an open channel. At high flowrates the method of
measuring the shear stress became inaccurate as the plate had a
tendency to oscillate and leave the vertical. A floating wall section

Table 1

$A = 0.30$ m
$B = 0.051, 0.076, 0.102, 0.28$ m
$C = 0.6, 1.2$ m
$D = 0.15, 0.20$ m
$H(\text{at } \varepsilon_o) = B + C + 0.08$ m

Sand: mean dia. 168 μm
      $U_o = 0.03$ m/s
      $\varepsilon_o = 0.47$
      $\rho_s = 2630$ kg/m^3

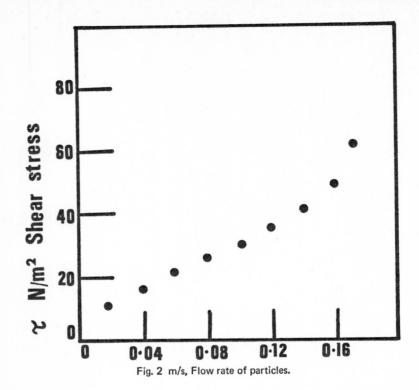

Fig. 2 m/s, Flow rate of particles.

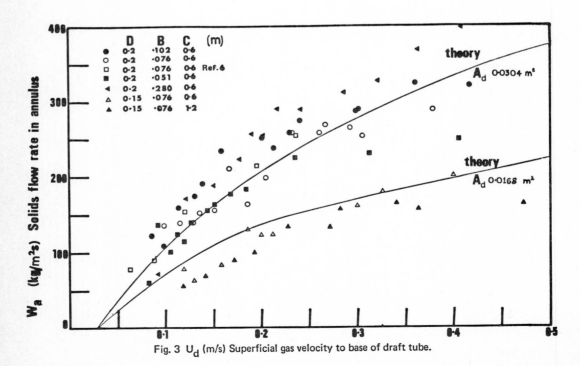

Fig. 3 U_d (m/s) Superficial gas velocity to base of draft tube.

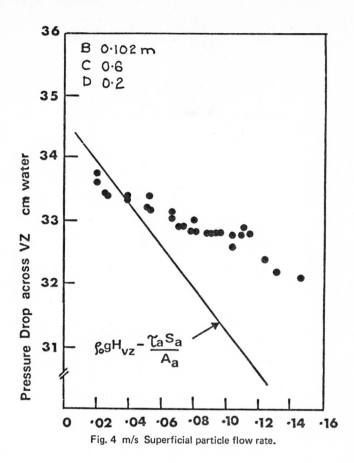

Fig. 4 m/s Superficial particle flow rate.

is being built to overcome these difficulties. However, the suspended
plate is expected to give a reasonable representation of the shear
stress as the velocity profiles in the annulus are reasonably flat and
different length plates gave similar results, indicating that the
boundary layer was established quickly.

(ii) Fig. 3 shows W_a against U_d for draft tube areas 0.0168 and
0.0304 m^2. The results show a steady increase in circulation rate with
U_d which the model predicts reasonably well. The more gradual increase
in W_a at higher air velocities is a reflection of increasing voidage in
the tube which is not matched by a corresponding increase in U_s.

Keairns et al[1] were able to correlate W_a in terms of (P_4-P_1). In
our analysis, based on earlier experiments[6], we assumed that (P_4-P_1) is
zero. Our experiments show that a pressure drop of a few cm of water
exists but we have so far been unable to relate this in the same

manner to the circulation rate. Keairns et al[1] were unfortunately
unable to vary A_a and the gap height B both of which would affect
(P_4-P_1). Our experiments (Fig.3) show that the gap height B influences the circulation rate. On the other hand increasing the height of
the tube at constant gap height does not alter the flowrate as both
head and wall area are increased proportionally.

The pressure drop across VZ in the annulus is compared with the
theoretical expression on Fig.4. The results indicate that the shear
stress has been over-estimated. For the draft tube we expect the
pressure drop to differ from the predictions as we have applied the
shear stress for a bubble-free bed to a slugging situation. Clearly
this is inaccurate and a much more difficult analysis of shear losses
in the draft tube will have to be made, but if one considers that the
wall area/volume is less in the tube the results of this assumption
may not be serious.

REFERENCES

1. Keairns, D. et al. Evaluation of the Fluid-Bed Combustion Process
         Vol.II EPA-650/2-73-048b Dec.1973

2. Rowe, P.N. Chem.Engng Sci. 1973, __28__, 979.

3. Matsen, J.M. Powder Technology 1973, __7__, 93.

4. Bessant, D.J., Ph.D. Thesis, University of Birmingham, U.K., 1973.

5. Mitchell, F.R.G., J.Sci.Instrum. 1969, __2__, 812.

6. Jarry, I.G. and Scott, A.M. Research Project, Dept.of Chem.Eng.
         University of Cambridge, 1971.

ADDITIONAL INFORMATION

(1) Expression for Bubble Voidage, ϵ_d

In the summary paper the expression

$$\epsilon_d = \frac{(U_t-U_o) - U_s \dfrac{\epsilon_o}{1-\epsilon_o}}{(U_t-U_o) + U_s + u_b}$$

was presented. This may be derived by considering a section of the
draft tube, Figure 5, in which there is a net upward superficial
volume flowrate of particles, U_s. The total gas superficial

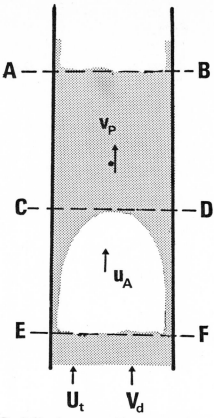

Fig. 5 Slugging system with net particle flow.

velocity U_t, including the gas entrained with the particles, is
given by

$$U_t = u_A \epsilon_d + U_o + \frac{U_s \epsilon_o}{1-\epsilon_o}$$

since $\frac{U_s}{1-\epsilon_o}$ is the particle velocity on the emulsion phase and
$\frac{U_s \epsilon_o}{1-\epsilon_o}$ the flow of gas entrained with it. u_A is the rising velocity
of the slug relative to the particle velocity, v_p, at its nose, i.e.
$u_A = v_p + u_b$. Now $v_p = U_t - U_o + U_s$, so by substituting for v_p
into the expression for u_A, and u_A into that for U_t, we obtain by
manipulation the above expression for ϵ_d.

(2) Direction of The Gas Flowing in The Annulus

The relative velocity between gas and particles in a fluid-ized bed is equal to the interstitial fluidizing velocity, u_o, that is $u-v_p = u_o$, where u_o is constant and positive in the upward direction. Depending on the magnitude and direction of v_p then u may either be positive or negative. If the downflowing particles exceed a velocity of u_o then gas will be dragged down with the particles, while if $v_p < u_o$ then the gas flows counter-current to the particles.

Merry and Davidson (7) demonstrated these two modes for gulf streaming in a large bed. By extension, the flow patterns for the draft tube circulating bed are illustrated in Figure 6. For the case where $v_p > u_o$ in the downcomer there is a region within the bed which forms a closed circulating loop of gas.

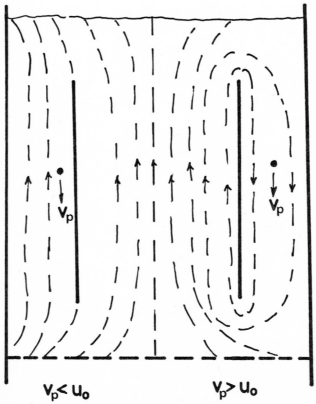

Fig. 6 Gas stream lines in a circulating fluidized bed.

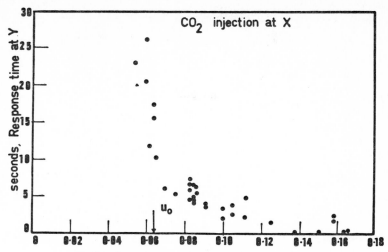

Fig. 7 m/s Downward superficial particle flow rate.

These considerations are important where two different fluidizing gases are being used for the riser and downcomer, in for example, some type of regeneration system. It would be necessary in those cases to minimise the gas interchange, which would place an upper constraint on the circulation velocity.

The direction of the flow of the interstitial gas was tested for particles flowing down the draft tube (i.e. the reverse direction in Figure 1) by injecting a tracer gas (CO_2) at point X (Figure 1) and observing the response of a katherometer at Y, 15 cm below X. Figure 7 shows the results which confirm that both regimes of gas flow occur and that the change in direction of the gas flow occurs close to a solids downward velocity of u_o as predicted.

When the particle velocity exceeds u_o and we expect a closed loop of gas, the probe registered excess CO_2 in the bed for a number of minutes after the tracer had been turned off. This indicates that the gas circulation patterns are fairly stable. Gas escapes from the loop by diffusion and bubble interchange in the riser. The gas interchange in the riser can be estimated from the theory of Davidson and Harrison (8) which predicts a value of 0.9 for the quantity X which represents the number of times a bubble is washed out by cross flow on its passage through the bed. This low

value indicates why tracer gas remains in the loop for a considerable period.

(3) Particle Circulation Rate

(i) Effect of gap between distributor and draft tube

Figure 8 plots some of the results of circulation rate
against superficial gas velocity in the draft tube in such a manner
as to emphasise the effect of the gap B. It is apparent that at
higher gas velocities the gap height limits the circulation rate.

By placing capacitance probes across the annulus, just above
the base of the tube, it was demonstrated that the particulate
phase flowing down the annulus remained at a voidage close to ε_o
and bubble free until at a certain value of U_d bubble activity
occurred. This leakage of gas up the annulus limited the particle
flowrate. A wider gap removes the tube from the distributor zone
and smooth flow again occurred with increased U_d until a new
critical gas velocity was reached. For a large gap of 28 cm no
limit to the circulation velocity was obtained, and little bubble

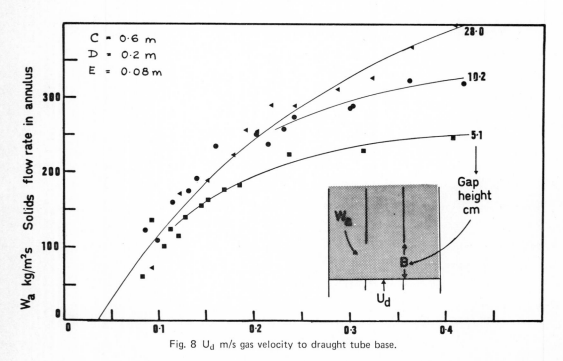

Fig. 8 U_d m/s gas velocity to draught tube base.

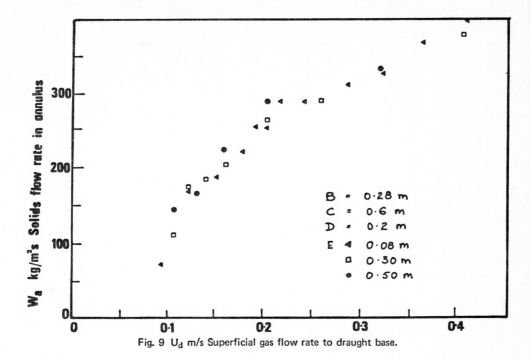

Fig. 9 U_d m/s Superficial gas flow rate to draught base.

activity in the annulus observed, for gas velocities in the draft
tube up to 0.4 m/s.

(ii) The effect of bed height above the draft tube

Figure 9 plots particle mass flux in the annulus
against draft tube gas velocity for differing depths of bed above
the draft tube. Although the particle flowrate in the annulus is
unaffected by the bed height it should be pointed out that the
mixing of particulate phase above the tube appeared to be poor.
Horsler and Thompson (9), who worked on a circulating gasifier for
the British Gas Corp.   reached a similar conclusion when they
observed a temperature non-uniformity in the deep section above the
draft tube while the temperature in the circulating section
remained uniform.

REFERENCES

7.   Merry J.M.D. and Davidson J.F.
          Trans.Instn Chem.Engrs, Lond. 1973, <u>51</u>, 361

8.   Davidson J.F. and Harrison D. 'Fluidised Particles"
          1963 (Cambridge: University Press)

9.   Horsler A.G. and Thompson B.H. 'Fluidisation'
          Proc.of Symp. Tripartite Chem.Engng Conf. Montreal,
          Canada, 1968 p.58 (Instn Chem.Engrs Lond.)

COCURRENT DOWNFLOW OF SUSPENSIONS IN STANDPIPES

L. S. LEUNG

INTRODUCTION

Particulate solids are often transferred downwards out of a fluidized bed in a standpipe. In a typical fluid bed catalytic cracking unit for instance (Matsen, 1973), spent alumina catalyst at a rate of several tons per minute flows from the reactor down a standpipe to the regenerator. The regenerated catalyst flows through another standpipe to be transported back through a riser to the reactor. For downflow of solids, two types of flow pattern are possible (Leung and Wilson, 1973): fluidized bed flow in which particles are in suspension; and packed bed flow (or moving bed flow) in which particles flow en bloc at the voidage of a packed bed with little relative motion between particles.

This paper is concerned with the possibility of coexistence of the two flow patterns in a standpipe; the transition of one flow pattern to another without change in mass flow rates; and how such transition can cause abnormal operations such as pressure reversal in standpipes.

FLOW PATTERNS

At a given set of mass flowrates, the slip velocity U_{sl} is related to voidage by

$$U_{sl} = \frac{U_f}{\varepsilon} - \frac{U_s}{1-\varepsilon} \tag{1}$$

where U_f = superficial fluid velocity (negative in downward
                 direction)

            U_s = superficial solid velocity (negative in downward
                 direction)

            ε = voidage in pipe

For cocurrent downflow, at a fixed set of U_f and U_s, slip
velocity increases monotonically with voidage approaching infinity
at ε = 1(firm line in Figure 1). The broken line in the figure
represents the fluidization expansion curve for the particular
fluid-solid system. Such a curve depends on the minimum fluidization
velocity, and in the case of a gas-solid mixture, on the average
bubble velocity. Equations for the fluidization expansion curve
have been reported by Davidson and Harrison (1963) for bubbling
fluidized beds; by Matsen et al (1969,1973) for slugging fluidized
beds; and by Richardson and Zaki (1954) for liquid-solid fluidized
beds. The fluidization expansion equations are applicable for
downflow of suspension provided slip velocity is used for
comparison (Lapidus and Elgin,1957).

Figure 1 indicates that for the fixed set of U_s and U_f considered,
two operating points are possible. At intersection A, the downward
flowing system can operate at a slip velocity greater than that at
minimum fluidization and at a voidage corresponding to that of a
fluidized bed with similar slip velocity. Point A represents
fluidized bed flow. At point B, the slip velocity is less than
that at minimum fluidization and the voidage corresponds to that

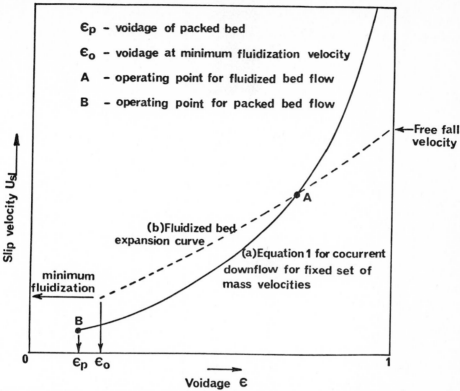

Fig. 1 Slip velocity — voidage relationship. (a) For cocurrent downward flow at given solid and fluid velocities (eqn. 1), continuous line; (b) for a nonflowing fluidized bed of the same solid-fluid system, broken line.

of a packed bed. Point B represents packed bed flow, or moving bed flow as it is often known.

Note that the firm line in Figure 1 represents a fixed set of U_f and U_s. For a different set of velocities, it is possible that the curve for equation (1) may lie above the fluidization expansion curve over the entire voidage range. In this case fluidized bed flow is not possible for the particular set of U_f and U_s.

The analysis of Figure 1 suggests that at a given set of mass flowrates of fluid and solid, the mixture may flow down a standpipe in fluidized bed flow, or in packed bed flow, or with

fluidized bed flow in some part of the standpipe and packed bed
flow in another part of the same pipe.

PRESSURE PROFILE

For fluidized bed downflow, wall frictional losses are generally insignificant compared with the gravity head. The pressure gradient in
a vertical standpipe in this case is approximately equal to $\rho_m g$
(Matsen, 1973) as illustrated in Figure 2a, where

$\qquad \rho_m$ = density of mixture in the standpipe

$\qquad g$ = gravitational acceleration.

For packed bed flow pressure gradient is caused mainly by
relative velocity between fluid and solid and can be estimated by a
modified Ergun equation using the correct slip velocity(Yoon and
Kunii 1970). Figure 2b illustrates possible pressure profiles in
a standpipe for packed bed flow for negative and positive slip
velocities.

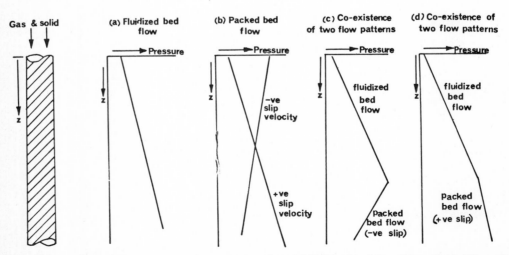

Fig. 2 Possible pressure profile for co-current gas-solid downflow. (a) Fluidized bed flow; (b) packed bed flow with
(i) positive slip velocity (ii) negative slip velocity; (c) & (d) fluidized bed flow & packed bed flow coexisting in pipe.

Finally if two flow patterns are to coexist in the same standpipe, the pressure profile may be represented by Figure 2c or Figure 2d depending on the direction of the slip velocity in the packed bed flow section. Discussion of the pressure profile and of the prediction of flow pattern in standpipe flow has been reported at length previously (Leung and Wilson, 1973).

EXPERIMENTAL

The aims of the current experimental program are to (i) verify the possibility of coexistence of two flow patterns in a standpipe; (ii) verify the existence of pressure profiles indicated in Figure 2c and 2d; and (iii) study the conditions for transition of fluidized bed flow to packed bed flow. In the experimental setup, metered sand and water at various flowrates are passed downwards through a 2 3/4 inch diameter vertical standpipe. Pressure gradient and flow patterns were observed in the experiments.

With no constrictions in the standpipe, stable fluidized bed flow was observed in the whole range of flowrates studied. Packed bed flow can be initiated if a constriction (such as an orifice) is placed in the standpipe. With a given constriction, fluidized bed flow is stable over a range of sand/water flowrates. However, as the water flowrate is reduced at a fixed sand mass flowrate, a critical flowrate was observed at which fluidized bed flow becomes unstable and the flow pattern in the tube changes to packed bed flow. By careful control of the flowrates it was possible to achieve coexistence of the two flow patterns in the standpipe. Packed bed flow was observed in the section immediately above the orifice while

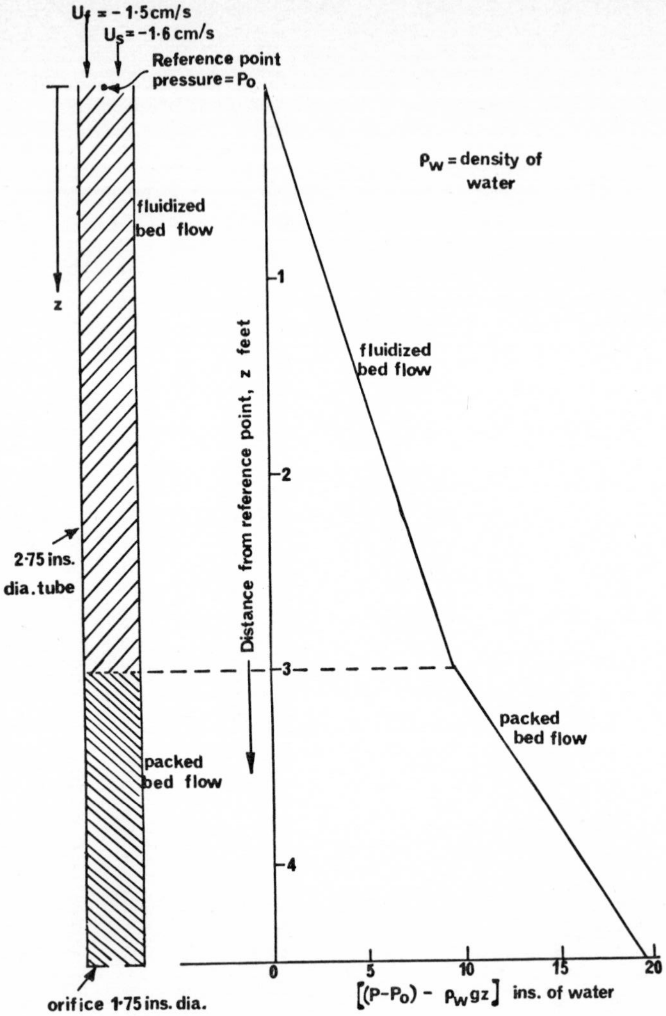

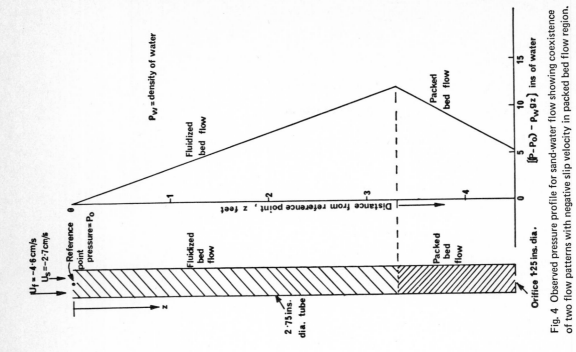

Fig. 4 Observed pressure profile for sand-water flow showing coexistence of two flow patterns with negative slip velocity in packed bed flow region.

Fig. 3 Observed pressure profile for sand-water flow showing coexistence of two flow patterns with positive slip velocity in packed bed flow region.

fluidized bed flow occurred in the next section above. Results of
such observation are given in Figures 3 and 4. In Figure 3, the
flowrates were such to give rise to a positive slip velocity in the
packed bed flow section, and the observed pressure profile is
similar to that given in Figure 2d. The results in Figure 4
correspond to a negative slip velocity giving rise to a pressure
profile similar to that described in Figure 2c.

Experiments have been carried out over a range of flowrates
using three different orifices (1¼, 1½ and $1^3/4$ inches diameter
openings). The flow transition velocities are dependent on size
of constriction in the standpipe. Experimental work is still in
progress and will be published later.

PRACTICAL IMPLICATIONS

Downflow of solids through a standpipe is often regulated by
the use of a slide valve at the bottom part of the standpipe. Often
in normal operation it is important to maintain fluidized bed flow
in the pipe to maintain a high pressure above the slide valve,
higher than the pressure below the slide valve (i.e. $P_2 > P_3$
in Figure 5). Pressure reversal in the system described in Figure
5 will result in the hazardous flow of hydrocarbon into the
regenerator through the slide valve.

The present analysis and experimental work suggest that rapid
loss of pressure above the slide valve can be triggered by a change
in flow pattern in the tube (from fluidized bed flow to packed
bed flow with pressure profile changing from Figure 2a to Figure
2c).This change in flow pattern can be caused by the presence

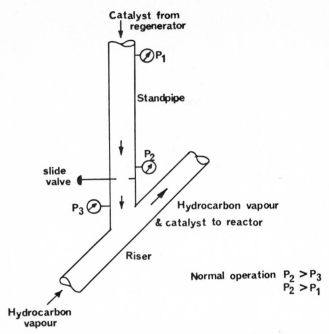

Fig. 5 Regenerator standpipe in fluid bed cracker unit.

of a constriction in the pipe, (corresponding to that offered by

a partially open slide valve),or by changing the flowrate and pressure

in the system. Our experiments suggest that the presence of a

constriction in the standpipe may promote packed bed flow. The

flow instability described by Matsen (1973) may be caused by

sudden change in fluidized bed flow in the tube to packed bed flow.

It is significant that the flow instability was observed only after

the replacement of a thermal joint (normally of the same diameter as

the standpipe) by an undersized thermal joint, creating a constriction

which may be the cause of flow pattern change. Matsen (1973)

postulated that the instability may be triggered when downwards

solid flow is just sufficient to hold bubbles stationary in the

tube.

In another catalytic cracker unit with which the author is familiar, sudden pressure loss at the bottom of a standpipe was sometimes observed. This occurred after a constriction was unadvertently installed at the bottom of the standpipe at a turnaround. The interesting feature here is that after occurrence of pressure loss above the slide valve, circulation of catalyst at the same rate (about 15 tons per minute) can be maintained and the plant can in fact be run under a new steady state. This observation is consistent with our conclusion that change in flow pattern can occur without any change in mass flowrate.

ACKNOWLEDGEMENT

Experimental work described here was carried out by W.K. Mapleston. The author wishes to acknowledge useful discussions with Mr. P.C. Brooks and Dr. R.H. Weiland.

REFERENCES

Davidson, J.F. and Harrison D., Fluidised Particles, Cambridge University Press, p.29, (1963).

Lapidus, L. and Elgin, J.C. , A.I.Ch.E. Journal, 3, 63 (1957).

Leung, L.S. and Wilson, L.A., Powder Technology 7, 343, (1973).

Matsen, J.M., Powder Technology 7, 93, (1973).

Matsen, J.M., Hovmana, S. and Davidson, J.F., Chem. Eng. Sci. 24, 1743, (1969).

Richardson, J.F. and Zaki, W.N., Trans. Inst. Chem. Engrs. 32, 35, (1954).

Yoon, S.M. and Kunii, D., Ind. Eng. Chem. Process Deisgn Develop. 9, 559 (1970).

SOME CHARACTERISTICS OF LARGE SOLIDS CIRCULATION SYSTEMS

J. M. MATSEN

Circulation of fluidized solids through standpipes and risers is an essential part of such fluid bed processes as catalytic cracking and fluid coking. Rigorous experiments in small scale circulation systems have been few. Therefore, a critical analysis of commercial data may prove useful in understanding the phenomena involved. The purpose of this paper is to provide data and insight into the behavior of commercial circulation systems and to explain some of the observations in terms of current theories of bubble behavior. The paper draws upon the normal and unusual operating experience of some 65 cat crackers and fluid cokers. The data cover standardpipe diameters up to 5 feet, heights 130 feet, circulation rates to 120 tons per minute, and pressure buildups of almost 3 atmospheres.

OBSERVED RATES

Circulation rates are usually determined by process considerations and overall system pressure balance, although flow limitations may occur in unusual circumstances. Most catalytic cracker standpipe circulation rates are in the range of W=120 to 300 pounds per square foot per second, with no strong differences apparent between small and large ones (1 to 5 feet diameter). Rates for fluid cokers tend to be lower in rate though 300 lbs/ft^2 sec has been achieved. Risers are usually the same or somewhat larger in diameter than the associated standpipes with a corresponding effect on mass flow rates.

Aeration

Although bubbles may flow either up or down in standpipes, net gas flow, including interstitial gas, is almost always down. Because this gas is being compressed as it is carried downwards, it is necessary to add aeration gas along the standpipe to maintain proper fluidization. Recent equations (3) relating standpipe density to gas and solids flow rates can be used to calculate the aeration requirements. Good agreement with empirical determinations of optimum aeration is usually obtained.

When net gas flow is upwards in a standpipe, gas expands as it rises. Thus, although the bottom of the standpipe may operate at minimum fluidization conditions, the top of the standpipe will be highly aerated and will give a low pressure buildup.

Lift gas rates in risers are usually adjusted to give pressure gradients of $0.5\rho_0$ and less, as required to obtain pressure balance. Slip factors of about 2 (i.e., average gas velocity is twice average particle velocity) are typical in riser operation.

Friction

Friction between particles and walls can sometimes be a significant part of transfer line pressure drop. However, few pertinent experimental data and theorical predictions are available. Some commercial friction data have been obtained by using the attenuation of gamma radiation by flowing solids to measure solids density. The difference with this density and the apparent density as indicated by the observed pressure gradient can be attributed to friction. Data from an 8 inch diameter test standpipe show that friction can approach zero below the minimum fluidization density but becomes very great as true density exceeds minimum fluidization density. Commercial units show similar behavior. Particles packed at densities greater than minimum fluidization density are not fluidized; and hence, the pressure gradient in gas percolating past these particles must be less than the bed density. Leung and Wilson (2) have examined flow under such conditions. Pressure gradients under packed bed flow are very sensitive to aeration, contributing greatly to the scatter of friction measurements.

Upper limits for friction in properly fluidized standpipes can be established by comparing maximum observed pressure gradients with densities at minimum fluidization conditions. In several commercial cat crackers and cokers, pressure gradients of 95% of minimum fluidization density have been measured across long standpipes. Since a long standpipe would be defluidized and/or aerated to a density less than ρ_0 over much of its length, friction debits in the fluidized zones are probably substantially less than 5% of ρ_0. Friction may therefore be neglected in most design work, and pressure gradients of 70% to 80% of ρ_0 may be selected to allow for operating flexibility and non-optimum aeration.

Pressure losses of 1 to 2 psi normally observed across the long radius "U-Bends" of a Model 4 cat cracker, are conventionally attributed to friction. Radiation measurements have shown, however, that actual densities can be much higher on the upflowing side than on the downflowing side. On a bend with a 12-1/2 ft radius, radiation densities of 39 and 20 lbs/ft^3 respectively were measured, giving rise to a 1.6 psi pressure drop even without friction. The density difference is due mainly to the fact that bubbles rise relative to catalyst and are moving slower than catalyst on the down leg and faster on the up leg of the bend. This slippage is accentuated by the tendency of bubbles to stratify along the upper surface of the bend.

Friction in a 12 inch diameter catalyst riser was measured at about 4 lb/ft^2 ft at radiation densities of 10 to 15 lbs/ft^3. Measurements in a 23 inch diameter coke riser showed much more scatter and lower average friction losses.

FLOW PATTERNS AND BUBBLE OBSERVATIONS

The simplest types of gas-solids flow are a uniform suspension of particles and a random dispersion of small bubbles in an emulsion. More complex flows have often been observed by radiographic measurements. Highly non-uniform flow patterns such as stratification and slug flow clearly affect the performance of standpipes and risers. Extensive measurements on a high velocity catalyst riser were recently reported by Saxton and Worley (4). In a 25 inch diameter line with gas velocities in excess of 35 ft/sec, areas having average densities of 3 to 60 lbs/ft^3 were observed in the same cross-section. This extreme segregation in solids flow illustrates the difficulty of predicting entrainment or pneumatic transport with models that assume uniform dispersion of particles in dilute phase flow.

Measurements in a 60 inch diameter catalyst standpipe, sloped at 30° to vertical showed a density of 52 lbs/ft^3 measured horizontally through the line diameter and 42 lbs/ft^3 measured through a chord near the upper surface of the line. Poor pressure buildup was a result of such stratification.

Radiation measurements have also shown the occasional presence of very large bubbles. Such bubbles were almost certainly wall to wall, Stewart and Davidson's (5) Type A slugs. Though not seeming to halt or obstruct solids circulation, the bubbles contributed to rough operation and poor pressure buildup in the standpipes. Bubbles of such size seem common with coke or with the coarse cracking catalysts in use 20 years ago. Their presence today in cat crackers is usually a sign of abnormally low fines content.

Bubble Problems

Large bubbles in standpipes which either rise or are carried downwards rapidly cause no real difficulty other than noticeable vibrations and short term oscillations in circulation rate. However, standpipe geometry and flow rate can be such that large bubbles are held stationary and can grow, restricting flow or causing upsets. One standpipe recently analyzed (3) had a 34-1/2 inch diameter constricted section above the 45-1/2 inch diameter main standpipe. The downward solids flow rate was such that a slug bubble would rise in the large diameter portion but was prevented from rising through the constriction by the higher solids velocity there. Once formed, such a bubble grew down in the standpipe, slowly reducing the pressure buildup and finally upsetting the circulation control system. Leung and Wilson (2) have noted that reduced pressure buildup can also occur in packed bed flow. However, considering the aeration patterns and transient responses in the case of (3), bubble formation rather than packed bed flow seems to be appropriate explanation in that particular instance.

U-Bends can provide a suitable environment for large bubbles evidenced in the radiographic data discussed above. Occasionally, a bubble is formed over a period of hours and will then persist for many hours. The loss in pressure buildup is moderate (limited by the size of the U-Bend) and can easily be compensated for by changes in circulation controls. However, the escape of such a bubble will be rapid and cause a sharp rise in pressure buildup and circulation rate. In one instance, normal pressure buildup in the 15-ft. high downflow portion of a 38-inch diameter bend was 2.9 psi, but with bubble formation, this would slowly decrease to about 1 psi. The bubble would be released once or twice a day causing a step change in circulation rate. Pressure buildup in the vertical standpipe and in the upflow side of the bend were not affected by bubble formation.

A set of three bubble problems was found in a recent standpipe study. The top 14 foot portion of the standpipe was 30 inches in diameter, then expanding to 36 inches and entering a 10 foot section sloped at 20° to vertical and returning to vertical for 105 feet before reaching a slide valve. The first phenomenon was "bridging", i.e., Stewart and Davidson's Type B slug, which would occur a short distance below the sloped portion of the line, and which was probably stabilized by the bend in the line. No catalyst flow would occur in this condition. The lower 100 ft of the standpipe was at constant pressure indicating presence of a large bubble. Across the bridged portion a pressure drop of 15 psi would occur between pressure taps only 8 feet apart. Above the bridge, the standpipe was filled and the pressure gradient was about 37 lbs/ft^3. This phenomenon would not occur once good flow once established.

A second phenomenon was the trapping growth of a Type A (5) slug just below the sloped section. The angled flow of solids leaving the sloped section seemed to form a roof for bubbles in the lower section. This was illustrated in a small model with flowing water. In the real standpipe, a series of differential pressure recorders was used to follow the formation of the bubble and its growth downwards to a length of perhaps 50 feet. The bubble typically would last 4 to 8 minutes; and upon release, its passage upwards could be followed by rise of a region of low density through the sloped section and into the regenerator. This behavior led to unstable circulation.

Operating personnel found that high aeration rates to the sloped section could stabilize circulation and give a constant, high pressure buildup in and above the sloped section. Mechanistically the high aeration was upsetting the solids flow pattern in the sloped portion and not allowing a roof to form for a bubble in the lower portion. Although operating was steady, circulation seemed to have a very definite upper limit of about 20 tons per minute. We believe that the high aeration rate caused upward gas flow into the regenerator, causing countercurrent choking. If slugs are rising even slowly in a pipe, the maximum downwards solids mass flow rate is given by $W/\rho_o = 0.35\sqrt{g\,D}$. With an upper line diameter D of 2.5 ft g = 32 ft/sec and $\rho_o \cong 42$ lbs/ft^3 a mass flow rate of 130 lbs/sec ft^2 corresponding to circulation of 19 tons per minute is possible. This choking phenomenon is similar to that observed in emptying a narrow necked bottle. As the survey of standpipe circulation rates has shown, mass flow rates W much higher than $0.35\,\rho_o\sqrt{g\,D}$ are usually possible, but in this case choking occurred because the aeration conditions forced gas to flow upwards.

CONCLUSIONS

Operating characteristics of a number of commercial fluid-solids circulation systems have been presented. Data include circulation rates, friction measurements, and radiographic flow pattern determinations. Current bubble theory is used to predict aeration requirements and to explain certain phenomena due to large bubbles.

APPENDIX

Commercial Circulation Rates

Figures 1 and 2 show typical solids mass flow rates in standpipes of fluid catalytic crackers and fluid cokers respectively. The rates shown were drawn from reports of long-term operating conditions and probably do not represent the extremes of circulation rates actually achieved on any of these units. Even maximum circulation rates are usually determined by overall operating constraints of the unit, such as regenerator air supply, catalyst activity, etc., rather than by flow limitations in standpipes or risers. The dashed lines on these figures represent the downwards solids mass flow rates which will hold a large bubble or slug stationary or stagnant. The data seem to indicate that slug stagnation does not generally place limits on solids flow rate in standpipes. Some unusual cases are cited later in this paper, however, in which slug formation and stagnation have caused circulation difficulties.

Aeration Requirements

In a normally operating standpipe there is a net downward flow of gas, and aeration is added in order to counteract gas compression. The basic equation for required aeration is: $R = AU \, (T_o/T) \, (1/P_o) \, (dP/dL)$. The total gas superficial velocity U is given by (3), $U = U_b \, (\rho_o/\rho - 1) + W$

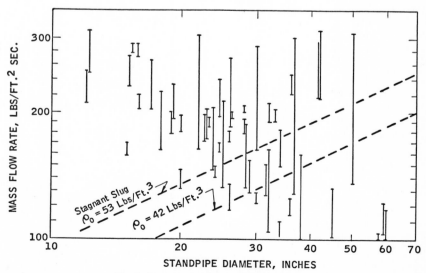

Fig. 1 Circulation rates in catalytic crackers.

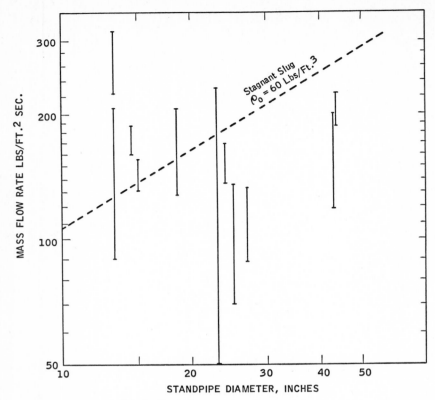

Fig. 2 Circulation rates in coker standpipes.

$(1/\rho - 1/\rho_s) + U_o$ and if friction loss is low - $(dP/dL) = \rho$. The aeration needed then is R = A $(T_o/T)(1/P_o) \left[U_b (\rho_o - \rho) + W (1 - \rho/\rho_s) + U_o\rho \right]$.

Early analysis of aeration did not allow for slip between gas and particles, due to rising bubbles and minimum fluidization gas, giving U = W$(1/\rho - 1/\rho_s)$. Aeration requirements calculated with the no-slip assumption are typically 20% to 30% greater than those assuming bubble slip. Actual aeration studies of several standpipes showed that the standpipes gave best pressure buildup and performance with 70% or 80% as much aeration as had been calculated using the no-slip assumption.

Friction Measurements

Figure 3 shows friction measurements on catalyst in an 8 inch diameter test standpipe. Friction is mainly a function of catalyst density and hence aeration, and the scatter seems to be due to measurement limits in radiation attenuation and pressure differences. Catalyst mass flow rate does not make a noticable contribution to scatter or friction. Simultaneous measurements at different elevations usually showed wide differences in true density and friction, despite the fact that solids mass flow rate was the same for all elevations. The catalyst used in these tests was quite compressible, having a minimum fluidization density of 41.0 lbs/ft^3, a loose settled density of 45.2 lbs/ft^3, and a packed density (estimated from other catalyst data) of 50 lbs/ft^3. The measured radiation densities above 50 lbs/ft^3 are suspect and probably indicate accuracy limitations of the radiation technique.

Table 1 shows results of some density and pressure gradient measurements on commercial units. These data certainly do not present a balanced picture since most of these measurements were taken precisely because the standpipes were operating with poor pressure buildup. The very last group of data in the table were taken for research rather

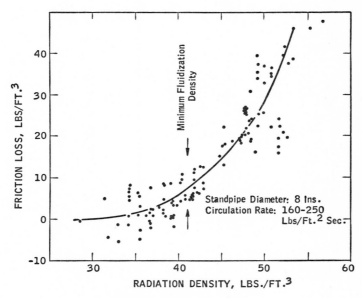

Fig. 3 Friction loss in standpipe.

Table 1 Commercial Data on Friction in Standpipes

Diameter Inches	Material	Mass Rate lbs/ft^2 sec	Radiation Density lbs/ft^3	$\Delta P/\Delta L$ lbs/ft^3	Friction lbs/ft^3
23	coke	75	61	31	30
23	coke	75	65	20	45
29	catalyst	145	42	38	4
29	catalyst	145	49	36	13
30	catalyst	160	54	39	15
33	coke	37	60	55	5
42	catalyst	115	47	41	6
42	catalyst	115	53	32	19
42	catalyst	115	55	21	34
50	catalyst	200	39		-1.5
50			42	{40.5±4.5}	1.5
50			37		-3.5

than troubleshooting purposes, and the standpipe was operating very well. These data were also taken several years later than the others, and they include any benefits due to improved catalysts.

Figure 4 gives results of some radiation density measurements on a 25 inch diameter U-bend with a catalyst mass flow rate of 180 lbs/sec ft^2. Pressure drop data are not available, but measurements clearly show a much higher density in the upflowing portions of the bend than in the downflowing legs. This density difference gives rise to much of the pressure drop across the bend that is sometimes attributed to friction.

Figure 5 shows some density and pressure drop measurements in a 12 inch diameter catalyst riser, with an apparent friction effect. Figure 6 shows similar data in a 23 inch diameter coke riser, and friction is much less obvious.

Figure 7 shows riser density contours reported by Saxton and Worley (4). Each contour plot is the result of radiation attenuation measurements across 18 separate chords covering the cross-section. Reduction from attenuation data to the contour plot was by the method of Bartholomew and Casagrande (1).

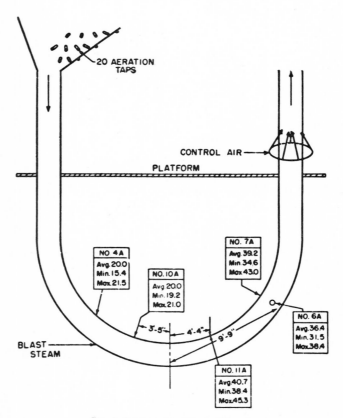

Fig. 4 Catalyst densities in U bend.

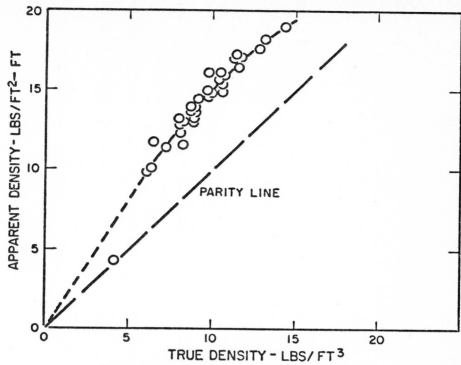

Fig. 5 True density, apparent density relationship in riser.

Table 2 gives some radiographic data on the existence of large, long lived bubbles which have been observed in commercial standpipes.

Bubble Problems

Figure 8 shows a previously reported (3) case of a large bubble or slug held stationary below a constriction by downwards flowing solids.

Figure 9 shows a pressure profile attributed to formation of a type B slug (5) or bridge of packed solids in a 36 inch diameter standpipe.

Figure 10a and 10b show a 1 inch diameter hydraulic model of a bend in the same standpipe. Water flows down the tube and an air bubble is injected just below the bend. At water velocities of 0.57 to 0.8 ft/sec (i.e., 1.0 to 1.4 times slug rise velocity of 0.35 $\sqrt{g\ D}$) the bubble is held stationary by the flow of water at the bend. If air is added continuously, the bubble grows downwards. A similar

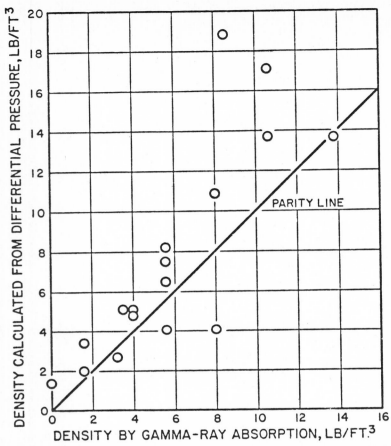

Fig. 6 Comparison of densities in cold coke riser calculated from differential pressure and from gamma-ray absorption.

phenomenon existed in the commercial standpipe, as indicated in Figure 11. Differential pressures across several sections of the standpipe were recorded on a high-speed multi-channel recorder. Starting at minute 1, a bubble forms in zone 4 just below the bend in the standpipe. The bubble soon fills zone 4 and travels downwards into zone 5. Starting at minute 3, zone 5 begins to fill again with catalyst and when zone 5 is full, zone 4 begins to fill as the bottom of the bubble continues to rise in the standpipe. As zone 4 starts to fill, the nose of the bubble apparently is released, and its passage up the standpipe can be followed by the sudden dip in pressure in zone 3, 2, and 1. Meanwhile the bottom of the bubble sequentially rises through zones 4, 3, 2, and 1. After the bubble has been released, all portions of the standpipe operate at high differential pressure for a short time, until at minute 5 a new bubble forms in zone 4 and the pattern repeats. This pattern would sometimes repeat itself for hours. In Figure 11, however, starting at minute 9 a high rate of blast aeration is added at taps between zones 2 and 3. This great amount of aeration creates a very low density region in zones 1, 2,

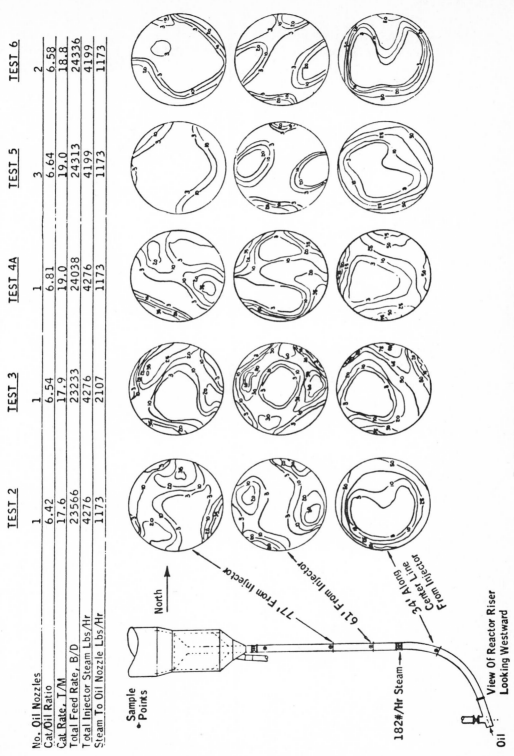

	TEST 2	TEST 3	TEST 4A	TEST 5	TEST 6
No. Oil Nozzles	1	1	1	3	2
Cat/Oil Ratio	6.42	6.54	6.81	6.64	6.58
Cat Rate, T/M	17.6	17.9	19.0	19.0	18.8
Total Feed Rate, B/D	23566	23233	24038	24313	24336
Total Injector Steam Lbs/Hr	4276	4276	4276	4199	4199
Steam To Oil Nozzle Lbs/Hr	1173	2107	1173	1173	1173

• Sample Points

North

7'7" From Injector

6'1" From Injector

3'4" Along Center Line From Injector

182#/Hr Steam

Oil

View Of Reactor Riser Looking Westward

Fig. 7 Density contours in reactor feed riser.

145

Table 2 Radiation Detection of Large Bubbles in Standpipes

Line Diameter	Material	Mass Rate	High Density	Low Density	Duration(1) of Low Density	Cycle Time(1)
Inches		lbs/ft^2 sec	lbs/ft^3	lbs/ft^3	min	min
22	catalyst	200	38.5	0.8	0.2	0.4
23	coke	70	55-65	0-6	1.2	2
42, Unit A	catalyst	86	40	8	2-6	8
42, Unit B	catalyst	105	45	5	up to 12 hrs.	

Note (1) Duration and frequency of low density (bubbles) was variable. These numbers
are representative of many bubble observations is each of these standpipes.

and 3, and the flow pattern of solids flowing downward in zone 3 does not
permit a bubble to be held stationary in zone 4. As a result, the standpipe can operate for many hours in a very stable manner, with the top at
low density and the bottom free of large bubbles and at high density.

Once a large bubble is formed in zone 2, the blast aeration
cannot all be carried downward, and some must travel up the standpipe
into the regenerator. This causes a choking phenomenon which limits
the countercurrent flow of catalyst. An hydraulic analogy of the choking is shown in Figure 12, a simplified approximation to the phenomenon
of emptying a narrow-neck bottle. Data on the apparatus show that the
emptying rate is very close to the rate at which liquid can flow around
a slug of the diameter of the outlet tube, i.e., volumetric discharge
rate = (tube areas) $(0.35\sqrt{g\,D})$. Although in this particular experiment
the upwards flow of gas must equal the downwards flow of liquid, consideration of the countercurrent flow phenomenon shows that very nearly
the same liquid downflow will result for a wide range of gas upflow
rates.

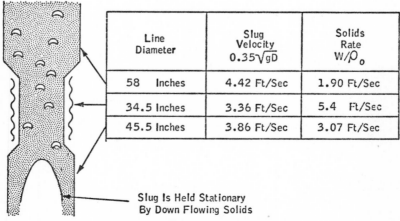

Fig. 8 Bubble stagnation below construction.

Line Diameter	Slug Velocity $0.35\sqrt{gD}$	Solids Rate W/ρ_0
58 Inches	4.42 Ft/Sec	1.90 Ft/Sec
34.5 Inches	3.36 Ft/Sec	5.4 Ft/Sec
45.5 Inches	3.86 Ft/Sec	3.07 Ft/Sec

Slug Is Held Stationary
By Down Flowing Solids

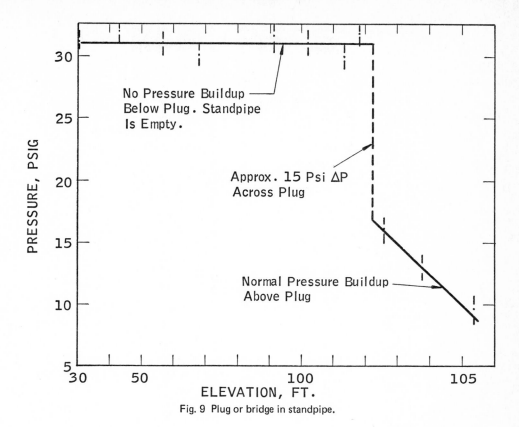

Fig. 9 Plug or bridge in standpipe.

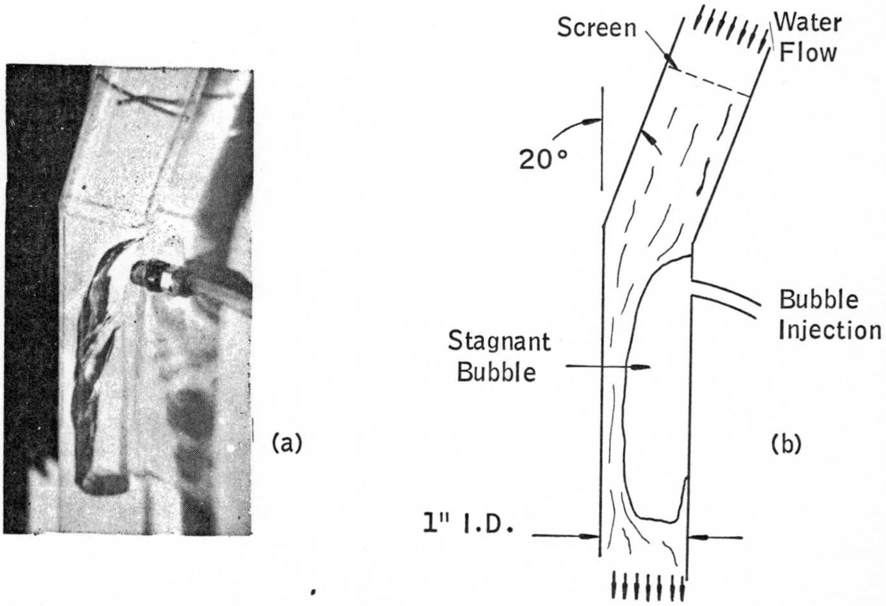

(a)   (b)

Fig. 10 Hydraulic model of stagnant bubble below bend in standpipe.

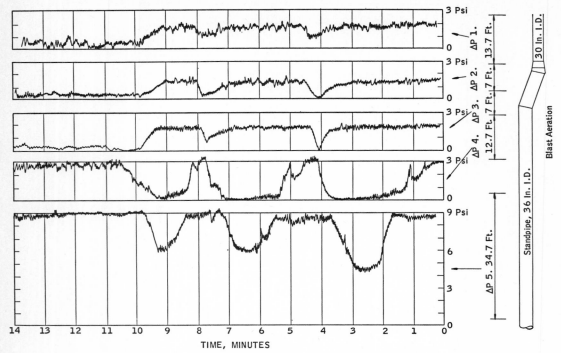

Fig. 11 Differential pressure recordings showing bubble formation.

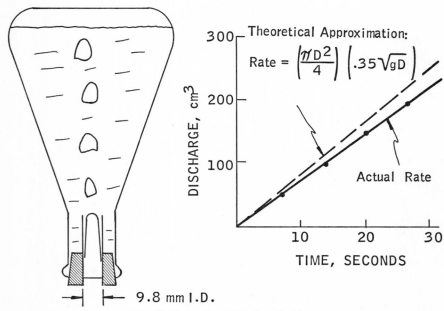

Theoretical Approximation:

$$\text{Rate} = \left(\frac{\pi D^2}{4}\right)\left(.35\sqrt{gD}\right)$$

Actual Rate

9.8 mm I.D.

Fig. 12 Hydraulic model of choked flow.

REFERENCES

(1) Bartholomew, R. N., and Casagrande, R. M., _Ind. Eng. Chem._, _49_, 428 (1957)

(2) Leung, L. S., and Wilson, L. A., _Powder Technology_, _7_, 343, (1973)

(3) Matsen, J. M., _Powder Technology_, _7_, 93 (1973)

(4) Saxton, A. L., and Worley. A. C., _Oil and Gas Journal_, _68_, No. 20, 84, (May 18, 1970)

(5) Stewart, P. S. B., and Davidson, J. F., _Powder Technology_, _1_, 61, (1967)

NOMENCLATURE

A = standpipe cross-sectional area, ft^2

D = diameter of standpipe or riser, ft

g = gravitational acceleration, 32.2 ft/sec^2

L = elevation in standpipe. dL is positive upwards

P = pressure, lbs/ft^2

P_o = standard pressure, 2110 lbs/ft^2

R = required aeration rate, std. ft^3/sec per ft of standpipe height

T = temperature, °R

T_o = temperature at standard conditions, 528°R

U = total superficial flow rate of gas in standpipe, and for upwards flow, ft/sec

U_b = single bubble rise velocity relative to stagnant solids, ft/sec

U_o = minimum fluidizing velocity

W = solids mass flow rate, lbs/ft^2 sec

ρ = fluidized solids density, lbs/ft^3

ρ_o = minimum fluidization density, lbs/ft^3

ρ_s = solids skeletal density (excludes voidage between particles and within open pores)

PARTICLE CONVEYING IN EXTRUSION FLOW

F. A. ZENZ and P. N. ROWE

Within all sectors of industry there is compelling interest in pneumatically handling powders at maximum density so that
less air (or other motive gas) and smaller pipes are needed for a
given solids transport rate. The particle loading of a given gas
stream cannot be varied continuously for discontinuities occur
equivalant to phase changes and only a very limited class of particles can be conveyed continuously at near maximum theoretical
density. In this article we describe a maximum theoretical density or extrusion type of solids flow and propose a simple test
for defining powders that will behave in this way.

CONVENTIONAL PNEUMATIC CONVEYING

The behaviour of a powder-laden gas flowing along a horizontal pipe is best understood by considering the variation in
pressure drop per unit length of pipe as the gas flow rate is
changed. Let W_S and W_G be respectively the solids and gas mass
flow rates per unit cross sectional area of the pipe (kg/s cm²).
Neglecting the usually very small decrease in density resulting
from pressure drop, W_G is equivalent to the gas superficial velocity in the pipe. In Figure 1 the gas flow rate is plotted as
abscissa, increasing on a logarithmic scale from left to right,
against the ordinate, pressure drop per unit length of pipe, increasing on a logarithmic scale from bottom to top.

In Figure 1, $W_S = 0$ refers to the flow of gas only through
a pipe. Other lines represent a positive added solids flow rate.
To the right of the vertical broken lines (which represent the so-
called saltation velocities, a function of the solids conveying
rate) the particles flow as a more or less uniform dispersion in

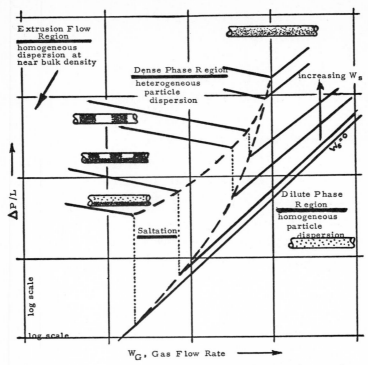

Fig. 1 Regimes of fluid — particle flow in horizontal pneumatic conveying.

the gas as indicated in the sketches. They travel like a spray
of bullets, a regime of flow uniformly referred to as "dilute
phase conveying".

Below the saltation velocity some particles settle out and
form an essentially stagnant layer at the bottom of the pipe.
Transport still occurs in the unobstructed upper part, but to
maintain a given solids flow with a reduced gas rate the pipe
becomes increasingly obstructed with stagnant particles thus increasing the local gas velocity to the necessary transport value
at the expense of an increased pressure drop.

If, after a layer has formed, the gas rate is further reduced, the open area may become so small that conveyance occurs
in a slug flow pattern where the slugs might occupy the entire
cross-section or "ride" over a non-moving salted layer. On the
other hand at very high solids flow rates the saltation point may
not permit a settling out into a non-moving layer simply because

there would be insufficient space above it to carry the necessary
solids flow and the layer is effectively dragged along the bottom
of the pipe.

The distinction between what is uniformly referred to as
dilute phase and what is commonly called dense phase conveying is
illustrated in Figure 1. There is no simple numerical division
between these regimes and no critical value of W_S/W_G. "Dense
phase conveying is essentially heterogeneous flow with variation
in particle concentration both in space and time. It would be
more orderly and logical to reserve this phrase for the limiting
and rather unusual case of flow at a density uniformly near to
the bulk density. However, common usage prevents this and a new
word must be found for which we propose "extrusion flow". It
occurs near the left-hand extremity of the curves in Figure 1.

EXTRUSION FLOW

The limiting case of maximum density of conveying would
occur when all solids in the pipe maintain bulk density and move
along as if a plastic material extruded through a pipe-like die.
If gas is to provide the driving force there must be a pressure
drop and, since the packed solids are porous to a greater or lesser
degree, there must be a net gas flow. Although W_G cannot be zero
it can be very small and as a result there will be less dusting
and a reduction in the commonly encountered problems of powder
containment. By no means are all powders capable of extrusion
flow under differential air pressure but those that are readily
maintain this mode of flow even through abrupt bends and restrictions and through pipes inclined at any angle. Suitable powders
are of small particle size (a few tens of microns) and of low bulk
density but these two properties are not of themselves sufficient.

In order that a powder shall take part in extrusion flow it
must possess a property that might be called "bulk deformability".
This is a property of all fluidized particles and in this way
there is little difficulty in bringing about extrusion flow in
vertical pipes. In non-vertical flow, however, it is not practical

to maintain a fluidizing gas flow in a pipe otherwise filled with ,
solids. Fine powders with very low fluidizing velocities may retain interstitial gas for some time after the supply has been dis-
continued and thus behave as if fluidized whilst, for example, in
horizontal flow, but this is a time dependent property of fairly
short duration. Such powders have a long settling or deaeration
time when left to stand in a vessel after vigorous agitation.
This test alone does not correctly identify powders suitable for
extrusion flow.

A SIMPLE TEST OF BULK DEFORMABILITY

A true liquid is capable of continuous deformation without
residual internal stress and it is this property of bulk deformability that is sought in a powder to be moved in extrusion flow.
Just as a liquid will allow a denser solid body to fall through
it, powders capable of extrusion flow display a similar continuous
yield or bulk viscosity. This is the basis of a very simple but
effective test. A 250 ml cylinder is filled with the powder to
be tested; a 1 cm diameter steel ball is released from just above
the surface and if it falls to the bottom, as if through a liquid,
the powder is capable of extrusion flow. The property must be permanent and not dependent on aeration immediately prior to the test.

BULK DEFORMABILITY TESTS OF SOME POWDERS

Rather than with a loose steel ball that is troublesome to
recover, the drop test described above can be carried out with a
solid on the end of a rigid stem such as the knobbed brass rod
drawn in Figure 2. The rod protrusion also indicates clearly that
the end has reached bottom. Several powders have been tested with
this simple apparatus. Finely powdered coal, all particles being less than 44 microns in diameter, readily passes this test
even after standing in the test cylinder for many days. If, however, the charge of material is tapped and compressed vigorously

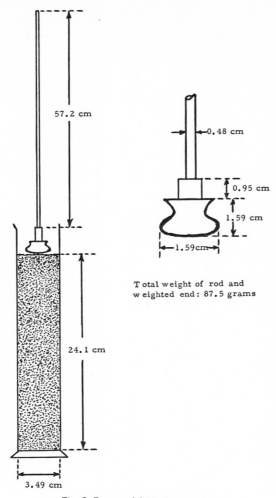

57.2 cm

0.48 cm

0.95 cm

1.59 cm

1.59cm

Total weight of rod and
weighted end: 87.5 grams

24.1 cm

3.49 cm

Fig. 2 Drop weight test apparatus.

the falling weight may only penetrate about two thirds of the
depth indicating that the property of bulk deformability can be
destroyed by strong compaction. This is a fairly obvious conclusion.

 The same coal powder can readily be conveyed through a
pipe in extrusion flow. Our conveying test equipment consisted
of a 14" diameter feed tank 48" high charged about one third full
with coal powder and subject to 5 psig top air pressure. The
outlet pipe was 1" diameter Plexiglas or vinyl about 16' in length,

the last 3' being vertical and the rest horizontal. In different
tests the change from horizontal to vertical was via a radiused
elbow, a 90° single miter elbow and a 90° tee. In all tests the
coal powder moved easily and rapidly in extrusion flow through
this system and issued as a coherent cylindrical jet of several
inches length before breaking up.

A selection of other conveniently available powders was
also similarly tested. Fine glass beads (120 microns), FCC
catalyst (43 microns) and finely divided TiO_2 (12 microns) showed
no signs of bulk deformability when tested. The dropped weight
would penetrate only a few millimeters. These materials could
not be induced to flow continuously in any manner through the
apparatus described above.

The addition of Cab-o-sil (finely divided SiO_2 of unagglo-
merated particle size less than one micron) can change the
physical properties of these powders and make them readily deformable. An equal volume added to FCC catalyst (a 4% by weight
addition) gives it a deformability comparable with the micronised
coal. The dropped weight fell readily and the mixture moved easily
in extrusion flow through the test pipe. Surprisingly and probably significantly the volumes of Cab-o-sil and catalyst were
roughly additive so that the density of the mix was almost half
that of the original catalyst. Under 50X magnification the
Cab-o-sil was seen as agglomerates of overall size about the
mean size of the catalyst particles suggesting that its role is
as interspersed deformable or "soft" particles amongst the normally rigid ones.

The addition of an equal volume of Cab-o-sil to rutile
(180 microns) and to the glass beads did not produce a mixture
through which the test weight could be dropped. In both these
cases there was a reduction in volume on mixing so that the resulting density was between 30 and 40% less than that of the denser component indicating some interstitial penetration by the
Cab-o-sil. Further details of the tests are given in Table 1.

Table 1 Summary of Exploratory Drop Test Results

Material	Wgt. gms	Vol. cc	Bulk Density gms/cc	Mix Bulk cc	Mix Bulk gms/cc	Drop Test
Glass Beads	262.4	180	1.46	-	-	Fail
FCC Catalyst	182.4	200	0.912	-	-	Fail
Ground Coal	106	210	0.505	-	-	Pass
Rutile	482	200	2.415	-	-	Fail
TiO_2*	55.2	180	0.306	-	-	Fail*
TiO_2	27.6	90	0.306	190	0.166	Pass
Cab-O-Sil	4.0	130	0.0336			
Rutile	241.5	100	2.415	155	1.58	Fail
Cab-O-Sil	3.8	120	0.0288			
Glass Beads	131.2	90	1.460	156	0.864	Fail
Cab-O-Sil	3.2	110	0.0291			
Ground Coal	51.4	100	0.514	210	0.262	Pass
Cab-O-Sil	3.6	110	0.0327	179	0.308**	Pass
				151	0.365**	Fail
FCC Catalyst	91.2	100	0.912	210	0.450	Pass
Cab-O-Sil	3.7	110	0.0308	176	0.536**	Fail
Ground Coal	53	105	0.505	196	0.72	Pass
FCC Catalyst	87.5	91	0.912			
Ground Coal	4.4	10	0.44	106	0.87***	Pass
FCC Catalyst	88	96	0.92			

 *Suspect, due to agglomeration by adsorbed? moisture (sticky material)
 **Tapped to a compacted density
***Passed only once; upon repeating test it failed

CONCLUDING REMARKS

These simple exploratory tests offer a clue as to the role
of fine particles in altering the rheological properties of a
powder. If the fines will agglomerate into "soft" particles dis-

persed amongst rigid bigger ones they make the whole more easily
sheared and give it a bulk deformability. The presence of this
property is readily measured by the simple drop test and the few
results so far indicate that such a material will easily move in
extrusion flow through pipes and bends under differential pressure
and with a minimum of net gas flow. The significance of this work
lies in the identification of powders suitable for a vast number
of commercial applications of extrusion flow over and above simple
pneumatic conveying. A prime example involves the charging of
powdered coal to blast furnace tuyeres where it is particularly
desirable to minimize the amount of diluting transport air and
where extrusion flow would afford essentially equal coal feed to
all tuyeres. The more uniform charging of coal to fluid bed gasifiers, combustors, boilers, or high pressure versions of Godel's
"ignifluid" grates, and the design of pressure let-down standpipes
are additional examples.

VERTICAL PNEUMATIC CONVEYING
OF BINARY PARTICLE MIXTURES

K. NAKAMURA and C. E. CAPES

INTRODUCTION

Pneumatic conveying has been applied not only as a
means of conveyance, but also in gas-solids dryers and reactors
in the chemical industry. Many studies have been reported on this
useful unit operation, but generally materials of uniform size have
been used for experiments and analyses. Zenz and Othmer (1960)
reported that saltation and choking velocities are almost equal for
uniform-size particles, while mixed-size materials give saltation
velocities three to six times as great as the choking velocities.
They recommended the use of the geometric weight mean particle
diameter to calculate choking velocity and pressure drop in vertical
conveying of material of non-uniform particle size when the gas
velocity is greater than the free-fall velocity of the largest
particles. Recently Hair and Smith (1972) reported on the behaviour
of mixed-size particles in a vertical conveying line and concluded
that the empirical dimensionless group equation of Vogt and White
(1948) may be applied to mixed-size particles if the geometric
weight mean particle is used in the equation of the pressure drop
as suggested by Zenz and Othmer. Muschelknautz (1959) made a more
theoretical study of pneumatic conveying of uniform- and mixed-size
particles. He used two additional parameters, one related to the

coefficient of repulsion and the other related to the effective relative velocity of the two colliding particles in order to calculate the pressure drop and the particle velocity in a line conveying mixed size particles.

Leung et al. (1971) analyzed the choking velocity for mixed-size particles with the assumption that the average voidage is 0.97 at the onset of choking. They predicted that the line would contain a higher fraction of the larger particles compared with the feed solids, although particle-to-particle collision was not considered in their rather simple analysis.

The present authors (Capes and Nakamura, 1973; Nakamura and Capes, 1973a) also studied vertical pneumatic conveying both theoretically and experimentally for uniform-size particles and derived the following relationship between gas and particle velocities:

$$v_F = b + m \frac{\epsilon}{1-\epsilon} v_p \tag{1}$$

A similar relationship applies reasonably well to the conveying of binary particle mixtures (Nakamura and Capes, 1973b) with the constants b_{mix} and m_{mix} given by a linear combination of the constants for the two types of particles:

$$b_{mix} = x_A b_A + x_B b_B \tag{2}$$

$$m_{mix} = X_A m_A + X_B m_b \tag{3}$$

The extent of particle segregation in the line predicted by the model, based on the assumption similar to that used by Leung et al. (1971) of no interaction between the two types of particles, was much greater than that found experimentally. In this present study, collisions between the particles flowing at different velocities are considered to explain more precisely the behaviour of non-uniformly sized particles in a vertical conveying line.

THEORETICAL CONSIDERATIONS

In this section we derive the equations of pressure drop and particle velocity which will be used later in the calculations for comparison of the model with the experimental data. Fig. 1 is a schematic representation of the momentum transport in a conveying line where mixed-size particles flow uniformly.

Equations of Motion

The equations of motion are derived here in the same way as in the uniform flow model describing the gaseous vertical

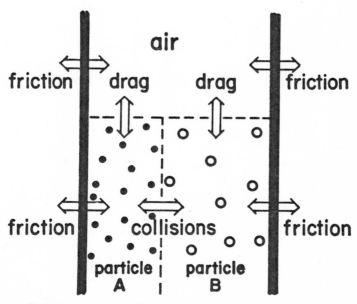

Fig. 1 Representation of momentum transport in a conveying line.

transport of mono-size particles (Nakamura and Capes, 1973a) with
two additional assumptions:

1) The same value of voidage can be used for the drag relationship
    between fluid and either type of particle "A" or "B".

2) Collisions between particles "A" and particles "B" do not
    influence the fluid-particle drag, the fluid-wall friction or
    the particle-wall friction relationships.

A momentum balance on the fluid for a small element of
volume with area equal to the pipe cross-section and increment of
vertical length, dZ, yields Equation (4) under the condition of
steady state and negligible gas expansion at low pressure drops:

$$-\varepsilon \frac{dp}{dZ} - x_A\beta_A \left\{ \frac{v_F}{\varepsilon} - \frac{v_A}{x_A(1-\varepsilon)} \right\}^2 - x_B\beta_B \left\{ \frac{v_F}{\varepsilon} - \frac{v_B}{x_B(1-\varepsilon)} \right\}^2 - \frac{4\tau_F}{D} = 0 \tag{4}$$

Momentum balances on the particles "A" and "B" give respectively:

$$-x_A(1-\varepsilon)\frac{dP}{dZ} + x_A\beta_A \left\{ \frac{\ddot{v}_F}{\varepsilon} - \frac{v_A}{x_A(1-\varepsilon)} \right\}^2 - \frac{4\tau_A}{D} - \rho_A x_A(1-\varepsilon)g - I_{AB} = 0 \tag{5}$$

$$-x_B(1-\varepsilon)\frac{dP}{dZ} + x_B\beta_B \left\{ \frac{v_F}{\varepsilon} - \frac{v_B}{x_B(1-\varepsilon)} \right\}^2 - \frac{4\tau_B}{D} - \rho_B x_B(1-\varepsilon)g + I_{AB} = 0 \tag{6}$$

Here x_A and x_B are the fractions of particle holdup, that is, the
ratios of each holdup to the total holdup in the line. I_{AB} is the
force the faster particles "A" transfer to the slower particles "B"
in a unit volume of the line due to collisions. The coefficients
related to fluid-particle drag, β_A and β_B, and the particle-wall
shear stresses, τ_A and τ_B, may be expressed as in eq. (7) to eq.
(10), taking assumptions (1) and (2) into account:

$$\beta_A = \frac{\rho_A(1-\varepsilon)g}{u_A^2 \varepsilon^{2n-3}} \qquad (7)$$

$$\beta_B = \frac{\rho_B(1-\varepsilon)g}{u_B^2 \varepsilon^{2n-3}} \qquad (8)$$

$$\tau_A = \left(\frac{C_A}{2}\right)\rho_A v_A \qquad (9)$$

$$\tau_B = \left(\frac{C_B}{2}\right)\rho_B v_B \qquad (10)$$

The particle-wall friction factor has been found to be almost inversely related to the solids velocity (Nakamura and Capes, 1973) and C is the constant for each kind of particle in this correlation. Several equations are available for calculation of the fluid-wall shear stress, τ_F.

The addition of eq. (4), (5), and (6) gives the following pressure drop equation:

$$-\frac{1}{g}\frac{dP}{dZ} = (\rho_A x_A + \rho_B x_B)(1-\varepsilon) + \frac{4}{gD}(\tau_F + \tau_A + \tau_B) \qquad (11)$$

It may be shown by substituting eq. (11) into eq. (5) and (6) that the pressure drop term is much less important than the other terms in eq. (5) and (6), and can be neglected since the voidage ε is close to 1 in dilute-phase transport. Eq. (12) and (13) then derive from (5) and (6) respectively.

$$x_A \beta_A \left\{\frac{v_F}{\varepsilon} - \frac{v_A}{x_A(1-\varepsilon)}\right\}^2 - \frac{4\tau_A}{D} - \rho_A x_A(1-\varepsilon)g - I_{AB} = 0 \qquad (12)$$

$$x_B \beta_B \left\{\frac{v_F}{\varepsilon} - \frac{v_B}{x_B(1-\varepsilon)}\right\}^2 - \frac{4\tau_B}{D} - \rho_B x_B(1-\varepsilon)g + I_{AB} = 0 \qquad (13)$$

Force of Collision I_{AB}

The force of collision I_{AB} cannot be neglected in the momentum equations when the particles flow in a line with different velocities. This force I_{AB} is given as the product of the number density of particles "B", the number of particles "A" colliding with the target particle "B" per unit time, and the average momentum change due to collision:

$$I_{AB} = n_B \cdot \frac{\pi(R_A + R_B)^2 \rho_A (v_A - v_B)}{m_A} \cdot M_B \cdot \overline{\Delta V_B} \qquad (14)$$

The change of velocity, ΔV_B, is easily derived from the law of momentum conservation and the equation defining the coefficient of restitution when a particle A collides <u>at the head of</u> a target particle B in the direction parallel to the pipe axis:

$$\Delta V_B = (1+e) \frac{m_A}{m_A + m_B} \left\{ \frac{v_A}{x_A(1-\varepsilon)} - \frac{v_B}{x_B(1-\varepsilon)} \right\} \qquad (15)$$

The number density n_B in eq. (14) is expressed as $\rho_B x_B(1-\varepsilon)/m_B$, and the masses m_A and m_B are equal to $\rho_A(\frac{4}{3})\pi R_A^3$ and $\rho_B(\frac{4}{3})\pi R_B^3$ respectively, so that eq. (14) can be rearranged into the following equation if $\overline{\Delta V_B}$ is approximately replaced with ΔV_B of eq. (15).

$$I_{AB} = \gamma \, x_A x_B (1-\varepsilon)^2 \left\{ \frac{v_A}{x_A(1-\varepsilon)} - \frac{v_B}{x_B(1-\varepsilon)} \right\}^2 \qquad (16)$$

where

$$\gamma = \frac{3}{4} \phi \, (1+e)\rho_A \cdot \frac{1}{R_B} \cdot \frac{(1+R_A/R_B)^2}{1+(\rho_A/\rho_B)(R_A/R_B)^3} \qquad (17)$$

The factor ϕ is introduced to compensate for such effects as multiplicity of scattering, fluid viscous motion causing particles to

avoid collision and impact other than the direct case assumed in
deriving eq. (15).

Soo (1967) gave the force acting on a single particle B in
a cloud of colliding particles A with the assumption of single
scattering. The coefficient of restitution does not appear in this
force, since he assumed a direction of collision parallel to the pipe
axis and specular reflection. The multiplication of Soo's force by
the number density of target particles B (see Appendex) results in
the same form of I_{AB} as eq. (16), but the coefficient is slightly
different from the one given by eq. (17), <u>viz</u>:

$$\gamma_S = \frac{3}{4} \eta \rho_A \cdot \frac{1}{R_B} \cdot \frac{(1+R_A/R_B)^2}{1 + (\rho_A/\rho_B)(R_A/R_B)^3} \tag{18}$$

A more refined analysis was reported by Muschelknautz (1959)
who considered the change of each particle velocity with the assumption
that the momentum is conserved not between m_A and m_B of a single
colliding particle A and a single target particle B , but between
their total masses existing in the line. In addition, Muschelknautz
took into consideration the unsteadiness of particle motion, that is
the fact that the finer particles before collision have larger
velocities than their average values while the reverse is true for the
larger particles. His result can be rearranged into the form of
eq. (16) (see Appendix) with a different coefficient:

$$\gamma_M = \frac{3}{4} \cdot 2\xi \cdot \frac{1+e}{1-e} \cdot \rho_A \cdot \frac{1}{R_B} \cdot \frac{(1 + R_A/R_B)^2}{1 + \rho_A x_A / \rho_B x_B} \tag{19}$$

where ξ is a parameter related to the effective relative velocity
between colliding particles and target particles which depends on the

deviation of impact from that producing the maximum momentum exchange. Muschelknautz assumed ξ and e to be 0.33~1.0 and 0.5 respectively when a mixture of 50 weight percent each of wheat and wheat grit was conveyed at a loading ratio of 0 to 4.5. It is evident from eq. (19) that the force of collision I_{AB} will become extremely large for materials with a coefficient of restitution close to 1 (e → 1) unless the velocities of the two types of particles are close to each other.

In the present analysis the force of collision I_{AB} was calculated from eq. (16) and (17) using appropriate assumed values of the combined coefficient, $\phi(1+e)$.

CALCULATION OF THEORETICAL RESULTS

In this section the particle momentum equations are used to calculate the particle velocity and holdup and the effects of the coefficients C and $\phi(1+e)$ are examined.

Rearrangement of the Particle Momentum Equations

The following dimensionless variables and Froude numbers are introduced here to simplify eq. (12) and (13):

$$\alpha_A \equiv \frac{v_A \, \epsilon}{v_F x_A (1-\epsilon)} \qquad (20) \qquad \alpha_B \equiv \frac{v_B \epsilon}{v_F x_B (1-\epsilon)} \qquad (21)$$

$$F_A \equiv \frac{u_A}{\sqrt{gD}} \qquad (22) \qquad F_B \equiv \frac{u_B}{\sqrt{gD}} \qquad (23)$$

$$F_F \equiv \frac{v_F}{\sqrt{gD}} \qquad (24) \qquad H_A \equiv \frac{C_A}{\sqrt{gD}} \qquad (25)$$

$$H_B \equiv \frac{c_B}{\sqrt{gD}} \tag{26}$$

The voidage ε is close to 1 in dilute-phase transport and eq. (7) and (8) can be approximated by:

$$\beta_A = \frac{\rho_A(1-\varepsilon)g}{u_A^2} \tag{27}$$

$$\beta_B = \frac{\rho_B(1-\varepsilon)g}{u_B^2} \tag{28}$$

This approximation leads to the use of the same fluid-particle drag relationship as in Muschelknautz's (1959) analysis and the excess pressure drop is derived from eq. (11), (12), and (13) as done by Muschelknautz:

$$-\frac{1}{g}\frac{dP}{dZ} - \frac{4}{gD}\tau_F = \frac{\rho_A x_A(1-\varepsilon)}{u_A^2} \cdot \left\{ \frac{v_F}{\varepsilon} - \frac{v_A}{x_A(1-\varepsilon)} \right\}^2$$

$$+ \frac{\rho_B x_B(1-\varepsilon)}{u_B^2} \left\{ \frac{v_F}{\varepsilon} - \frac{v_B}{x_B(1-\varepsilon)} \right\}^2 \tag{29}$$

Eq. (12) and (13) may be rewritten with the above dimensionless variables after eq. (16) and (17) are substituted into them:

$$(1-\alpha_A)^2 - 2\frac{H_A F_A^2 \alpha_A}{F_F} - \left(\frac{F_A}{F_F}\right)^2 - \left(\frac{\gamma_S D}{\rho_A}\right) \cdot \frac{F_B F_A^2}{F_F} \cdot \frac{(\alpha_A - \alpha_B)^2}{\alpha_B} = 0 \tag{30}$$

$$(1-\alpha_B)^2 - 2\frac{H_B F_B^2 \alpha_B}{F_F} - \left(\frac{F_B}{F_F}\right)^2 + \left(\frac{\gamma_S D}{\rho_B}\right) \cdot \frac{F_A F_B^2}{F_F} \cdot \frac{(\alpha_A - \alpha_B)^2}{\alpha_A} = 0 \tag{31}$$

Then eq. (30) and (31) can be solved for either α_A or α_B:

$$\alpha_A = \frac{1}{\{1-(\gamma_S D/\rho_A)(F_A^2 F_B/F_F \alpha_B)\}} \left[1 + \frac{F_A^2 H_A}{F_F} - \left(\frac{\gamma_S D}{\rho_A}\right)\left(\frac{F_A^2 F_B}{F_F}\right) \right.$$

$$\left. - \sqrt{\left\{1 + \frac{F_A^2 H_A}{F_F} - \left(\frac{\gamma_S D}{\rho_A}\right)\left(\frac{F_A^2 F_B}{F_F}\right)\right\}^2 - \left\{1 - \left(\frac{\gamma_S D}{\rho_A}\right)\left(\frac{F_A^2 F_B}{F_F \alpha_B}\right)\right\}\left\{1 - \left(\frac{F_A}{F_F}\right)^{-2} - \left(\frac{\gamma_S D}{\rho_A}\right)\frac{(F_A^2 F_B \alpha_B)}{F_F}\right\}} \right]$$

$$(32)$$

$$\alpha_B = \frac{1}{\{1 - (\gamma_S D/\rho_B)(F_A F_B^2/F_F \alpha_A)\}} \left[1 + \frac{F_B^2 H_B}{F_F} + \left(\frac{\gamma_S D}{\rho_B}\right)\left(\frac{F_A F_B^2}{F_F}\right) \right.$$

$$\left. - \sqrt{\left\{1 + \frac{H_B F_B^2}{F_F} + \left(\frac{\gamma_S D}{\rho_B}\right)\left(\frac{F_A F_B^2}{F_F}\right)\right\} - \left\{1 + \left(\frac{\gamma_S D}{\rho_B}\right)\left(\frac{F_A F_B^2}{F_F \alpha_A}\right)\right\}\left\{1 - \left(\frac{F_B}{F_F}\right)^2 + \left(\frac{\gamma_S D}{\rho_B}\right)\left(\frac{F_A F_B^2 \alpha_A}{F_F}\right)\right\}} \right]$$

$$(33)$$

An initial value of either α_A or α_B is assumed, for example $\alpha_A = 1 - \frac{u_A}{v_F}$, and iterative calculations involving eq. (32) and (33) lead to converging solutions of α_A and α_B for gas velocities above a minimum value. The fractions of particle holdup, x_A and x_B, and the voidage, ε, can be calculated by substitution of the solutions, α_A and α_B, and the relationship $x_A + x_B = 1$ into Equations (20) and (21):

$$x_A = \frac{\alpha_B v_B}{\alpha_A v_B + \alpha_B v_A} \qquad (34)$$

$$x_B = \frac{\alpha_A v_B}{\alpha_A v_B + \alpha_B v_A} \qquad (35)$$

$$\varepsilon = \frac{\alpha_A \alpha_B}{\alpha_A \alpha_B + \alpha_A (F_B/F_F) + \alpha_B (F_A/F_F)} \qquad (36)$$

Effect of the Coefficients C and ϕ (1 + e)

The relationships between gas velocity, particle velocity and the extent of segregation in the line were calculated by the procedure outlined above. These relationships depend not only on the coefficients C and $\phi(1+e)$ but also on such operating variables as particle feed rate, feed ratio of the two particle types, etc. The effects of C and $\phi(1+e)$ will be examined for the case of a mixture of 1.083 mm and 2.90 mm glass beads conveyed in a vertical pipe of 3 in. diameter with feed rate, S, of 3 kg/min and feed ratio, X_A, of 0.2, 0.5, and 0.8.

Fig. 2 and Fig. 3 show the effect of the coefficients C and $\phi(1+e)$, respectively, on the velocity relationship. It can be seen that the relationship is non-linear particularly with larger feed fractions of the finer particles. This is contrary to the result obtained in the conveying of mono-size particles in which a straight line was found which was characterized simply by the slope m and the intercept b. (Capes and Nakamura, 1973). In Fig. 2 for the binary mixture, however, the slope does become virtually constant at higher velocities and increases with the coefficient C as in the conveying of uniformly-sized particles where the relationship $m = 1 + Cu/gD$ was derived and found to apply quite well (Nakamura and Capes, 1973a). As shown in Fig. 3, the coefficient $\phi(1+e)$ does not influence the slope as much as does the value of C, but the lines move downwards almost in parallel as $\phi(1+e)$ is increased. It should be noted that the intercepts of the theoretical curves will differ from those found by least squares fitting of straight lines by Nakamura and Capes (1973b) due to the non-linearity of the theoretical relationships.

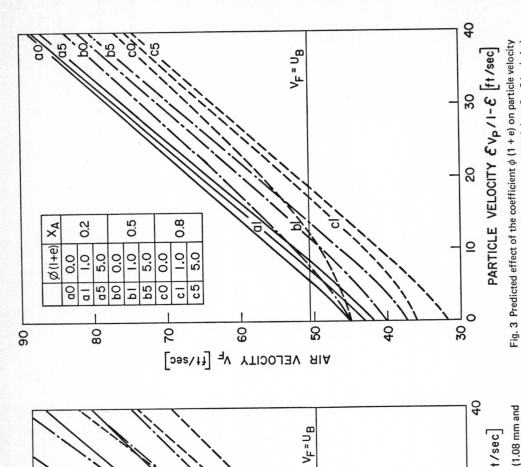

Fig. 3 Predicted effect of the coefficient ϕ (1 + e) on particle velocity (1.08 mm and 2.90 mm glass beads, C = 0.0328 ft/sec, S = 3 kg./min.)

Fig. 2 Predicted effect of coefficient V on particle velocity (1.08 mm and 2.90 mm glass beads, ϕ (1 + e) = 1.0, S = kg./min.)

Fig. 4 and Fig. 5 show the effect of the value of C and $\phi(1+e)$ on the ratio of holdup fraction to feed fraction, x_A/X_A. The coefficient C does not influence this ratio greatly while the coefficient $\phi(1+e)$ shows a very strong influence indicating that particle collisions may not be ignored in analyzing the behaviour of binary mixtures.

COMPARISON WITH EXPERIMENTAL DATA

Experimental Details

The vertical conveying line used in the experiments was the same as that described previously (Capes and Nakamura, 1973).

Three different binary mixtures were studied, two employing steel shot and one employing glass beads. For each mixture several different compositions were examined. For each composition two sets of experiments at two different solids feed rates were investigated. The operating conditions were set and the system was allowed to come to steady state for approximately one hour before the pressure gradient up the riser and the solids feed rate were recorded. The four valves were fired simultaneously and the solids trapped in the lower part of the column up to the holdup valve (the acceleration zone) were blown out through the tee at the bottom of the line. The holdup valve was then opened and the solids which had been trapped in the test section were removed in the same fashion. All four samples, the bottom feed sample from the lower diverting valve, the top feed sample from the upper diverting valve, the acceleration fraction from the acceleration zone and the holdup fraction from the test section were

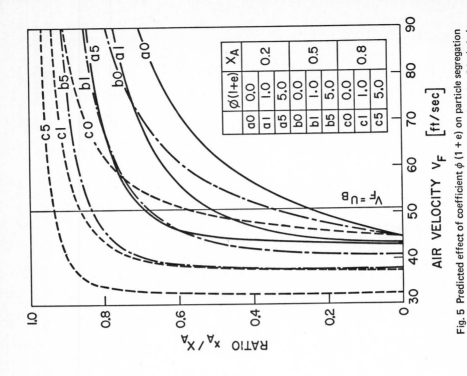

Fig. 5 Predicted effect of coefficient $\phi (1 + e)$ on particle segregation (1.08 mm and 2.90 mm glass beads, C = 0.0328 ft/sec, S = 3 kg/min.)

	$\phi(1+e)$	X_A
		0.2
a0	0.0	
a1	1.0	
a5	5.0	
		0.5
b0	0.0	
b1	1.0	
b5	5.0	
		0.8
c0	0.0	
c1	1.0	
c5	5.0	

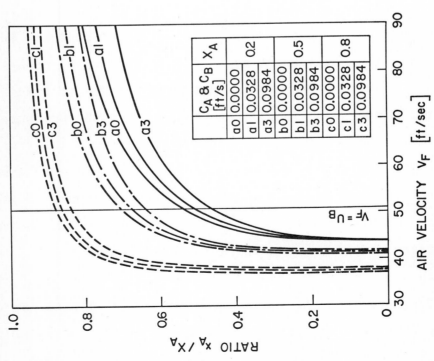

Fig. 4 Predicted effect of coefficient C on particle segregation (1.08 mm and 2.90 mm glass beads, $\phi (1 + e)$ = 1.0, S = 3 kg/min.)

	C_A & C_B [ft/s]	X_A
		0.2
a0	0.0000	
a1	0.0328	
a3	0.0984	
		0.5
b0	0.0000	
b1	0.0328	
b3	0.0984	
		0.8
c0	0.0000	
c1	0.0328	
c3	0.0984	

Table 1 Properties of Particles Used

Material	Particle Density (g/cm³)	Diameter (mm)	Calculated Terminal Velocity (ft/sec)	Calculated Coefficient C‡ (ft/sec)
Glass	2.90	1.08	26.1	0.0123
	2.86	2.90	50.6	0.0351
Steel	7.85	0.535	27.1	0.0344
	7.70	1.20	48.9	0.0705
	7.70	2.34	75.2	0.0587

‡
Calculated from eq. (12) of Nakamura and Capes (1973a).

weighed and hand-screened to determine the extent of particle segregation in the riser. The air velocity was then changed to the next desired level and the procedure was repeated, the solids feed rate remaining at the same level. Once measurements at a suitable number of air velocities (usually five or six) had been taken, the solids feed rate was changed to a different level. When the set of experiments using this new solids feed rate was completed, the composition of the mixture was changed and the whole procedure was repeated.

A summary of the properties of the particles used is presented in Table I.

RESULTS AND DISCUSSION

The experimental data points and the analytically predicted lines of the relationships between gas and particle velocities are shown in Fig. 6, 7 and 8, while the corresponding segregation data are given in Fig. 9, 10, and 11, respectively. The coefficients C

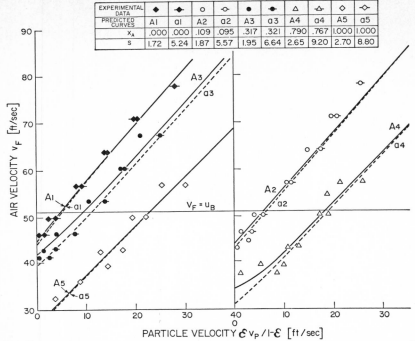

EXPERIMENTAL DATA	◆	◆—	○	-○-	●	●—	△	△—	◇	◇—
PREDICTED CURVES	A1	a1	A2	a2	A3	a3	A4	a4	A5	a5
X_A	.000	.000	.109	.095	.317	.321	.790	.767	1.000	1.000
s	1.72	5.24	1.87	5.57	1.95	6.64	2.65	9.20	2.70	8.80

Fig. 6 Relationship between gas and particle velocities for 2.08 mm and 2.90 mm glass beads mixture [ϕ (1 + e) = 2.0 for predicted curves].

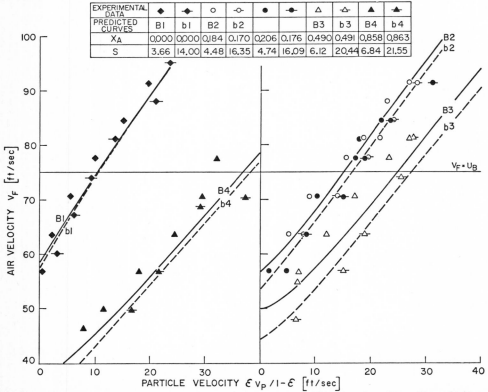

EXPERIMENTAL DATA	◆	◆—	○	-○-	●	●—	△	△—	▲	▲—
PREDICTED CURVES	B1	b1	B2	b2			B3	b3	B4	b4
X_A	0.000	0.000	0.184	0.170	0.206	0.176	0.490	0.491	0.858	0.863
S	3.66	14.00	4.48	16.35	4.74	16.09	6.12	20.44	6.84	21.55

Fig. 7 Relationship between gas and particle velocities for 0.54 mm and 2.34 mm steel particles mixture [ϕ (1 + e) = 2.0 for predicted curves].

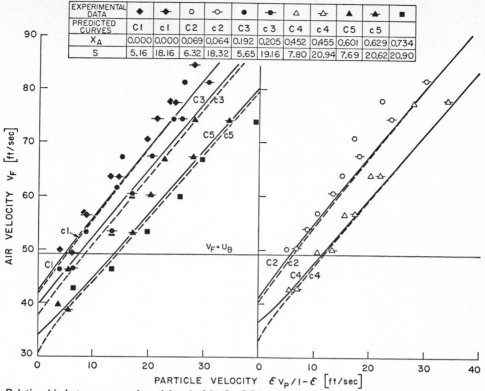

EXPERIMENTAL DATA	◆	◆	○	-○-	●	-●-	△	-△-	▲	-▲-	■
PREDICTED CURVES	C1	c1	C2	c2	C3	c3	C4	c4	C5	c5	
X_A	0.000	0.000	0.069	0.064	0.192	0.205	0.452	0.455	0.601	0.629	0.734
S	5.16	18.16	6.32	18.32	5.65	19.16	7.80	20.94	7.69	20.62	20.90

Fig. 8 Relationship between gas and particle velocities for 0.54 mm and 1.20 mm steel particles mixture [$\phi (1 + e) = 6.0$ for predicted curves].

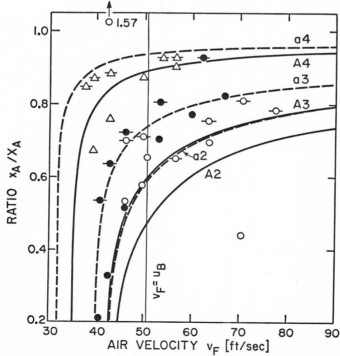

Fig. 9 Particle segregation for 1.08 mm and 2.90 mm glass beads mixture [$\phi (1 + e) = 2.0$; see Fig. 6 for legend].

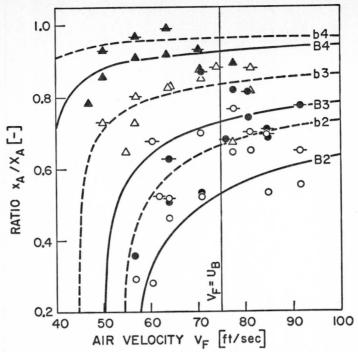

Fig. 10 Particle segregation for 0.54 mm and 2.34 mm steel particles mixture [ϕ (1 + e) = 2.0; see Fig. 7 for legend].

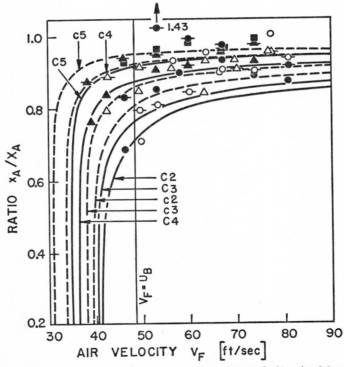

Fig. 11 Particle segregation for 0.54 mm and 1.20 mm steel particles mixture [ϕ (1 + e) = 6.0; see Fig. 8 for legend].

used in the calculations were those obtained by substitution in the
relationship, $m = 1 + Cu/gD$, using the values of m for the uniformly
sized particles making up the mixtures. The coefficient $\phi(1+e)$ for
each mixture was chosen by visual inspection to obtain the best fit
of the experimental points.

The velocity data in Fig. 6, 7, and 8 show considerable
scatter but the calculated and experimental results are in agreement
to the extent that there is an almost linear relationship between
air and particle velocities at each end of the concentration range
with a curved portion in evidence at intermediate concentrations
(see Fig. 6 especially). Thus our earlier assumption of a linear
relationship (Nakamura and Capes, 1973b) is not strictly true in
the case of mixtures of different particle sizes and densities. The
experimental data also agree with the prediction of greater particle
velocities as the feed rate is increased although the effect of
particle feed rate is not large.

The model presented here is most successful in fitting
the segregation results in Fig. 9, 10, and 11 where the fraction of
fines in the column to that in the feed, x_A/X_A, is shown as a function
of gas velocity. The inclusion of particle collisions in the model
through the factor $\phi(1+e)$ allows the analytical relationships to be
fitted quite well to the experimental points. As was noted earlier
and may be seen in Fig. 5 with the curves for $\phi(1+e) = 0$, the
assumption of no particle interaction leads to the prediction of more
particle segregation than is actually observed. Particle collisions
slow down the finer particles and accelerate the coarser ones leading
to greater fines concentrations and fewer coarse particles,
respectively, in the column than would be the case if collisions did

not occur. The model also predicts the observed trends of greater accumulation of coarse particles in the line as the gas velocity is reduced and the decrease in choking velocity as the concentration of finer particles increases.

The values of the coefficient $\phi(1+e)$ found to best represent the experimental data are:

1.08 mm and 2.90 mm glass beads: $\phi(1+e) = 2.0$

0.54 mm and 2.34 mm steel shot: $\phi(1+e) = 2.0$

0.54 mm and 1.20 mm steel shot: $\phi(1+e) = 6.0$

These values suggest that ϕ is always greater than unity since the coefficient of restitution, e, has been reported (Iinoya, 1965) to be 1.0 for glass particles and 0.2 to 0.6 for metal particles. Some of the factors affecting ϕ are the fraction of particles undergoing impaction, the effective relative particle velocity and multiple collisions. Recalling that it was assumed that a given particle never returns to the same target particle for another collision in deriving the force, I_{AB}, it is seen that multiple collisions would have the effect of making $\phi > 1$. The other factors which are considered to affect ϕ would tend to decrease its value below unity. It is possible therefore, especially with binary mixtures of similar amounts of the two particles, that the finer particles are accelerated to collide repeatedly with the same larger particle resulting in values of ϕ greater than unity. Further experiments and analyses are required to elucidate this point.

Fig. 12 shows some experimental pressure drop data compared with theoretical curves. The following equation, obtained by substitution of eq. (9) and (10) into eq. (11), was used to predict the pressure drop relationship:

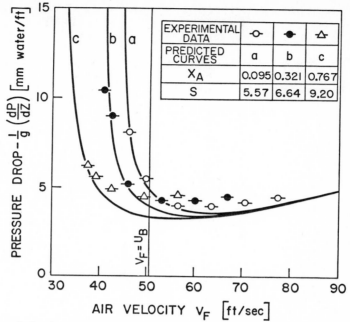

Fig. 12 Pressure drop in line for 1.08 mm and 2.90 mm glass beads mixture [ϕ (1 + e) = 2.0 for predicted curves].

$$-\frac{1}{g}\frac{dP}{dZ} = (\rho_A x_A + \rho_B x_B)(1-\varepsilon) + \frac{2}{gD}(\rho_A x_A c_A + \rho_B x_B c_B)v_p + \frac{4}{gD}\tau_F$$

It should be noted that the feed rate, S, was not constant
in these experiments but varied with the feed fraction, x_A. The
predicted curves agree with experiment at low gas velocities in which
the pressure drop is larger for lower feed fractions of fines due
to the accumulation of coarse particles in the line. This relationship
appears to be reversed in the experimental data at higher gas rates.
This may be ascribed in the present experiments to differences in
particle feed rate, although a similar phenomenon was observed by Hair
and Smith (1972) at constant feed rate conditions.

CONCLUSION

The theoretical model presented here, which takes into
account particle-particle collisions during the vertical pneumatic

conveying of mixtures, indicates a number of interesting features, including:

(1) The particle-gas velocity relationship is non-linear with a curved portion at lower gas velocities. The relationship can, however, be represented by a straight line at higher gas velocities.

(2) The constant related to the particle-wall friction factor, C, influences the slope of the velocity relationship while the coefficient related to particle-particle collisions, $\phi(1+e)$ moves the relationship upward or downward in parallel.

(3) The extent of particle segregation is affected considerably by the coefficient $\phi(1+e)$ and less by the constant C. Particle segregation also depends on the feed rate, S, more so than does the velocity relationship.

These features are also observed in the experimental data, to a greater or lesser extent. The values of the coefficient $\phi(1+e)$ which provide the best fit of the segregation data depend on the properties of the three mixtures studied. Momentum transfer due to collision appears to be much larger for the 0.535 and 1.20 mm steel particles mixture than for either the 1.083 and 2.90 mm glass beads or the 0.535 and 2.34 mm steel shot mixtures.

APPENDIX

Derivation of Equations (18) and (19)

Soo (1967) gave the force acting on a single particle B due to collision with particles A by the equation (cf. page 200 in his book):

$$f_{AB} = \eta_{AB} n_A \# (R_A + R_B)^2 \frac{m_A m_B}{m_A + m_B} \left\{ \frac{v_A}{x_A(1-\epsilon)} - \frac{v_B}{x_B(1-\epsilon)} \right\}^2 \quad \text{(A-1)}$$

The force of collision per unit volume of line can be expressed as the product of the above force, f_{AB}, and the number density of particles B, n_B:

$$I_{AB} = n_B f_{AB} = \eta_{AB} n_A n_B \pi (R_A + R_B)^2 \cdot \frac{m_A m_B}{m_A + m_B} \left\{ \frac{v_A}{x_A(1-\epsilon)} - \frac{v_B}{x_B(1-\epsilon)} \right\}^2 \quad \text{(A-2)}$$

The number densities, n_A and n_B, and the particle masses, m_A and m_B are given by the following equations:

$$n_A = \frac{\rho_A x_A (1-\epsilon)}{m_A} \quad \text{(A-3)}$$

$$n_B = \frac{\rho_B x_B (1-\epsilon)}{m_B} \quad \text{(A-4)}$$

$$m_A = \rho_A \cdot \frac{4}{3} \pi R_A^3 \quad \text{(A-5)}$$

$$m_B = \rho_B \cdot \frac{4}{3} \pi R_B^3 \quad \text{(A-6)}$$

Eq. (A-3) to (A-6) are substituted into eq. (A-2) and the force I_{AB} from Soo's analysis is derived in the form of eq. (16) with γ_S given by eq. (18).

Muschelknautz (1959) gave the force acting on particles B existing in a section of line of length ΔZ due to collision of particles A (cf. page 14-16 in his paper):

$$\Delta W_S \leqq F_o \Delta Z \ I_{AB} = \frac{\Delta \bar{M}_B}{m_B} \cdot \frac{(R_A + R_B)^2 \pi}{F_o} \cdot M_A \cdot \left\{ \frac{v_A}{x_A(1-\varepsilon)} - \frac{v_B}{x_B(1-\varepsilon)} \right\}$$

$$\cdot \frac{x_A(1-\varepsilon)}{v_A} \times 2\xi \ \frac{1+e}{1-e} \cdot \left\{ \frac{v_A}{x_A(1-\varepsilon)} - \frac{v_B}{x_B(1-\varepsilon)} \right\} / \left\{ 1 + \frac{\mu_A}{\mu_B} \cdot \frac{x_A v_B}{x_B v_A} \right\}$$

$$(A-7)$$

where

$$\Delta \bar{M}_B = \rho_B x_B (1-\varepsilon) \ F_o \ \Delta Z \qquad\qquad (A-8)$$

$$m_B = \frac{4}{3}\pi \rho_B R_B^{\ 3} \qquad\qquad (A-9)$$

$$F_o = \frac{\pi D^2}{4} \qquad\qquad (A-10)$$

$$M_A = \rho_A v_A F_o \qquad\qquad (A-11)$$

$$\mu_A = \rho_A v_A / \rho_F v_F \qquad\qquad (A-12)$$

$$\mu_B = \rho_B v_B / \rho_F v_F \qquad\qquad (A-13)$$

Eq. (A-8) to (A-13) are substituted into eq. (A-7) and the force I_{AB} from Muschelknautz's analysis is derived in the form of eq. (16) with γ_M given by eq. (19).

NOMENCLATURE

b = constant in eq. (1), ft./sec.

C = constant related to the particle-wall shear stress, ft./sec.

D = pipe diameter, ft.

e = coefficient of restitution

F = dimensionless number [cf. eq. (22), (23) and (25)]

g = acceleration due to gravity, ft./sec.^{2}

H = dimensionless number [cf. eq. (24) and (26)]

$I_{AB}=$ force of collision per unit volume of line, $lb/ft^2 sec^2$

m = constant in eq. (1), ft./sec. or particle mass, lb.

n = exponent in Richardson – Zaki equation [cf. eq. (7) and (8)] or number density of particles, ft^{-3}.

P = gas pressure, $lb./ft.sec^2$.

R = particle radius, ft.

S = particle feed rate, kg./min.

u = free falling terminal velocity, ft./sec.

v = superficial velocity, ft/sec.

ΔV = change of particle velocity due to collision, ft/sec.

$\Delta \bar{V}$ = average value of ΔV, ft/sec.

x = volume fraction of given type of particle in the line.

X = volume fraction of given type of particle in the feed.

Z = vertical coordinate, ft.

Greek Letters

α = ratio of particle to gas velocity [cf. eq. (20) and (21)].

β = constant in fluid-particle drag relationship, lb/ft^4.

γ = coefficient in eq. (16) for force of collision, lb/ft^4.

γ_M = coefficient γ derived from Muschelknautz (1959) analysis.

γ_S = coefficient γ derived from Soo (1967) analysis.

ε = void fraction.

η = fraction of impaction.

ξ = coefficient related to the effective relative velocity of collision [cf. eq. (19)].

ρ = density, lb/ft^3.

τ = shear stress, lb/ft.sec^2

ϕ = parameter introduced into equations for force of collision
     [cf. eq. (17)].

Subscripts

A = refers to the finer particles of a binary mixture

B = refers to the coarser particles of a binary mixture

F = refers to the fluid

mix = refers to the particle mixture

p = refers to a particle

LITERATURE CITED

Capes, C.E., Nakamura, K., Can. J. Chem. Eng., 51, 31 (1973).

Hair, A.R., Smith, K.L., Mech. Chem. Eng. Trans., 19 (May, 1972).

Iinoya, K. (ed.), "Handbook of Powder Technology", p. 127, Asakura
     Book Pub. Co., Tokyo, Japan, 1965 (in Japanese).

Muschelknautz, E., V.D.I. (Ver.Deut.Ing.) - Forschungsh. 476 (1959).

Nakamura, K., Capes, C.E., Can. J. Chem. Eng., 51, 39 (1973a).

Nakamura, K., Capes, C.E., J. Res. Assoc. Powder Tech., 10, 603
     (1973b) (in Japanese).

Leung, L.S., Wiles, R.J., Nicklin, D.J., Ind. Eng. Chem. Process Des.
     Develop., 10, 183 (1971).

Soo, S.L., "Fluid Dynamics of Multiphase Systems", p. 200, Blaisdell,
     Waltham, Mass., 1967.

Vogt, E.G., White, R.A., Ind. Eng. Chem., 40, 173 (1948).

Zenz, F.A., Othmer, D.F., "Fluidization and Fluid-Particle Systems",
     pp. 322-332, Reinhold, New York, N.Y., 1960.

SOLIDS MIXING IN SLUGGING FLUIDIZED BEDS

O. E. POTTER and W. THIEL

SUMMARY.   Studies of mixing in 2 in. dia., 4 in. dia. and 9 in.
dia. fluidized beds of high aspect ratio are reported. Slugs may
be round-nosed[1] or square[2]. Square-nosed slugs were found in
the 9 in. dia. as well as the smaller diameter beds, depending on
the material. Materials like cracking catalyst gave round-nosed
slugs in all beds. The mixing obtained in the case of round-nosed
slugs was correlated by a simple model.

EXPERIMENTAL DETAILS.   The beds were 5.1 cm dia. x 360 cm height,
10.2 cm x 470 cm height and 21.8 cm by 690 cm height. The powders
employed were fresh cracking catalyst, aluminium powder, mixtures
of the two, a coarse fraction of fresh cracking catalyst, copper
powder, ballotini, talc and a narrow range and broad range equilibrium cracking catalyst. Heat was transferred to the bed at the
bottom end and removed at the top end, leaving a working section
in the middle which was thermally insulated. The axial temperature
gradient was measured in the working section. Rise velocities and
lengths of slugs and interslug material were measured.

EXPERIMENTAL RESULTS. (i) A notable result was that 'square-
nosed' or 'raining' slugs appeared normally in all materials other
than cracking catalyst or aluminium powder (U < 15 cm s^{-1}) or
small diameter ballotini (U$_{mf}$ = 0.25 cm s^{-1}). Vibration, applied
in the two smaller beds only, had little effect on axial diffusivity when round-nosed slugs were present but in some cases helped
to convert a system from the square-nosed slug to round-nosed slugs
or half slugs. Presumably if sufficient vibration energy could be
applied, round-nosed slugs could be produced always. (ii) With

cracking catalyst, as velocity was increased, the flow regime
changed to the 'turbulent' regime described by Davidson *et al*[4][5].
(iii) Rise velocities of slugs compared reasonably well with the
findings of Davidson *et al*, as also did slug-lengths. Because the
beds employed were rather deep, coalescence was found to continue
until the length of interslug material was in some cases 6-8 bed
diameters. (iv) Mixing rates observed were very high when round-
nosed slugs were employed, e.g. in the 9 in. dia. bed with
cracking catalyst, a value of $E_{SA} \approx 1000$ cm^2 s^{-1} was observed.
In Figure 1 data of this and other work is presented and compared
with the model. The data relate only to round-nosed slugs (a
presumption for the data of other workers). It will be noted that
the model prediction varies with the length of interslug material
(expressed as T, the number of column diameters), which is small,
say 2-3 for shallow beds and up to 6-8 for deep beds in the case
of beds of diameter 5-10 cm.. For beds of larger diameter T
becomes smaller but the increase in value with depth persists.

SIMPLE MODEL FOR MIXING WITH ROUND-NOSED SLUGS. The model
initially proposed that the interslug material be considered well-
mixed so that the system reduced to a series of CSTR vessels with
the vessels moving in the direction of flow. The flow from vessel
to vessel is the material which flows around the slug. This
model gave too large an axial dispersion. It was then found that
if the interslug region was divided in two, one part (71% v/v)
being well-mixed and the remainder being in piston-flow, that a
good representation of the mixing data was obtained; see Figure 1.

FORMATION OF SQUARE-NOSED SLUGS. Measurements of the angles of
internal friction of packed beds of powder order the powders with
respect to the ease of forming round-nosed bubbles. Thus cracking
catalysts (tan β of order 6-8, where β is the angle of internal
fraction) forms round-nosed slugs persistently, glass spheroids
(tan β = 5.9) forms round-nosed slugs more readily than glass
spheroids (tan β = 3.9) and sand (tan β = 3.4).

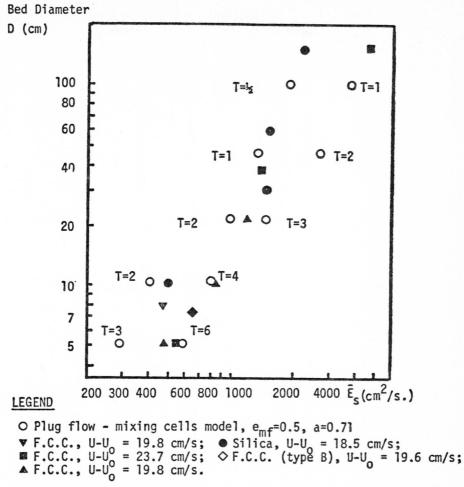

Fig. 1 Comparison of the plug flow-mixing cells model with all the available experimental data.

LEGEND

○ Plug flow - mixing cells model, e_{mf}=0.5, a=0.71
▽ F.C.C., $U-U_o$ = 19.8 cm/s;   ● Silica, $U-U_o$ = 18.5 cm/s;
■ F.C.C., $U-U_o$ = 23.7 cm/s;   ◇ F.C.C. (type B), $U-U_o$ = 19.6 cm/s;
▲ F.C.C., $U-U_o$ = 19.8 cm/s.

NOMENCLATURE

E_{SA} = axial diffusion coefficient

β = angle of internal friction.

REFERENCES

(1)   Stewart, P.S.B., and Davidson, J.F. *Powder Technology*, *1*, 61,
      (1967); Kehoe, P.K.W. and Davidson, J.F. "Chemeca '70",
      The Institution of Chem. Engrs. Symp. Ser. No. 33,

Butterworth's of Australia, (1970), p.1-97; Hovmand, S.,
Freedman, W. and Davidson, J.F. *Trans. Inst. Chem. Engrs.*
49, 149, (1971); Hovmand, S., and Davidson, J.F.
"Fluidization" ed. Davidson, J.F. and Harrison, D.,
Academic Press, London and New York (1971), p.195.

(2) Gelperin, N.I., Ainshtein, V.G. and Sikhanova, L.I. *Theor.*
Found. Chem. Eng. 4, 123, (1970); Gelperin, N.I.,
Ainshtein, V.G., Sulkanova, L.I. and Pavlinko, I.V., *ibid*
4, 426 and 515, (1970).

SUPPLEMENTARY INFORMATION

Figure 1 shows the equipment employed in the study, more
precisely the rig 10.2 cm. dia. Table 1 shows the general dimensions
of each of the three rigs. The two smaller beds could be vibrated by
an electromagnetic vibrator attached to the framework of the rig.
The heat flux could be determined from the heat removed in the cooling
water. The presence of gas slugs in the bed was detected by semi-
circular capacitance plates attached round the outside of the glass
experimental section. The 21.8 cm. dia. bed was operated with two
different cooling sections. Initially the cooler was constructed
with internal cooling tubes but it was observed that with this cooler
materials which would otherwise yield round-nosed slugs began to form
square-nosed slugs. Hence a water-jacket only was employed to avoid
obstructions in the bed itself. Measured heat losses from the
experimental section on the smallest bed were found to be about 1%
of the axial heat flow.

Since the formation of square-nosed slugs appears fairly
normal, some attempts were made to see if there was a critical length
to diameter as expressed in the angle of internal friction. Figure 2
illustrates the conventional packed bed measurement (A) and equipment
devised for the fluidized bed (B) in which the distributor consisted
of a sliding chamber to which air was supplied, the upper surface being
à porous distribution plate. However no locking force could be

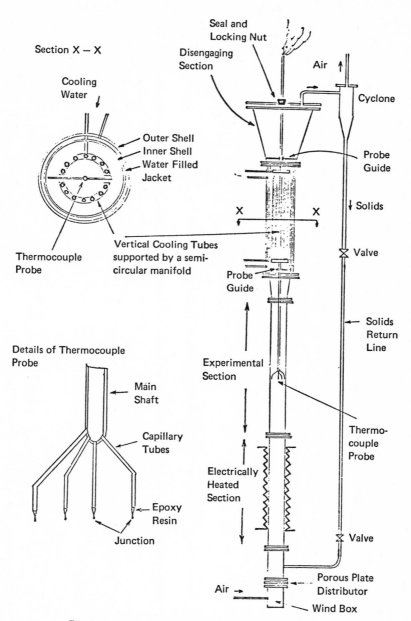

Fig. 1 Diagrammatic sketch of the 10.2 cm. diameter bed.

Table 1 Details of the Experimental Apparatus

	Bed Diameter (cm)		
Distributor type	5.1	10.2	21.8
	Porous Bronze Plate	Porous Bronze Plate	Single Bubble Cap
Overall height of bed	12 feet	15½ feet	22½ feet
Height of the cooling section	2½ feet	3 feet	5½ feet
Height of the experimental section	4 feet	6 feet	6 feet
Height of the heated section	1 foot	3½ feet	6 feet
Height of the disengaging zone	1 foot	2 feet	2½ feet
Internal diameter of the disengaging zone	4 inches	1 foot	2 feet
Electric heaters	1 kW.	3 kW.	15 kW.
Details of the cooling section	Tight wound spiral coil on the outside of the bed wall	Water jacket and a single bank of immersed tubes in an annular shape	1.Water jacket and a double bank of immersed tubes in concentric annuli (Cooler a) 2.Water Jacket only (Cooler b)

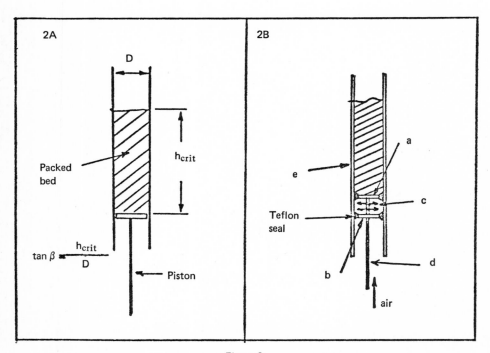

Figure 2.

Table 2 Slug Flow Regimes formed by the Different Solids

| SOLID | 5.1 cm. Diameter Rig | | 10.2 cm. Diameter Rig | | 21.8cm. Diameter Rig |
	Vibrated	No Vibration	Vibrated	No Vibration	No Vibration
F.C.C.	U ≤ 23 cm/s. a U ≥ 23 cm/s. d	a	U ≤ 11 cm/s. a U ≥ 11 cm/s. d	U ≤ 20 cm/s. a U ≥ 20 cm/s. d	U ≤ 2.5 cm/s. a U ≥ 2.5 cm/s. d
F.C.C.(+200)	a	a	-	-	-
F.C.C.-A1	U ≤ 20 cm/s. a U ≥ 20 cm/s. a+c	U ≤ 20 cm/s. a U ≥ 20 cm/s. a+c	a	U ≤ 15 cm/s. a U ≥ 15 cm/s. a+c	-
A1	U ≤ 10 cm/s. a+c 10 ≤ U ≤ 17 cm/s. c U ≥ 17 cm/s. b+c	b + c	U ≤ 15 cm/s. a U ≥ 15 cm/s. b+c	U ≤ 15 cm/s. a+c U ≥ 15 cm/s. b+c	-
E.F.C.C.	a	a	a	a	a
Cu	a + b + c	b	a + b + c	b + c	-
MS/XL	U ≤ 5 cm/s. a 5 ≤ U ≤ 12 a+c U ≥ 12 cm/s. c	b + c	U ≤ 6 cm/s. a U ≥ 6 cm/s. b+c	b + c	U ≤ 9 cm/s. a U ≥ 9 cm/s. b+c
MS/II	b + c	b	b + c	b + c	b + c
350μ glass	b + c	b	b + c	b + c	-
SAND	b + c	b	b + c	b + c	b + c
TALCUM	b + c + e	b + c + e	-	-	a + f

Legend

a - smoothly slugging bed   c - half slugging bed   e - gas channeling

b - raining slugging bed   d - turbulent flow regime   f - fissure structure in the bed

measured when the bed was operated just above incipient fluidization and sliding piston moved.

Table 2 shows the slug flow regimes with the different solids.

MIXING CELLS MODEL

The cells are the portions of the bed in between gas-slugs, see Figure 3. In the first form of the model it was assumed that all the material between gas-slugs was well-mixed. If slugs were can-shaped then

$$E_S \doteqdot 0.5(1 - \varepsilon_{mf})TD[\overline{U - U_{mf}} + 0.35(gD)^{\frac{1}{2}}] \qquad \cdots (1)$$

By comparison with the experimental results it was established that this predicted value was considerably too high. The insertion of a piston-flow section - which is physically more realistic was next attempted, and proved satisfactory when the mixing cell was supposed to constitute about 71 per cent of the inter-slug material, the remainder being a piston-flow region. This is illustrated in Figure 4 which shows the state of the bed at intervals.

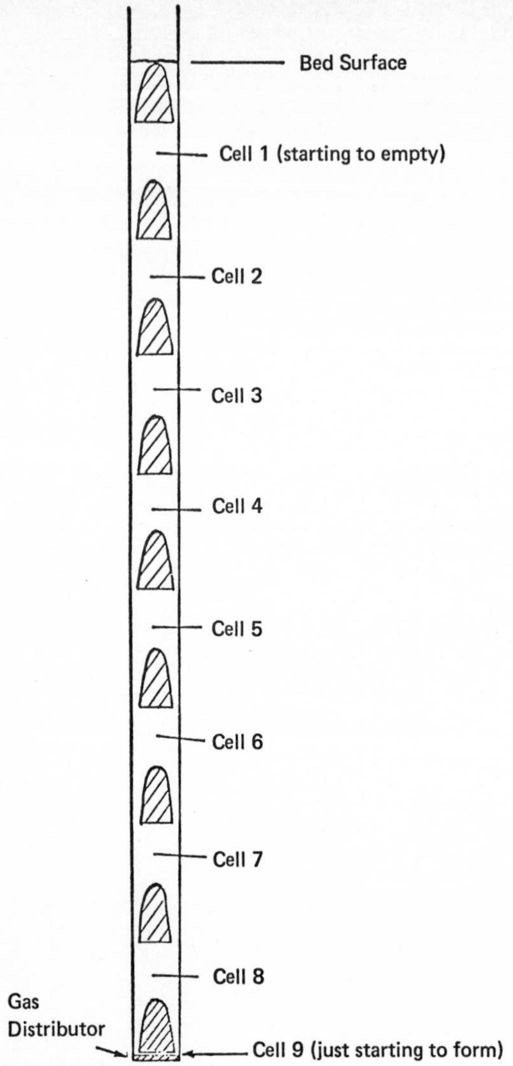

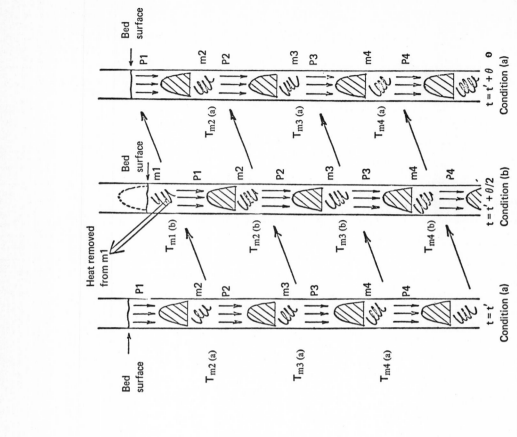

Fig. 4 The plug flow — mixing cells model.

Fig. 3 Simulation of the mixing of a tracer solid, the bed at $t^a = 0$.

ADSORPTION EFFECTS IN FLUIDIZED BEDS

H. V. NGUYEN and O. E. POTTER

<u>INTRODUCTION</u>. Previous adsorption studies[1] in fluidized beds
have not been sufficiently extensive to characterize the main
features. This paper is an extension of a previous report[2] on
back-mixing experiments with adsorption in fluidized beds. Tracer
gases were uniformly and continuously injected at the top of a
fluidized bed operated at $U > U_{CR}$, where U_{CR} is the critical
velocity for back-mixing, i.e. the velocity at which back-mixing
commences for non-adsorbed gases. Gas samples were withdrawn and
analysed for tracer concentration at different levels below the
injecting plane. A porous adsorptive solid was used with helium -
which is not adsorbed -, Freon-12 and mercury vapour - which are
adsorbed - for the evaluation of the effect of adsorption on
mixing. Further details of experimental equipment and procedures
are available in[2].

<u>COUNTER-CURRENT BACK-MIXING MODEL WITH ADSORPTIVE TRANSFER</u>. The
experimental data were interpreted in terms of the three-phase,
counter-current back-mixing model. The overall gas transfer
coefficient K_{BP} is calculated from the individual coefficients
K_{BC} (between bubble and cloud-wake phases) and K_{CP} (between cloud-
wake and dense phases), the forms of which are as proposed by
Davidson and Harrison[3] and Kunii and Levenspiel[4]. Comparison
of measured K_{BP} with K_{BP} thus calculated is made in terms of α
where :

$$\alpha = {K_{BP}}_{measured} / {K_{BP}}_{calculated}.$$

However, there is an additional mechanism for gas exchange
between the cloud-wake and dense phases. As the solids are
exchanged between these phase, the gas associated with this
quantity of solids may also be exchanged. The additional gas

transfer term is therefore equal to $K_S \epsilon_{mf}$ where K_S is the solids exchange coefficient. The inclusion of $K_S \epsilon_{mf}$ into the model equations reduces the value of α.

Equations allowing the calculation of concentration profiles and their solutions are presented in the supplement appended to this paper.

RESULTS. Experimental back-mixing data are presented in Figure 1 for particles of average size of 85 μm. Previously the profiles of log (concentration) versus depth below tracer injection level have been presented as linear for depths up to 8 inches. When measurement is continued to greater depths approaching that of the

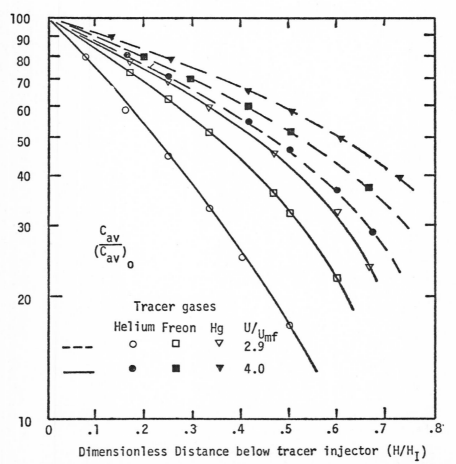

Fig. 1 Back-mixing profiles (particle size = 85 μm level of tracer infection H_I 12 in. above distributor.)

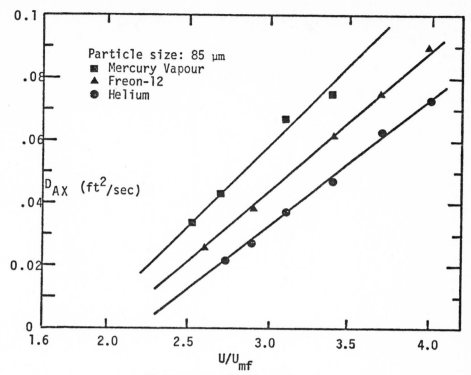

Fig. 2 Apparent diffusivity coefficient.

distributor it is found that the profiles are concave downwards.
This is believed to be due to the greater gas exchange rate of
smaller bubbles. From the helium runs, the ratio of cloud-wake
to bubble volume f_w is determined by evaluating the apparent
diffusivity coefficient D_{AX} (Figure 2) and α is calculated as a
function of fluidizing velocity for both cases: (i) when $K_S \epsilon_{mf}$
is included and (ii) when $K_S \epsilon_{mf}$ is neglected. Measured values of
K_{BP} can be used in the model to determine concentration profiles
for various operating conditions and the apparent coefficient of
adsorption [2] assuming adsorption processes are at equilibrium.
The apparent adsorption coefficient m was found to be much less
than the true value.

DISCUSSION. (i) Gas-mixing is enhanced by adsorption even if the
adsorption processes are not at equilibrium. (ii) The introduction
of the additional gas transfer represented by $K_S \epsilon_{mf}$ into the model

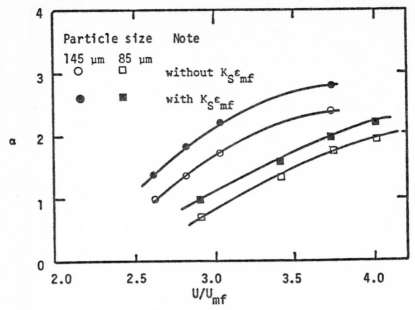

Fig. 3 α as a function of U/U_{mf}.

brought the prediction for overall gas transfer coefficient K_{BP} closer to the experimental value. (iii) Bubble size is important in modelling fluidized beds. A selection of an effective mean bubble size ignores the variation of axial mixing with bubble size. The influence of adsorption on mixing means that the behaviour of a reactor may depend on the state of fluidization. A particle as it moves from wake to dense phase and then back again into a wake undergoes a dynamic variation in the concentration of reactant and in the magnitudes of adsorbed species. Catalysts may exhibit chemisorptive behaviour or may have selective adsorption of a certain reacting species. Unfortunately our knowledge of the adsorbed species for the reactions applied industrially is negligible.

REFERENCES

(1)   Miyauchi, T., Kaji, H., Saito, K., *J. Chem. Eng. Japan* *1*, 72 (1968); Sagetong, P., Gilbert, H., Angelino, H., *Chimie et industrie*, *105*, 1825 (1973); Yoshida, K., Kunii, D., Levenspiel, O., *Ind. Eng. Chem. Funds. 8*, 402 (1969);

Zalewski, W.C., Hanesian, D., A.I. Ch. E. Symp. Ser. 69, No. 128, 58, (1972).

(2)   Nguyen, H.V., Potter, O.E. *Advances in Chem.* Series, *133*, 290 (1974).

(3)   Davidson, J.F., Harrison, D., "Fluidized Particles" CUP (1963).

(4)   Kunii, D., Levenspiel, O. "Fluidization Engineering", Wiley, New York (1969).

APPENDIX

Solution of the Three-Phase Back Mixing Model

Equation describing the material balances for each phase are as follows:

Bubble phase

$$U_{GB}\frac{dC_B}{dh} + K_{BC}\varepsilon_B(C_B - C_C) = 0 \qquad (A.1)$$

Cloud-wake phase

$$U_{GC}\frac{dC_C}{dh} + K_{BC}\varepsilon_B(C_C - C_B) + K_{CP}\varepsilon_B(C_C - C_P)$$

$$+ \varepsilon_{mf}K_S (C_C - C_P) + (1 - \varepsilon_{mf})K_S m(C_C - C_P)$$

$$+ (1 - \varepsilon_{mf})f_w U_{GB} m \frac{dC_C}{dh} = 0 \qquad (A.2)$$

Dense phase

$$U_{GP}\frac{dC_P}{dh} + K_{CP}\varepsilon_B(C_P - C_C) + (1 - \varepsilon_{mf})K_S m(C_P - C_C)$$

$$+ \varepsilon_{mf}K_S(C_P - C_C) - (1 - \varepsilon_{mf})f_w U_{GB} m \frac{dC_C}{dh} = 0 \qquad (A.3)$$

In equations (A.2) and (A.3), the terms involving $\varepsilon_{mf}K_S$ represent the additional gas transfer due to the exchange of solids between the wake and the dense phases.

Using dimensionless variables, viz :

$$C_1 = \frac{C_B}{C_{av}} , \; C_2 = \frac{C_C}{C_{av}} , \; C_3 = \frac{C_P}{C_{av}} , \; z = \frac{h}{H} \qquad (A.4)$$

where C_{av} = average concentration at the level of tracer injection.
$\quad\;\; H$ = level of injection.

After rearrangement, (A.1), (A.2), (A.3) become :

$$\frac{dC_1}{dz} = A_1C_1 + A_2C_2 \qquad\qquad (A.5)$$

$$\frac{dC_2}{dz} = A_3C_1 + A_4C_2 + A_5C_3 \qquad\qquad (A.6)$$

$$\frac{dC_3}{dz} = A_6C_2 + A_7C_3 \qquad\qquad (A.7)$$

where :

$$A_1 = \frac{-HK_{BC}\varepsilon_B}{U_{GB}}$$

$$A_2 = \frac{HK_{BC}\varepsilon_B}{U_{GB}}$$

$$A_3 = HK_{BC}\varepsilon_B / [U_{GC} + (1 - \varepsilon_{mf})f_w U_{GB} \, m]$$

$$A_4 = \frac{-H\{\varepsilon_B(K_{BC} + K_{CP}) + K_S[m(1 - \varepsilon_{mf}) + \varepsilon_{mf}]\}}{U_{GC} + (1 - \varepsilon_{mf})f_w \, m \, U_{GB}}$$

$$A_5 = \frac{HK_S[m(1 - \varepsilon_{mf}) + \varepsilon_{mf}]}{[U_{GC} + (1 - \varepsilon_{mf}) f_w \, m \, U_{GB}]}$$

$$A_6 = \frac{H\{K_{CP}\varepsilon_B + K_S[(1 - \varepsilon_{mf})m + \varepsilon_{mf}]\}}{[U_{GP} - U_{GB}(1 - \varepsilon_{mf})f_w m]}$$

$$A_7 = -A_6.$$

Boundary conditions

@ $Z = 1.0$

$$C_1 = C_{1H}, \quad C_2 = C_{2H}, \quad C_3 = C_{3H} \qquad (A.8)$$

@ $Z = 0$

At the distributor the entering fresh gas would have a tracer concentration C_i and it is reasonable to assume that the bubble gas would be entirely made up by fresh gas, viz:

$$C_{1i} = C_i \qquad (A.9)$$

Balance on total gas at the distributor :

$$U = U_{GB} + U_{GC} + U_{GP} \qquad (A.10)$$

Balance on tracer gas at the distributor:

$$UC_i = U_{GB}C_{1i} + U_{GC}C_{2i} + U_{GP}C_{3i}$$

$$+ (1 - \varepsilon_{mf})f_w U_{GB} m(C_{3i} - C_{2i}) \qquad (A.11)$$

The last term in (A.11) accounts for the gas adsorbed on the solids moving downwards in the dense phase and on the same volumetric rate which moves upwards with the forming wake.

If entering gas contains no tracer then :

$$C_i = 0 \qquad (A.12)$$

thus

$$C_{2i} = C_{3i} \qquad (A.13)$$

with $\quad k = \dfrac{-U_{GP} + (1 - \varepsilon_{mf})f_w U_{GB}\, m\, C_{3i}}{U_{GC} + (1 - \varepsilon_{mf})f_w U_{GB}\, m.}$

The system of equations (A.5), (A.6), (A.7) now can be solved.

An analytical solution is possible. However, since all bed properties change with bubble size, the analytical solution is not applicable to the whole bed. It is possible, however, starting from the distributor, to divide the bed into a number of stages within each bubble size is assumed to be constant, and use the analytical solution to obtain the necessary tracer concentration for each individual phase. Values of these concentrations at the top of the nth stage then will be used as the boundary condition for the $(n + 1)$th stage. By this marching forward method, one can satisfy the top boundary condition and solve for the axial concentration profiles.

However, in this case, the abovementioned approach was not working very well even when the bed was divided into 40 stages.

It was much simpler to solve the equations by Runge Kutta method with Gill's coefficients. The solution converges to the set top boundary condition after 2 or 3 iterations in most cases with the bed divided into 25 stages.

MIXING OF AN EXPANDED MATERIAL
WITH FINE DENSE POWDERS BY FLUIDIZATION

J. P. SLONINA, J. GELLON, P. VERLHAC, D. VIEL, M. ARMBRUSTER, and P. RENARD

INTRODUCTION

Many authors, specially J.C. GUICHARD and J.L. MAGNE, like
D. ROLAND and D. GELDART, tried to specify the electrostatic phenomena
linked with silent discharges inside fluidized industrial beds of
mineral particles, and with surface triboelectricity. Our industrial
experience got us to mixing dielectric very fine mineral particles, with
flakes of expanded graphite, i. e. an expanded of very great specific
surface.

1 Revealing the Triboelectricity of the Exfoliated Graphite Surfaces

1.1 Formation of exfoliated graphite from natural graphite

Graphite crystallizes in the hexagonal system with Angstroem
lattice parameters :

$$c = 6,708 \overset{o}{A}$$
$$a = 2,461 \overset{o}{A}$$

Rather perfect single crystals naturally exist as flakes of
several tenths of millimeter thickness and from half a millimeter to
several millimeters diameter. These single crystals were formed by a
combination of high temperature and high pressure.

Artificial or natural graphites can be lattice modified
by chemical agents such as halide vapours of some acids, specially
oxidizing acids. We experimented complexing lamellar single crystal
flakes of graphite with a mixure of nitric and sulfuric acids. At
insertion operation, parameter $c/2$ increases from 3, 354 to 7, 98 $\overset{o}{A}$.
It seems that this structure modification is accompanied by a sulfuric
acid ions and molecules concentration in the lattice defects,

especially in the twinings. Called "acid sulfate of graphite",
the laminar complex has the chemical formula C_{24}^{+} HSO_4,$2H_2$,$SO_4$and can. $\vdash$
degraded by desorption or water washing, to a residual compound.
This degradation brinks back the lattice structure of the original
graphite, but with about 10 % acid inserted in the graphite structure.
A brutal thermic shock can exfoliate that residual compound normally
to its lattice planes. The exfoliated graphite has an apparent
volume two hundred or three hundred times greater than the original
volume, and its specific surface lies between fifty and a hundred
square meters for one gramme. The lenticular separation between
packets of lattice planes, is produced by the thermic expansion
of gases occluded as complete layers or local concentrations at
lattice defects.

To evaluate exfoliation kinetics, the phenomenon was filmed
with a very fast camera allowing three thousands images per second.
Slow at low temperatures, the formation of lentils in the c-axis
direction, starts from 160 ° C, the exfoliation speed being directly
linked with the temperature of the exfoliation oven. When the thermic
shock is brutal, for instance by the residual compound entering
an oven maintained at 2 000 °C, the residual compound explodes.
Cut out of the 16 mm film, photographs show the time evolution of the
exfoliated crystal.

1.2 Triboelectricity appearing at exfoliation

Exfoliation develops electrostatic charges from tribo-
electrization. To avoid parasitical phenomena, we tried to take all
precautions, as indicated by J.C. GUICHARD and J.L. MAGNE.
Figure 4 shows the scheme of the assembly used to measure the triboelectricity generated at the moment of natural graphite flake
expansion.

The assembly comprises : a heating resistance enclosed
in a Faraday cage ; an electrode travelling with a railed carriage ;
an electrometer measuring the charges developed at exfoliation ; and
a recorder.

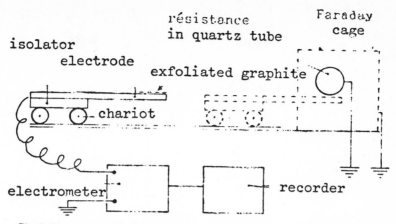

Fig. 1 Measurement of triboelectricity during exfoliation. Experimental set-up.

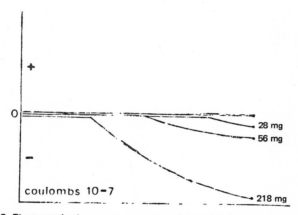

Fig. 2 Electrostatic charges measurements time residual compound masses.

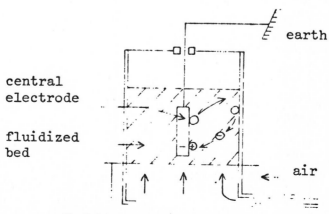

Fig. 3 Schematic of the fluidized bed pile.

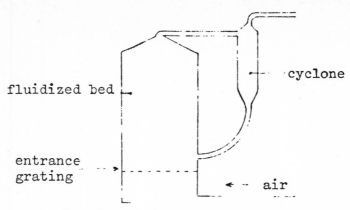

Fig. 4 Schematic of the industrial fluidized bed.

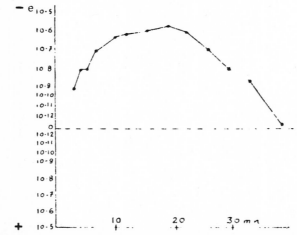

Fig. 5 Triboelectricity charges development in fluidized bed of expanded graphite only.

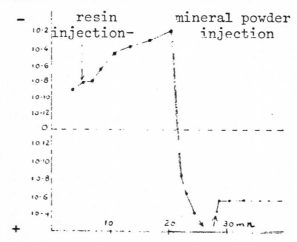

Fig. 6 Triboelectricity charges development after pulverized resin and mineral powder injections.

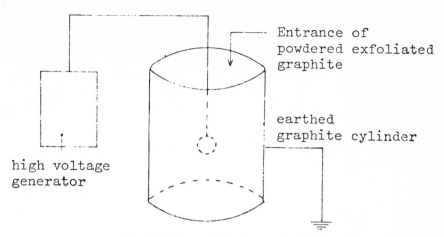

high voltage
generator

Entrance of
powdered exfoliated
graphite

earthed
graphite cylinder

Fig. 7 Powder extraction by corona effect.

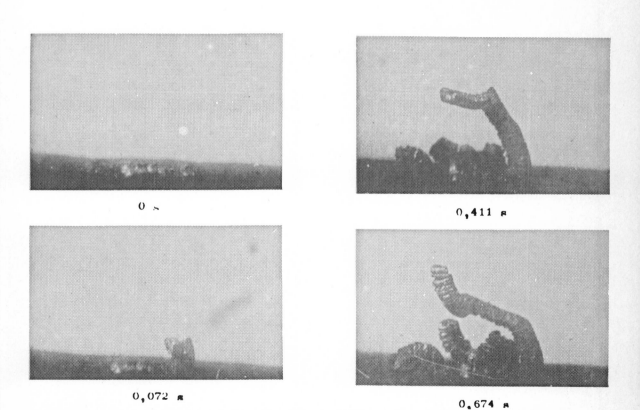

0 s

0,411 s

0,072 s

0,674 s

Fig. I As a time function, exfoliation of a residual compound flake. Magnification X2. From film shot at three thousand images per second with very fast camera.

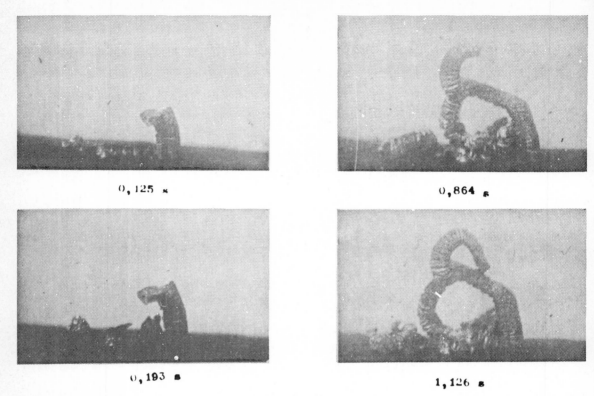

0,125 s   0,864 s

0,193 s   1,126 s

Figure I (Cont.).

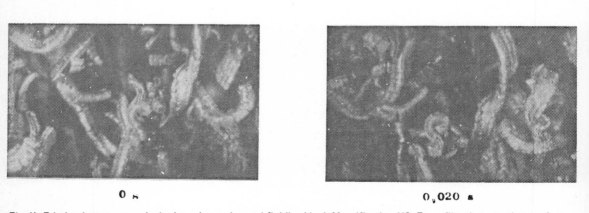

0 s   0,020 s

Fig. II Friction between vermicules in and experimental fluidized bed. Magnification X2. From film shot at a thousand images per second with very fast camera.

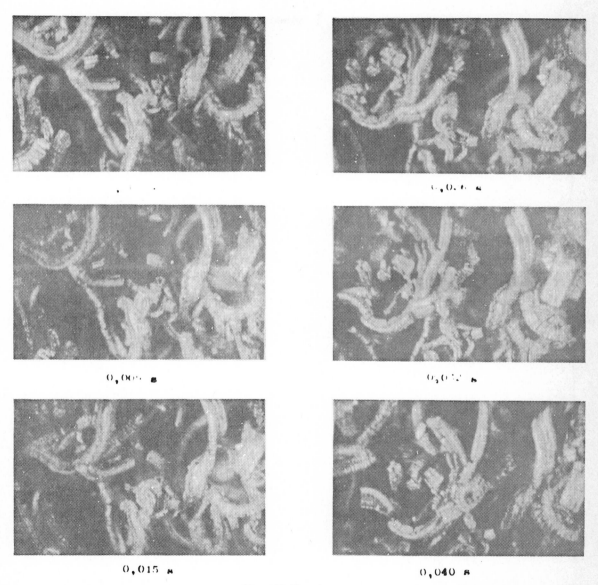

0,015 s   0,040 s

Figure II (Cont.).

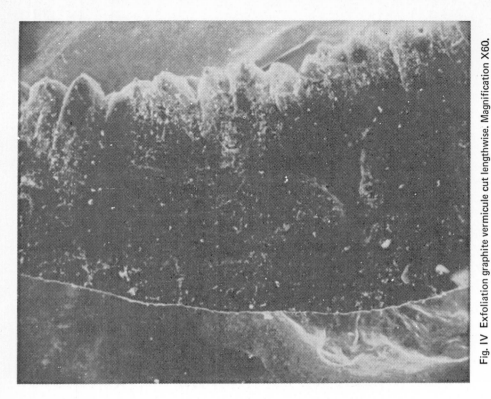

Fig. IV Exfoliation graphite vermicule cut lengthwise. Magnification X60.

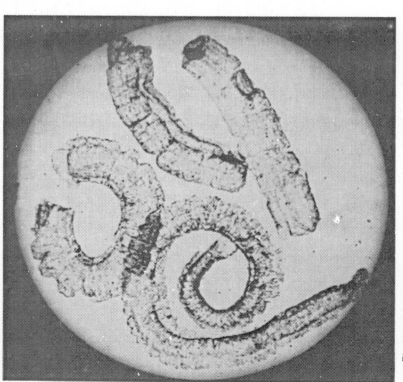

Fig. III Microradiography of exfoliation graphite vermicule after mixing. Magnification X10.

208

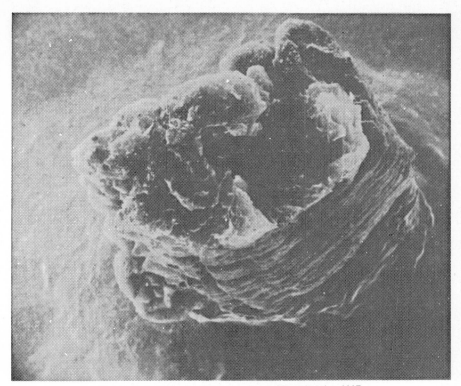

Fig. V Exfoliation graphite vermicule. Magnification X45.

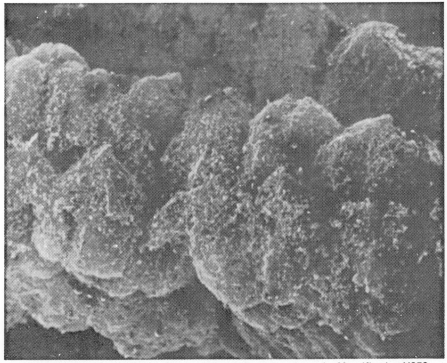

Fig. VI Exfoliation graphite surface loaded with fine mineral powder.Magnification X950.

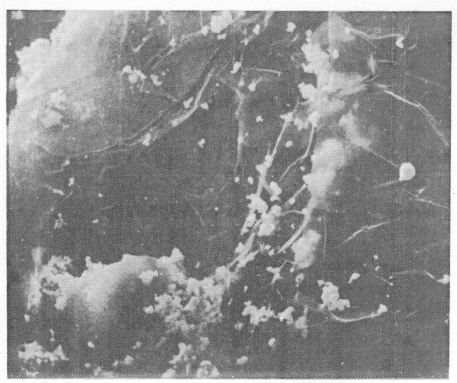

Fig VII Mineral powder distribution on the exfoliated graphite surfaces. Magnification X3000.

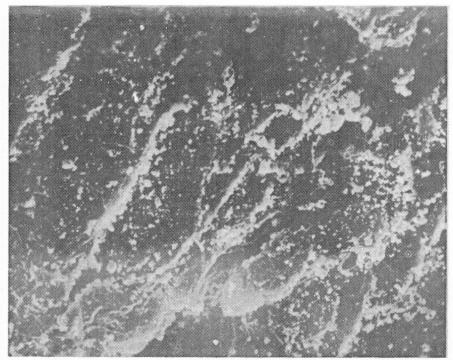

Fig. VIII Mineral powder distribution with greater concentrations on the layer edges. Magnification X300.

The graphite flake expansion takes place directly on the
electrometer electrode. Travelling with the railway carriage,
that electrode is electrically insulated from the carriage.
The assembly allows us to introduce the flakes to be expanded in the
Faraday cage, under the heating resistance, in a position always
identically the same at all tests.

Preliminary measures showed no electrometer variation when only
the resistance grew hotter, or when the material put on the heating
resistance is no more under electric tension, a brutal deviation is
recorded. At graphite flake expansion, the sign of developed charges
is always negative.

As we could verify, the size of the evolved charge was proportional to the quantity of prepared flakes. Figure 2 gives measures
recorded for different masses of residual compound. Because
of inevitable contacts, interferences between flakes are also
probable.

2 Experimental Results Linked with the Influence of Fluidization on the Triboelectrization of Exfoliated Graphite Vermicules

2.1 The "Fluidized Bed Pile"

Adopting from J.C. GUICHARD and J.L. MAGNE, the so called
"Fluidized Bed Pile" assembly, the graphite vermicules get a permanent
electrostatic charge after some time. On figure 3, a central electrode
can partially discharge the fluidized bed. That discharging will make
a compensating current appear on the gases introduction grating, as
the fluidized bed continuously conduces to reestablish a static
equilibrium charge.

The electrometer measured current corresponds to the discharge
current from the central electrode assimilating such a population
of exfoliated graphite vermicules to a distribution of brownian moveme

particles, we can evaluate collisions between vermicules and with
the containing walls. At each wall collision, the vermicules get the
same charge, whatever their first electric state. Every ulterior
contact of a vermicule having just touched the wall, with a vermicule
not having yet touched the wall, will equalize their charges
(supposing there is no triboelectric phenomenon at vermicule
separation). Thanks to the enclosing wall, a stationary equilibrium
charge will present after some time, equal charges of same sign on all
vermicules, and at the limit, the central electrode can be considered
to discharge all colliding vermicules. Discharged directly or by contact with neighbouring particles, a particle will retake a charge
by liberating an opposite sign charge on the container wall.

Figure 4 shows the classical cyclone fluidized bed used at our
industrial operations.

2.2 Experimental observations

When only vermicules of expanded graphite are fluidized, the
beginning of fluidization gives on the record, a succession of
both positive and negative impulses. These impulses correspond
to residual charges got by vermicules at exfoliation and from
friction (Figure 5). One must wait five minutes to equalize all
charges. From that moment, the electrometer records only negative
flow. The first recorded flow is very weak (3×10^{-8} Coulomb)
and tends to increase. If fluididation is continued, the flow
is greatest after 25 minutes (4×10^{-6} Coulomb) and begins to
decrease. That maximum effect corresponds to a notable proportion
of mechanically crushed vermicules. When all the graphite is crushed,
practically no additional current is recorded. Cut from a 16 mm film
shot at three thousand images per second, the photographs show the
nature of contacts between vermicules.

To strenghten and amplify the triboelectric phenomenon, one
method introduce pulverized resin in the fluidized bed of

exfoliated graphite (Figure 6). Remaining negative, the recorded
charge is tenfold multiplied. A prolongation of fluidization allows
a more intense current to overload the electrometer in about
twenty minutes. But injecting a fine mineral powder in
the mixure having already acquired the maximum electrostatic charge
provokes a brutal change of charge sign. The current saturates the
electrometer and decreases to an equilibrium point at injection
ending. If the mixing continues, the current increases to a
maximum and begins to decrease after crushing of some proportion
of graphite flakes.

3 Examining the Fine Powder Contract with the Graphite Flakes

3.1 Stereoscan examination

Stereoscan examination of a vermicule shows the powder distribution
all round the flake. But we ascertain a slight preference of
the clinging powder for the folds of the accordion like vermicule.
That is rather normal in fluidization mixing, when electrostatic forc·
developed in the bed, add the mechanical forces applied by the injector to the fine particles. Arriving with some force in the
fluidized bed, of expanded graphite, the powder is more easily
stopped by the vermicule indentations.

The powder distribution is very regular on the inside of the
vermicule spongy texture. At flake level, these electrostatic
charges make point effects on the layers of graphite partly loosened
at expansion. In consequence, during the fluidized bed mixing, the
powder clings preferentially to the edges of the graphite folds.

Such point effects certainly explain the good distribution
of the powder at flake level.

3.2 Corona effect powder extraction

We used the assembly of Figure 7, comprising a central
"coronade" enclosed in an earthed conductive cylinder of

polycrystalline graphite. When the charged vermicules are
thrown into the surrounding cylinder, fine powder particles are
electrically projected against the cylinder wall. After ten
passages through the system at 72 000 volts, the mixture lost
about fifty per cent of the original powder. For weaker applied
tensions, and always for a powder granulometry of the micron
order, no powder impoverishment of the mixture is noted.

CONCLUSION

From the phenomena quantitative measures, we can establish
observations about the preexistence of electric charges on the
surface of the exfoliated lamellar material. In an experimental system,
the fluidization of exfoliated graphite allowed the revelation of triboelectric phenomena from friction between graphite vermicules, and
from vermicule -bed separating wall friction. Revealed also was the
triboelectric effec reinforcement by preliminary pulverization
of first crushed phenolic resin, in the exfoliated graphite bed.
Finally, stereo scan examination of the bonding between extremely fine
 powder and the oxidized graphite surfaces, allows to recognize
preferential positioning of fine powder on some places. In the
fluidized bed, corona effect utilization allowed us also to attempt,
and even to obtain, partial separation of powder-graphite bonding.

SCREENING OF PARTICLES BY A PACKED FLUIDIZED PARTICLE CLASSIFIER

KUNIO KATO, YOSHIHIRO SHIROTA, and UTARO ITO

ABSTRACT

The nonuniform size particles are fed to the middle part
of the tower packed with open-end cylindrical screen packing
and are classified into two parts by flowing air upward in the
tower. This apparatus is defined as packed fluidized particle
classifier. The characteristics of particle classification by
this classifier is expressed by the partial separation efficiency. Glass beads from 35 mesh to 400 mesh and silica sand
from 60 mesh to 200 mesh were classified with this classifier
under the different operating conditions. The characteristics
of particle classification of this apparatus was much better
than that of centrifugal or gravitational particle classifier.
The partial separation efficiency can be calculated from the
operating conditions by establishing experimentally the relation between the distribution function of gas flow in the apparatus and operating conditions.

INTRODUCTION

Solid particles are dealt with by many industries very
often. Usually, solid particles are not uniform in size and
shape. Classification of nonuniform size particles to the
same size particles or selection of the particles consisting
of the same kind of material from the particles of the several
 different kinds of materials are a important unit operation
in many industries. Many different particle classifiers,

such as gravitational, centrifugal and other types of particle
classifiers are used for the classification of particles.

In this paper, the characteristics of particle classification by a packed fluidized particle classifier is investigated. The particles are classified by this classifier with
the following method. Nonuniform size particles are fed to the
middle part of the tower packed with open-end cylindrical
screen packings and are classified into two part by flowing air
upward in the tower. The characteristics of particle classification by this classifier is expressed by partial separation
efficiency. At first, the effect of the apparatus conditions,
such as the packing size, the distance between the feed point
of particles and lower particle withdrawal point or tower
height, upon the characteristics of particle classification
of this classifier was investigated with two different size
particles. The relation between the characteristics of
particle classification and operating conditions was investigated in the region where the characteristics of particle classification was not affected by the apparatus conditions.

The partial separation efficiency of this classifier is
calculated from the apparent terminal velocity of particles
and distribution function of gas flow in the apparatus. In
this experiment, glass beads from 35 mesh to 400 mesh size and
silica sand from 60 mesh to 250 mesh were classified with this
classifier under the different operating conditions and the
partial separation efficiencies of these particles were obtained. The relation between the apparant terminal velocity and
operating conditions and that between the distribution function
of gas flow in this classifier and operating conditions are
obtained from the partial separation efficiency. From these
relations, the partial separation efficiency of this classifer
can be calculated from operating conditions.

EXPERIMENTAL APPARATUS AND EXPERIMENTAL PROCEDURE

A schematic of experimental apparatus is shown in Figure 1.
The main part of this classifier was the tower with 8cm I.D.

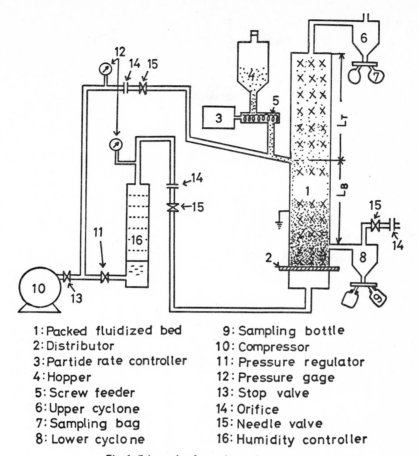

Fig. 1 Schematic of experimental apparatus.

1: Packed fluidized bed	9: Sampling bottle
2: Distributor	10: Compressor
3: Partide rate controller	11: Pressure regulator
4: Hopper	12: Pressure gage
5: Screw feeder	13: Stop valve
6: Upper cyclone	14: Orifice
7: Sampling bag	15: Needle valve
8: Lower cyclone	16: Humidity controller

and 150cm long in which the open-end cylindrical screens were packed. Table 1 shows the packings used in this experiment. Porous plate with 2mm thickness and 20μ pore size was used as gas distributor. Particles were fed to the middle part of the tower by screw feeder at constant feed rate. Particles clas-

Table 1 Packings Used in the Experiment

packings	sign	taylor mesh	d(cm)	D_p(cm)	ε (−)
	PN−1	16	1.15	1.29	0.960
	PN−2	16	2.60	3.01	0.990
cylindrical	PN−3	10	1.43	1.59	0.967
wire net	PN−4	32	1.22	1.39	0.975
	PN−5	16	1.80	2.07	0.975

sified by this classifier were collected into the sampling bag
below the uper cyclone and the sampling bottle below the lower
cyclone.

The experimental method was as follows ; at first, air of
which flow rate was measured by orifice was fed to the tower
at constant flow rate. The particles of which size distribution was measured were fed to the tower from hopper at constant feed rate. The size distribution of particles used in
this experiment is shown in table 2. The particle feed rate
of this experiment was 80~550 (g/min). The particle
feed rate was obtained from measuring the weight of particles
collected by the bag below the uper cyclone and bottle below
the lower cyclone under the constant time interval. A part
of particles in the bag and bottle was screened by J.I.S.
standard sieves and the size distribution of particles was
measured. To prevent static charging of the particles, packings are connected to an earthed line.

Table 2 Particle Distribution

	multi component glass particle			silica sand
system	weight fraction (-)			
J.I.S.mesh	A	B	C	D
−35+42	—	0.220	—	—
−42+48	—	0.213	—	—
−48+60	0.070	0.197	—	—
−60+80	0.270	0.153	—	0.313
−80+100	0.190	0.217	—	0.311
−100+115	0.200	—	—	0.227
−115+150	0.100	—	0.315	0.062
−150+170	0.100	—	0.138	0.041
−170+200	0.060	—	0.174	} 0.046
−200+250	0.010	—	0.079	
−250+270	—	—	0.066	—
−270+325	—	—	0.046	—
−325+400	—	—	0.362	—

EXPERIMENTAL RESULTS

To investigate the characteristics of particle classification of this classifier, the following experiments are performed. At first the effect of the apparatus conditions, such
as the packing size, tower height, the distance between the
feed point of particles and lower particle withdrawal point
upon the characteristics of particle classification was investigated with two different size particles. After that,
the relation between the characteristics of particle classification and operating conditions was investigated in the region
where the characteristics of particle classification was not
affected by the apparatus conditions.

Classification of Two Different Size Particles

To investigate the effect of the apparatus conditions
upon the characteristics of particle classification, the particles consisting of 50 wt% 80∿100 mesh glass beads and 50 wt%
48∿60 mesh glass beads are classified by this classifier.
Newton's classification efficiency η_N is used to express
the characteristics of this classifier. Fig.2 shows the re-

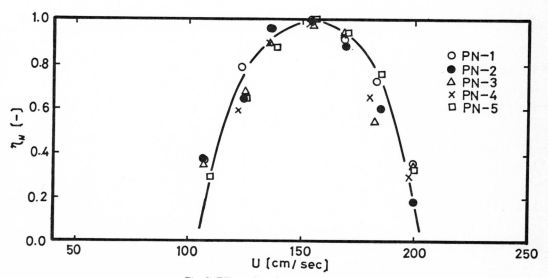

Fig. 2 Effect of packing size upon η_N.

lation between U and η_N for different packing size in
table 1. From Fig.2, if packing size is smaller than 3cm,
η_N is not affected by the packing size. To investigate the
distance L_T between the feed point of particles and the top
of the tower upon the characteristics of particle classification, particles were classified under different length of L_T.
From these experiments, if L_T is longer than 20cm, η_N is
not affected by L_T. On the other hand, if the distance L_B
between the feed point of particles and the lower particle
withdrawal point is longer than 40cm, η_N is not affected
by L_B.

Classification of Multi Component Particles

From the above experimental results, the multi component
particles are classified by the tower packed with PN-5 packings in table 1 with 150cm long. The partial separation efficiency of i component particles is

$$\beta_i = \frac{AY_i}{FX_i} \quad - - - - - \quad (1)$$

Fig.3 and Fig.4 show the relation between the partial separation efficiency and particle diameter with the operating gas
velocity as parameter where particles A (glass bead) and particle D (silica sand) are classified by this classifier respectively. From Fig.3, the characteristics of particle classification of this classifier is much better than that of a
centrifugal or graviational particle classifier. Fig.5 and
Fig.6 are cross plot of Fig.3 and Fig.4, that is, Fig.5 and
Fig.6 show the relation between the partial separation efficiency and operating gas velocity with the particle size as
parameter where the particles A (glass bead) and D (silica
sand) are classified with this classifier respectively.
The operating gas velocity when the partial separation efficiency of i-component particles become 0.5 is defined as the

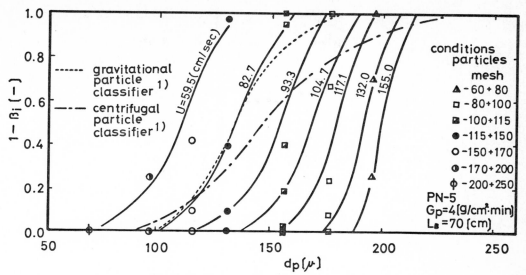

Fig. 3 Relation between d_p and $1 - \beta_i$ in the case of particles A.

appararent terminal velocity U_{Ti} of i component particles.
Fig.7 and Fig.8 show the relation between U/U_{Ti} and β_i in
the case of particles B (glass beads) and particles C (~~silica
beads.~~ ~~sand~~) being classified with this classifier respectively.
The characteristics of particle classification of this classifier is roughly grasped from Fig.7 and Fig.8.   From Fig.7 and

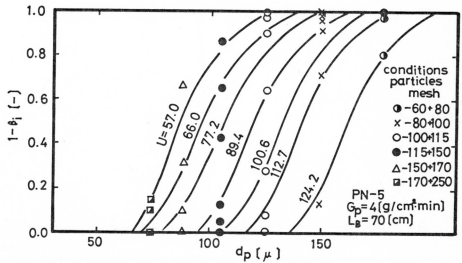

Fig. 4 Relation between d_p and $1 - \beta_i$ in the case of silica sand.

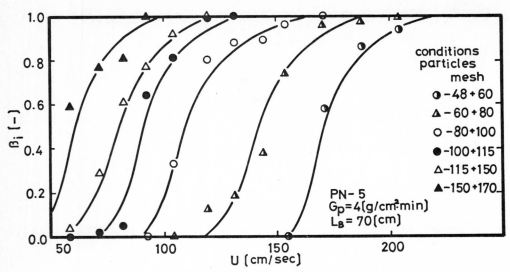

Fig. 5 Relation between U and β_i in the case of glass beads A.

Fig.8 the fractional recovery curve of coarse particles is
steeper than that of fine one.

ANALYSIS OF EXPERIMENTAL RESULTS

To calculate the partial separation efficiency of this
classifier, the quantitative relationship between the apparent

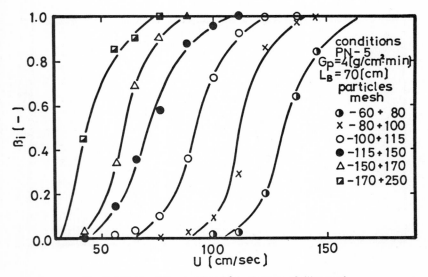

Fig. 6 Relation between U and β_i in the case of silica sand.

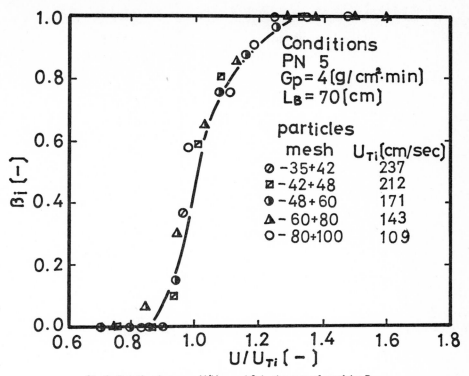

Fig. 7 Relation between U/U_{Ti} and β_i in the case of particles B.

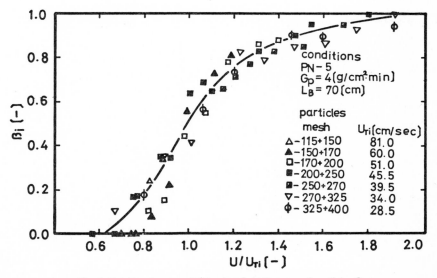

Fig. 8 Relation between U/U_{Ti} and β_i in the case of particles C.

terminal velocity of i-component particles and operating condition and also the quantitative relationship between distribution function of gas flow in the tower and operating condition must be obtained.

Relation Between the Apparent Terminal Velocity and Operating Condition

To investigate the effect of feed rate of particles per
unit cross sectional area of the tower G_p upon the partial
separation efficiency, particles D was classified by different feed rates of particles. Fig.9 shows the relation
between partial separation efficiency and particles size with
feed rate of particles as parameter. From Fig.9 the apparent
terminal velocity of each size of particles increases as the
feed rate of particles increases, that is, when the feed rate
of particles increases under constant gas velocity, the fraction of particles taken out from lower particle withdrawal
point increases. Fig.10 shows the relation between U_{Ti} and G_p
with the size of particles as parameter. When G_p is zero,
the value of U_{Ti} is almost equal to the terminal velocity of
particles calculated from the average diameter of i-component

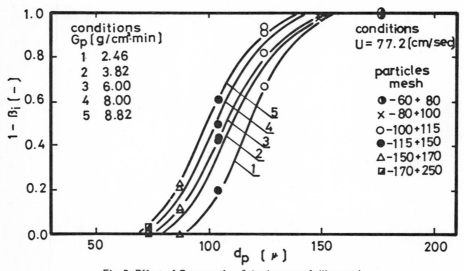

Fig. 9 Effect of G_p upon $1 - \beta_i$ in the case of silica sand.

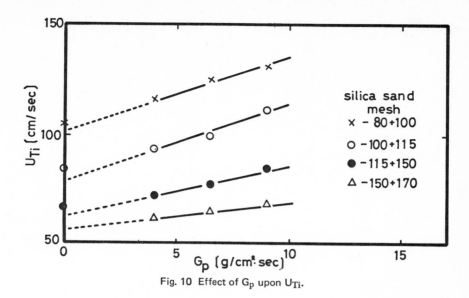

Fig. 10 Effect of G_p upon U_{Ti}.

particles. Fig.11 shows the relation between U_{Ti}/U_{Tio} and G_p. From Fig.11, the following empirical equation is obtained.

$$U_{Ti} = U_{Tio} (0.032G_p + 1.0) - - - - - (2)$$

where U_{Ti}, U_{Tio} (cm/sec), G_p (g/cm^{2}min)
Equation (2) is applicable when G_p is 0.5∼12.0 (g/cm^{2}min).

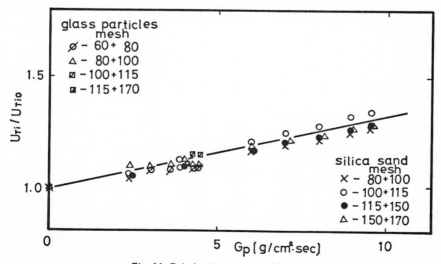

Fig. 11 Relation between U_{Ti}/U_{Ti0} and G_p.

Relation Between the Partial Separation Efficiency and Distribution Function of Gas Flow

If the distribution function of gas flow in this classifier
is expressed as F(u), the partial separation efficiency of
i-component particles is calculated as.

$$\beta_{ical} = \int_{U_{Ti}}^{\infty} F(u) \, du \quad - - - - - \quad (3)$$

If the distribution function of gas flow in this classifier
is assumed to be the standard distribution function, F(u) is
expressed as

$$F(u) = \frac{1}{\sqrt{2}} \frac{1}{\delta} \exp \left\{ -\frac{(u-U)^2}{2\delta^2} \right\} \quad - - - \quad (4)$$

From equations (3) and (4), the partial separation efficiency
is calculated from the operating gas velocity with δ as parameter. The relation between δ and operating gas velocity is
searched in order for measured partial separation efficiency
to well agree with calculated partial separation efficiency
β_{ical}. The following empirical equation is obtained.

$$\delta = \frac{U}{(0.0411U + 1.8)} \quad - - - - - \quad (5)$$

where δ and U is in (cm/sec). Equation (5) is applicable
in the range of operating gas velocity being in 20∿300 (cm/sec)
Fig.12 compares experimental partial separation efficiency with
the calculated partial separation efficiency. From Fig.12, the
experimental partial separation efficiency well agrees with the
calculated one.

From the above results, the apparant terminal velocity of
any particles in this classifier is calculated by equation (2)

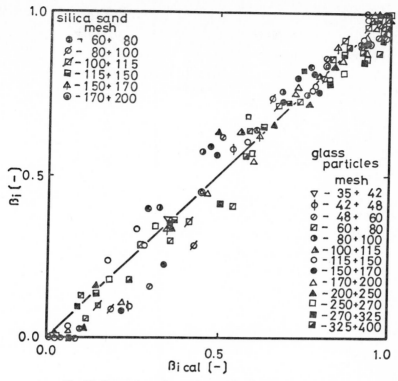

Fig. 12 Comparison of experimental β_i with calculated one.

and the partial separation efficiency can be calculated by equations (3), (4), (5) from operating conditions.

DISCUSSION

The particles form the dispersed phase in this classifier. In any place in this classifier, if the terminal velocity of particles is smaller than gas velocity in that place, the particles go up in the tower. In the case where the terminal velocity of particle is larger than gas velocity, the particles go down in the tower. Such a phenomenon happens in this classifier repeatedly and feed particles are classified into two parts. Open-end cylindrical screen packings make gas velocity rearrange in the tower and ~~and~~ make gas velocity uniform in radial direction. In this experiment, if size of packings is

in the range of 1~3cm, the characteristics of particle classification by this classifier is not affected by the packing
size. This fact means that packing size used in this experiment make gas velocity uniform well in the radial direction.
In this experiment, the characteristics of particle classification is not affected by the length of L_T and L_B if
L_T is longer than 20cm and L_B is longer than 40cm. This
fact means that packings used in this experiment make gas
velocity uniform effectively.

If large size particles are classified in this classifier,
operating gas velocity becomes very large. On the other hand,
if very small particles are classified in this classifier, the
characteristics of particle classification becomes bad owing
to the cohesion of particles to the packings and bed wall.
The particle size which is suitable for this classifier is
about 20~450 mesh.

The particles taken out from the upper cyclone of the
first tower is fed to the second tower which is almost the
same structure as the first tower and the particles are classified into two part by flowing gas in the second tower.
By this method, particles are classified into three parts.
If the particle distribution of feed particles is known, the
particle distribution of three outlet points can be calculated
from the operating gas velocity. By the same way mentioned
above, the particles are classified into 4 or 5 parts by using
3 or 4 towers.

CONCLUSION

1. The particles from 600μ to 30μ can be continuously classified by a packed fluidized particle classifier with simple
   operation.
2. The characteristics of the particle classification of this
   classifier is not affected by the apparatus conditions, if
   L_T is longer than 20cm, and L_B is longer than 40cm and
   size of packing is 1~3cm.

3. The partial seperation efficiency of particles can be
    calculated from operating conditions with equations (2),
    (3), (4), (5).

NOMENCLATURE

D_p = effecitve diameter of cylindrical wire net packing [cm]

d = length cylindrical wire net packing [cm]

d_p = particle diameter [cm]

G_p = mass velocity of feed particles based upon effective
             cross sectional area of particle classification
             tower [g/cm^{2}sec]

L_B = distance between the feed point of particles and
             lower particle withdrawal point [cm]

L_T = distance between the feed point of particles and
             the top of particle classification tower [cm]

U = superficial gas velocity based upon effective cross
             sectional area of the tower [cm/sec]

U_{Ti} = apparent terminal velocity of i component
             particles [cm/sec]

u = gas velocity [cm/sec]

β_i = partial separation efficiency of i component
             particles [-]

β_{ical} = calculated partial separation efficiency of i
             component particles [-]

η_N = Newton's classification efficiency [-]

ε = packing voidage [-]

δ = standard deviation of gas velocity distribution [cm/sec]

LITERATURE CITED

1) T. Hayakawa, S. Huchigami and S. Fujita ;
   Preprint for the 32th annual meeting of the Soc. of
   Chem. Engrs., Japan p.243
2) " Chemical Engineering Hand book. ",
   Soc. Chem. Eng. Japan.

VELOCITY MEASUREMENTS
AND ISOKINETIC SAMPLING
IN A TWO PHASE, SOLIDS IN GAS FLOW

T. E. BASE, M. A. BERGOUGNOU, C. G. J. BAKER, and P. E. MILLS

EXTENDED ABSTRACT

This paper discusses the measurement of velocity in dilute two phase systems with small solid particles or aerosols in gas suspension, using a new low cost rugged fluidic velocimeter. The paper also introduces an isokinetic sampling probe that utilizes hot wire anemometry to determine the correct sampling suction rate.

Development work on flow sensors with a jet immersed in a parallel flowing stream was investigated theoretically by Squire and Trouncer [1] and more recently N.A.S.A. have developed actual probes for aircraft operations [2,3,4]. A more complete summary of available unbounded turbulent jet devices used as transducer elements, including sensors being developed by The National Aeronautical Establishment (Canada), Lowspeed Aerodynamics Laboratory, is presented by Tanney [5].

In two phase flow systems the more conventional method of flow measurement such as pitot static tube, even with the assumption of the flow being steady and irrotational and the fluid incompressible and inviscid, give erroneous readings. For instance, the dynamic pressure of the discrete particle

phase in vacua is twice the dynamic pressure for a simple
single fluid phase flow for the same specific weights and
mean velocities. Such devices also tend to get "clogged" or
damaged by the solid phase.

The principle employed in the velocity sensor discussed
in this paper was based on the fact that when a jet of fluid
is exhausted from an orifice across a gap towards a receiving
port then most of the flow is collected at the receiving port.
However, if the probe is placed normal to a free stream or
placed with the jet parallel to the free stream, with the exhaust jet pointing upstream, then the mass transfer from the
exhaust jet to the receiving port is changed.

Figure 1 shows the layout of the velocimeter set up for
calibration in a wind tunnel. Compressed air, or any other
gas for that matter, was supplied to the airjet of the sensor
which was placed in the working section of a wind tunnel.
The air jet passed across the small gap and depending on
whether the probe was in normal or parallel mode, for a
particular plenum pressure, a certain percentage of the air
stream was collected at the receiving port. The received
air was first amplified as shown in Figure 1 by a momentum
amplifier and then passed to an open ended tube in which a
hot wire probe was placed and finally exhausted to the atmosphere. In this device the change of output voltage from
the hot-wire anemometer was proportional to the mass flow
across the velocimeter gap which was collected at the receiving
port. Figure 2 shows the variation of output voltage with
wind tunnel speed for the case of the airjet normal to the

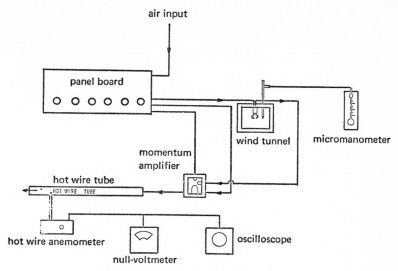

Fig. 1 Circuit of the fluidic velocimeter.

free stream and Figure 3 presents similar data with the probe
in parallel mode but with no momentum amplifier in the circuit.

In conclusion, the "fluidic velocimeter" shows considerable promise as a device to predict the flow velocity in two
phase flows (air and solid or liquid aerosol). The parallel

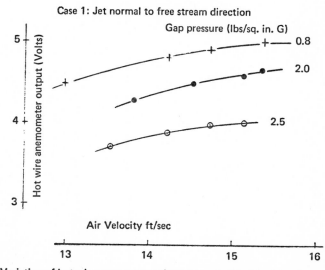

Fig. 2 Variation of hot wire anemomter voltage output with airspeed and gap pressure.

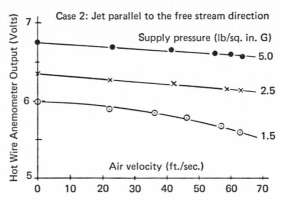

Fig. 3 Variation of the hot wire anemometer voltage output with airspeed and supply pressure.

flow case, where the airjet was projected in the upstream direction of the main airflow was the most preferred method since the backflushing of the downstream probe prevented clogging of the orifice.

At present, tests are being conducted to determine the effect of volume fraction, or solids to gas ratio, on the fluidic velocimeter calibration curve.

Another important flow property to be measured in a pneumatic transport line or in two phase (solids-gas) fluid-ized beds is the measurement of mass concentrations. Several papers have been published describing isokinetic sampling techniques and Mills [6] described several methods previously employed. However, since most of these probes are based on fluid dynamical pressure, errors occur at low velocities and such devices tend to get obstructed by the solid phase. Boothroyd [7] investigated an isokinetic probe using a single hot-wire anemometer as the flow measuring device and errors less than 2% were reported. This original work was the inspiration for the isokinetic anemometric probe discussed in this paper which used two hot wires, an inner wire to measure the flow entering the probe and another wire which measured the flow in the main gas stream.

Figure 4 shows two views of the anemometric isokinetic probe which shows the inner and outer hot wires made of tungsten which were constructed to be the same length and resistance. In order to achieve isokinetic flow, calibration curves were first obtained relating hot wire anemometer output voltages to the inner and outer stream velocity. From the two hot wire voltage outputs it was then possible to achieve isokinetic sampling. Figure 5 shows the calibration of the anemometric probe for two wind tunnel speeds for a range of suction rates.

Finally, further studies are being made to determine the effect of the solid phase on the heat transfer from the hot wires, the change of calibration with change of mass fraction and the maximum velocity that a particular tungsten hot wire of particular diameter could withstand in a two phase flow.

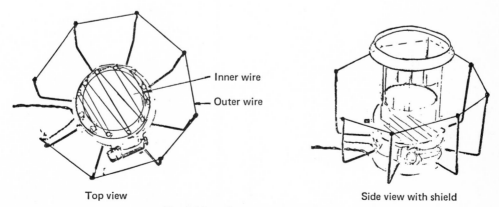

Inner wire
Outer wire

Top view   Side view with shield

Fig. 4 View of anemometer probe tip.

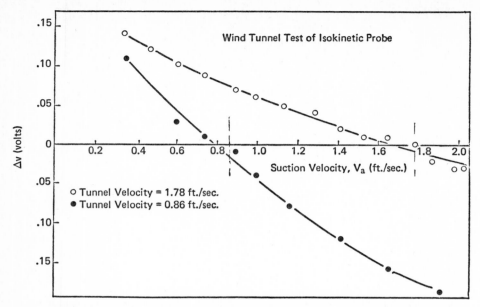

Fig. 5   Variation of differential hot wire anemometer output voltage with probe suction velocity.

REFERENCES

1. Squire, H.B. and Trouncer, J. *"Round Jets in a General Stream"*, Ministry of Aircraft Prod., Aero. Res. Com. R and M 1974, January, 1944.

2. Neradka, V.F. and Turek, R.F. *"Fluidic Low Speed Wind Sensor Research Study"*, N.A.S.A. CR86352. N.A.S.A. Electronics Research Center, December 1969.

3. Miner, R.J. *"A Fluidic Low Speed Air Speed Indicator"*, A.I.A.A. 2nd Aircraft Design and Operations Meeting, Los Angeles, July 20-22, 1970. A.I.A.A. Paper 70-906.

4. Neradka, V.F. *"Fluidic Wind Sensor Research Leading to a Flight Test Model"*, N.A.S.A. CR-111808, N.A.S.A./ Langley Research Center, December 1970.

5. Tanney, J.W. *"The Unbounded Turbulent Jet as a Transducer Element"*, Report No. DME/NAE 1971(2) Low Speed Aerodynamics Laboratory, National Aeronautical Establishment, National Research Council Quarterly Bulletin, April 1971.

6. Mills, P.E., M.E.Sc. Thesis, Faculty of Engineering
       Science, The University of Western Ontario, 1972.

7. Boothroyd, R.G. *"An Anemometric Isokinetic Sampling Probe
       for Aerosols"*, J. Sci. Instrum. <u>44</u> (1967)
       (pp. 249-253).

ACKNOWLEDGEMENTS

The authors would like to express their sincere thanks to the following undergraduate students, Messrs. P.J. Kwandt, G. Kallio and S. Lota who did some of the initial testing of the fluidic velocimeter and to fellow faculty member Dr. C.E. Livingstone and also Mr. M.J. Curry for assistance in the design and construction of the anemometric isokinetic probe. Finally, but not least, the research was funded by a National Research Council grant awarded to Dr. M.A. Bergougnou for the support of Mr. P.E. Mills.

BULK SOLIDS EFFLUX CAPACITY IN FLOODED AND STREAMING GRAVITY FLOW

F. A. ZENZ

ABSTRACT

The fundamental dimensionless relationship for gravity flow of catalyst from a flooded hopper whose conical sides are sloped at an angle shallower than the bulk solids angle of internal friction can be written as

$$W_s = \frac{\sqrt{g}}{\sqrt{\tan \alpha}} \, (\rho_p - \rho_f)(1 - \epsilon) \, D_e^{1/2}$$

Streaming flow, which more closely parallels the motion of collected solids within a cyclone as they swirl into the dipleg, occurs at significantly higher rates. In the case of FCC catalyst this can be at least 2.25 times that from the flooded hopper.

In the early days of FCC plant design it was common practice to size cyclone dipleg diameters on the basis of a fixed solids capacity usually expressed in terms of catalyst weight per unit time per unit of dipleg cross sectional area. Such figures as 125 lbs. of catalyst/sec x sq.ft. were considered safely applicable and yielded apparently sufficiently large pipes, proven able in practice, to handle the recovered entrained solids. As operating superficial velocities increased over the years, so did the entrainment rates and hence the design dipleg diameters. Fortunately interest in bulk solids flow within other areas of industry reached a significant level of sophistication before the continued use of 125 lbs/sec x sq.ft. would have led to unbelievably enormous dipleg diameters. Very recent work in this field has now led to what is believed to be the fundamental relationship for the maximum bulk solids capacity of any restriction, and particularly of cyclone diplegs.

The capacity of an open dipleg is limited by the ability to feed solids into its mouth, not the ability to discharge its contents. For example, if in the arrangements of Figure 1, at equal dimensions D_O, the solids valve in case (a) were abruptly fully opened the resulting efflux rate would show W(a) orders of magnitude greater than W(b). Past experimental studies of bulk solids flow therefore concentrated on the limits of arrangements such as in Figure 1(b). This is obviously the limiting case for design of storage bin hopper drawoff nozzles. It is important to establish the basic gravity flow equations for case (b) in order to appreciate the relative capacity of a situation such as in arrangement (c) where the solids have not "flooded" the inlet cone, as would more correctly apply in a properly operating cyclone cone-to-dipleg transition.

THE RAUSCH-BROWN STUDIES

Aside from the hour glass and other miscellaneous applications the process industries' first attempt to place the gravity flow of solids within a scientific frame of reference was the support of an investigation under the direction of Professor Joseph Elgin at Princeton University in the late 1940's financed by what was then the Socony-Vacuum Oil Co. The resulting thesis of Dr. John Rausch, completed in 1948, contained a proposed semi empirical correlation[1] which appeared to satisfy all possible conditions and found wide application for many years. Rausch's correlation which pertained solely to arrangements such as in Figure 1(b) could be expressed analytically in the form:

$$W_s = \left[\frac{0.23 \sqrt{g} \, C_o \, C_w}{\sqrt{\tan \beta}} \left(\frac{D_o}{D_p} \right)^{0.2} \right] \beta \, D_o^{1/2} \qquad (1)$$

In 1960 Dr. R. L. Brown[2] of the British Coal Utilization Research Association published results of some closer observations of the mechanism of the particles' movements in the vicinity of an orifice demonstrating a time-average blockage of the efflux port

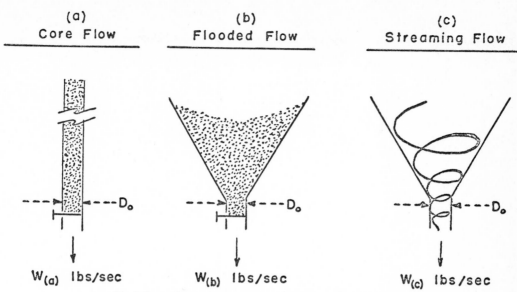

Fig. 1 Standpipe, bin, and cyclonic modes of solids gravity flow.

area at its periphery. An article appearing on pages 159-168 of
the February 1962 issue of Hydrocarbon Processing and Petroleum
Refiner reviewed the Rausch-Brown developments in detail and demonstrated that Brown's observations permitted equation (1) to be
greatly simplified and its resultant agreement with data substantially improved. This basic relationship derived from combining
the Rausch-Brown investigations gave:

$$W_S = \frac{1}{\sqrt{\tan \beta}}\, \rho_B\, D_e^{1/2} \tag{2}$$

where W_S = (lbs of solids/sec)/$[(\pi D_e^2)/(4 \times 144)]$

D_e = effective opening diameter, $(D_o - 1.5\, D_p)$
 in inches

D_o = actual diameter of opening, inches

D_p = particle diameter, inches

ρ_B = particles bulk density, lbs/cu.ft.

β = particles drained angle of repose, degrees

Aside from its simplicity there have always been several
disturbing aspects to equation (2):

1) It is dimensionally inconsistent, since W_s is expressed
on a per square foot basis while D_e is in inches. This can ob-

viously be circumvented by a numerical coefficient to handle the
conversion; however, a relationship without constants should be inherently sounder, if not simpler to recall, and hence there is re-
luctance to introduce such a conversion constant.

2) Equation (1) was rooted in a force balance on the particles
in the vicinity of the efflux port resulting in the $\sqrt{g}$ term which
in practical applications might simply be considered a constant.
Nevertheless, the question arises as to whether fundamentally it
should correctly appear. Would the efflux rate differ by $\sqrt{g}$ in
an acceleration field different than sea level ambient?

3) Following Rausch's lead, β in equations (1) and (2) is
the solids' drained angle of repose, which it would appear should
have little, if any, effect on the rate of solids efflux. From
the simple observations of solids flow in a two-dimensional Plexiglas
"slice" model of a bin[3], as illustrated in Figure 2, it is quite
evident that the shear plane which predominantly reflects the net
solids force above the efflux port is represented by the solids'
characteristic angle of internal friction, α , and not the drained
angle of repose, β , which represents solely a plane of equilibrium
between draining bulk solids and the surrounding fluid medium.

In the apparatus in which Rausch carried out his experiments
he was unable to observe the angles of internal friction of his
solids. Looking down onto the surface of the draining solids he
could at best only observe β; α is an internal shear plane between flowing bulk solids and the surrounding static bulk solids.

THE DIMENSIONLESS FLOW RELATIONSHIP

Rausch's tests were carried out with very free-flowing
solids exhibiting angles of repose on the average of 35°. Such
meager data as exist for free flowing solids suggest that α may
in general be approximated as β + 27. If the diameter term in
equation (2) were expressed in units of feet, thereby requiring a
conversion constant equal to the square root of 12, and if simultaneously tan β were replaced by tan α , and the gravitational
field introduced as suggested by equation (1) this would result in:

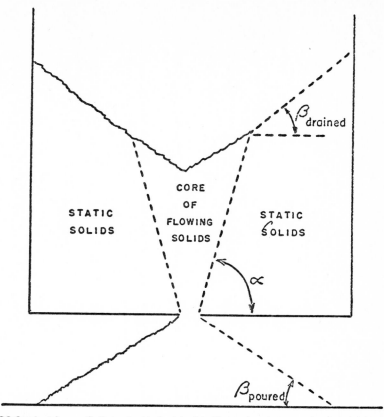

Fig. 2 Flow of FCC Catalyst from a 2-dimensional plexiglas bin illustrating the poured and drained angles of repose and the angle of internal friction.

$$W_s = \frac{\sqrt{g}}{\sqrt{32.2}} \frac{\sqrt{\tan \alpha}}{\sqrt{\tan \alpha}} \frac{1}{\sqrt{\tan \beta}} \rho_B D_e^{1/2} \sqrt{12} \qquad (3)$$

$$= \frac{\sqrt{12}}{\sqrt{32.2}} \frac{\sqrt{\tan \alpha}}{\sqrt{\tan \beta}} \frac{\sqrt{g}}{\sqrt{\tan \alpha}} \rho_B D_e^{1/2} \qquad (4)$$

Substituting average values of the angular properties of Rausch's solids:

$$W_s = \frac{\sqrt{12}}{\sqrt{32.2}} \frac{\sqrt{\tan (35+27)}}{\sqrt{\tan 35}} \frac{\sqrt{g}}{\sqrt{\tan \alpha}} \rho_B D_e^{1/2} \qquad (5)$$

but:

$$\frac{\sqrt{12}}{\sqrt{32.2}} \quad \frac{\sqrt{\tan 62}}{\sqrt{\tan 35}} \approx 1.0$$

so that equation (2) may be written in dimensionless form as:

$$W_s = \frac{\sqrt{g}}{\sqrt{\tan \alpha}} \, \rho_B \, D_e^{1/2} \tag{6}$$

where W_s = solids efflux rate, lbs/sec x ft^2 of effective
                        area
          g = acceleration field, ft/sec x sec (32.2 at
                        ambient sea level)
          α = solids angle of internal friction within their
                        surrounding interstitial medium, degrees
          ρ_B = solids bulk density in ambient air, lbs/cu.ft.
          D_e = effective opening diameter, in feet

Equation (6) is now dimensionless and bears no necessary
correlation or conversion coefficient. It correlates the data of
Rausch and of LaForge and Hatcher[4] more exactly than equations
(1) and (2), and it properly places the angle of internal friction
in the position of the principal flow determining variable where
it can now be clearly related to hopper configuration, bin design,
and the so-called mass flow[5] geometries. In media other than ambient air the term ρ_B should be replaced by:

$$\rho_B = \left[\rho_p (1-\epsilon) + \rho_f \epsilon \right] - \rho_f = (\rho_p - \rho_f)(1-\epsilon)$$

so that (6) becomes:

$$W_s = \frac{\sqrt{g}}{\sqrt{\tan \alpha}} \, (\rho_p - \rho_f)(1-\epsilon) D_e^{1/2} \tag{7}$$

where: ρ_p = particle density, lbs/cu.ft.
          ϵ = interparticle void fraction in bulk solids,
                        cu.ft./cu.ft.
          ρ_f = density of surrounding medium, lbs/cu.ft.

This density correction is significant only in very high pressure
gaseous media or in liquid systems.

RECTANGULAR PORTS

Equation (7) can be applied to efflux through a rectangular port of narrowest dimension w and longest dimension ℓ by substituting $w - 1.5D_p$ for D_e and $(\ell - 1.5D_p)(w - 1.5D_p)$ for the area in W_s.

EFFLUX IN AN ACCELERATION FIELD

The effect of the surrounding acceleration field as represented by the $\sqrt{g}$ term in equation (7) has been corroborated by means of a relatively simple experiment depicted schematically in Figure 3. A small can fitted with a corked hole of known diameter in its bottom and filled with a known charge of solids was slung around by A over his head through a known radius fixed by the length of the rope suspending the can. Upon a signal from B, at which time

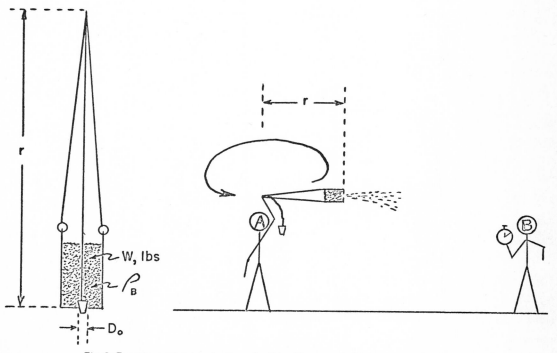

Fig. 3 Experimental determination of solids efflux rates in higher acceleration fields.

B began recording time on a stop watch, A pulled the cork from the
hole in the can (while maintaining rotation) and also began counting
revolutions. The moment B saw solids efflux cease he stopped his
watch and signalled A to stop counting revolutions. Such experiments yielded solids rates, W_s and acceleration fields, (peripheral
velocity)^2/r, derivable from the measured rpm, the radius, the
charge, and the time interval. The results, carried out up to 12
times ambient acceleration (e.g. up to 385 ft/sec x sec), showed
excellent agreement with equation (7).

EFFECT OF HOPPER ANGLE

Equation (7) applies to configurations of Figure 1(b) in
which the angle between the cone wall and the horizontal ranges
from $0°$ to α (or in other words as long as the core of effluxing
solids must shear against its own specie), and in which the ratio
of effective opening diameter to particle diameter exceeds a numerical value of about 12, which corresponds to the commonly accepted
minimum below which random particle interferences can lead to occasional faltering, if not complete blockage of flow at the opening.

When the hopper angle equals α , the peripheral flow area
loss, represented by 1.5 D_p reduction in effective opening diameter,
is reduced to 0.215 D_P which represents the fractional free area between spheres on a square pitch. This represents at most a 33% increase at the minimum design opening diameter of 12 D_P, diminishing
to less than 10% increase when D_o/D_P exceeds 34, and to less than a
2% increase when D_o/D_P exceeds 160. In view of the accuracy of the
experimental data this is of negligible consequence in the practical
range of dipleg design since 4"/60 microns represents a minimum
D_o/D_P of 1693.

When the hopper angle exceeds α, the efflux rate exceeds
that calculable from equation (7). This increase, shown in Figure
4, is based on the data of Rausch plus four experimental measurements, with glass beads in a series of relatively smooth sheet metal
conical hoppers, summarized in Table 1. The curves of Figure 4,
when θ exceeds α , will be a function of the smoothness of the

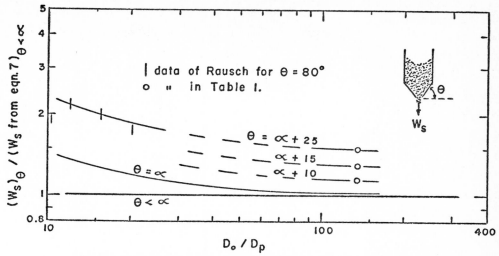

Fig. 4 Effect of hopper angle on solids efflux rates from smooth steel bins.

hopper walls and should therefore be applied with caution. In view of the relatively high values of α reported[6.] for FCC catalysts, a hopper of sufficient slope to materially effect the efflux rate, in excess of that given by equation (7), is rarely encountered and generally impractical.

STREAMING GRAVITY FLOW

If a coherent mass of solids were already accelerated to some velocity greater than zero upon entering the efflux port, it would certainly be expected that a higher effective efflux rate would be achieved than if the mass had to start from the port at

Table 1 Effect of Hopper Angle on Efflux Rate of 118 Micron Glass Beads

Hopper Angle, θ	Efflux Rate, lbs/secxsq.ft.	$(W_s)_\theta / (W_s)_{\theta < \alpha}$
$55°$	85.69	1.00
$\alpha + 5°$	87.40	1.02
$\alpha + 10°$	98.54	1.15
$\alpha + 15°$	111.40	1.30
$\alpha + 25°$	127.68	1.49

zero initial velocity. A series of comparative tests were therefore
carried out to evaluate the relative rate of bulk solids efflux,
or throughput, between arrangements such as in Figure 1(b) and 1(c).
In effect this constituted an attempt to determine to what extent a
streaming cyclone feed might be capable of exceeding a flooded gravity feed.

       The apparatus employed in these exploratory tests is
illustrated schematically in Figure 5. Four materials of widely
varying density, angle of internal friction and particle size were
investigated in arrangements (a) and (b). The 65° sloped cone was
simply made of galvanized sheet metal and was not in any way
polished or rubbed excessively to any desirably smooth finish.
Visual observation of the streaming flow tests left a suspicion
that metal surface finish and possibly hopper angle might have a
detectable, though probably not substantial, effect. The 65°
sloped cone was considered a reasonably close approximation to the
average of most FCC cyclone configurations.

       The properties of the test materials and the maximum
flow rates achieved are summarized in Table 2. At best it would
appear that streaming flow is consistently more effective with
decreasing particle size. However, it is not conclusive whether
the material's angle of internal friction is also significant as
a dependent or independent variable. In addition the smaller the
particle size the greater the tendency for powder to coat the cone
surface as a thin film. The lead powder, which exhibited such
coating most pronouncedly, had also somewhat of a lubricating
characteristic and hence leaves open to question whether this
property might have significantly affected the results. It may
also be argued that in actual cyclone operation the streaming flow
of collected solids might reach the dipleg entry at an even greater
velocity than was achieved in the simple tests of Figure 5 and
Table 2. It is obvious that a complete resolution of all these
variables would involve a relatively tedious and comprehensive
program which could likely be more academic than practical.

       At any rate, it would seem reasonable to accept 77° as
an average angle of internal friction for FCC catalyst as well as
a streaming flow factor of 2.25 in excess of flooded gravity flow

(a) Streaming Gravity Flow   (b) Flooded Gravity Flow

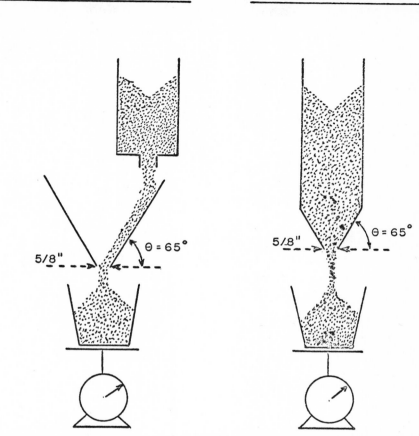

Fig. 5 Apparatus for exploratory comparative determination of streaming and flooded gravity flow rates.

Table 2 Maximum Streaming Flow Rates Through the Hopper and Port of Fig. 5

Material	α	β_B	Mean D_p Microns	Flooded Flow $(W_s)_{\theta < \alpha}$	Streaming Flow $(W_s)_{\theta = 65°}$	$\dfrac{(W_s)_{streaming}}{(W_s)_{flooded}}$
Lead	88°	413	< 44	99.94	253.85	2.54
FCC Catalyst	77°	59	60	36.12	81.63	2.26
Glass Beads	60°	90	118	85.69	161.95	1.89
Rutile	65°	145	150	127.64	181.25	1.42

thus leading to the suggested design criteria:

Maximum Capacity in Streaming Gravity Flow
(basis for minimum FCCU dipleg diameter):   (8)

$$W_s = 1.77 \, \beta_B \, D^{1/2}$$

Maximum Capacity in Flooded Gravity Flow
(basis for safe maximum necessary FCCU dipleg diameter):   (9)

$$W_s = 0.787 \, \beta_B \, D^{1/2}$$

where in both (8) and (9) D is the dipleg diameter in inches,
β_B, the catalyst bulk density in lbs/cu.ft., and W_s the dipleg
capacity in lbs of catalyst per second per square foot of dipleg
cross section.

The constant in (8) is based on a 60 micron mean particle size (from Table 2) hence it could be reasoned that second
and third stage diplegs collecting finer particles might justify
an even higher constant? The only justification for (9) would
lie in designing for the unlikely possibility that in a violently
slugging dipleg solids might occasionally be thrown back up into
the cyclone cone (if insufficient free space had been allowed
between the apex of the cone and the back-up height of solids in
the dipleg) whereupon they would then drain at the rate given by
(9); if the cyclone normally collected at a faster rate it might
then take a considerable time to clear. The loading to the
plugged cyclone would meanwhile be somewhat relieved because of
its resultant higher pressure drop.

It is interesting to note that the parallel relationship
to (8) and (9), derived on the erroneous basis of angle of repose
as opposed to angle of internal friction, resulted in a constant
of 1.2[7] which is fortuitously almost exactly the mean of the constants in (8) and (9).

REFERENCES CITED

1. Zenz, F. A., Petroleum Refiner 36, No. 10, 162-170 (1957)

2. Brown, R. L., Trans. Instn. of Chem. Engrs. (London) 38,
       243-256 (1960)

3. Zenz, F. A., Petroleum Refiner 36, No. 4, 173-178 (1957)

4. LaForge, R. M., Hatcher, S. T., Univ. of Tennessee, paper
        presented to the ASME (see reference 7.)

5. Jenike, A. W., ASME paper No. 63-WA-52

6. Zenz, F. A. and Othmer, D. F., "Fluidization and Fluid-
        Particle Systems", pp. 86-89, Reinhold, 1960

7. Zenz, F. A., Hydrocarbon Processing & Petroleum Refiner 41,
        No. 2, 159-168 (1962)

THE DETERMINATION OF GAS-SOLIDS PRESSURE DROP AND CHOKING VELOCITY AS A FUNCTION OF GAS DENSITY IN A VERTICAL PNEUMATIC CONVEYING LINE

T. M. KNOWLTON and D. M. BACHOVCHIN

INTRODUCTION

Pneumatic conveying of particulate solids in vertical lift lines is a widely employed transport technique. This technique has been extensively studied,[1-11] primarily at pressures near atmospheric. However, very little information exists in the literature on how high gas densities affect pneumatic conveying parameters and, thus, the design of high-pressure conveying equipment.

Because of the increasing number of coal gasification and related processes which operate at elevated pressures and utilize vertical lift systems, an investigation was undertaken to determine the effect of gas density on various lift parameters in a large-scale, cold-flow, high-pressure research facility. This facility was originally constructed to provide high-pressure fluidization and solids transport data to help improve the design and operation of the HYGAS[®] coal gasification pilot plant and is believed to be the only such large-scale facility in existence. Vertical lift-line pressure drops were studied over a gas density range of 0.14 to 3.4 lb/cu ft, while choking velocities were determined for a gas density range of 0.33 to 3.4 lb/cu ft.

EQUIPMENT AND MATERIAL

Equipment

A flow sheet of the equipment used in the vertical lift-line study is shown in Figure 1. Basically, the equipment consisted of a gas-recycle compressor, a solids recirculation loop, and gas-solids cleaning equipment.

A nonlubricated, reciprocating compressor was used to circulate nitrogen gas through the system. A bypass loop recirculated the gas in excess of that needed for conveying the solids back into the compressor inlet. A cooler removed the

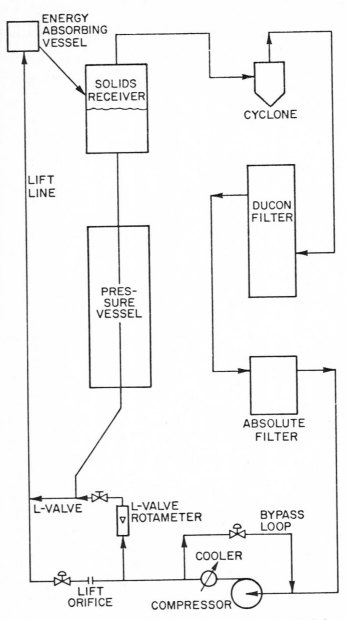

Fig. 1 Flow sheet of high-pressure-fluidization unit used for solids-recirculation studies.

heat of compression of the gas to prevent temperature buildup in the system. Following the recycle compressor, the gas was routed through a variable-area orifice meter and then passed through a valve which controlled the amount of gas fed to the lift line. The lift line was constructed of 3-inch Schedule 80 carbon steel pipe approximately 50 feet long. Pressure taps were installed approximately every 10 feet in the lift line to analyze the pressure drop in the lift line segmentally (Figure 2).

The gas-solids mixture in the lift line passed through an energy-absorbing vessel and then into the solids receiver where the solids were deposited. The solids receiver was constructed of 24-inch NPS Schedule 60 carbon steel pipe. The lift gas then passed out of the solids receiver into a cyclone and Ducon granular bed filter, through an absolute filter, and back to the compressor.

The solids flowed out of the solids receiver in gravity flow via a 3-inch Schedule 80 downcomer line to the pressure vessel. The pressure vessel was constructed of 24-inch NPS Schedule 60 carbon steel pipe approximately 16 feet long. Grayloc closures were provided at each end of the test vessel to facilitate the changing of internals.

The downcomer line passed through the middle of the pressure vessel as shown in Figure 3. A Plexiglas tube was installed in the downcomer line inside the pressure vessel so that solids motion in the downcomer could be observed. Ten sight ports, each constructed of 2-inch-thick Pyrex glass with a 5-5/8 inch viewing diameter, were positioned over the outside of the pressure vessel so that solids movement in the line could be visually monitored.

The solids then passed down through the remaining length of the downcomer to the L-valve, also constructed of 3-inch Schedule 80 pipe. The solids flow rate around the recirculation loop was controlled by the amount of aeration fed to the L-valve.

Material

The solids investigated were siderite ore and Montana lignite, having bulk densities of approximately 150 lb/cu ft and 47 lb/cu ft, respectively. The lignite material was that used in the IGT HYGAS coal gasification pilot plant in Chicago. The siderite material used was the iron ore to be used in the steam-iron hydrogen generating pilot plant to be operated at the same location. Both were wide-particle-

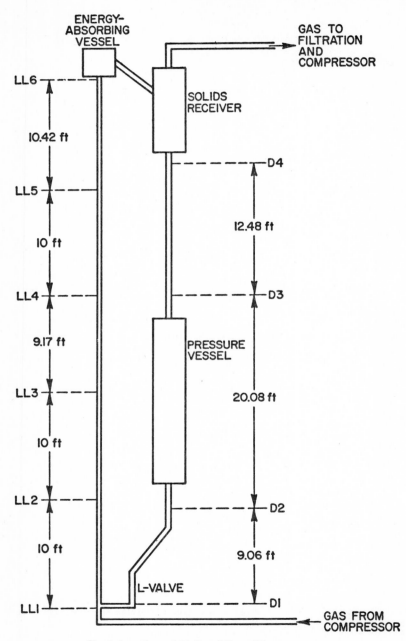

Fig. 2 Locations of lift-line differential pressure taps.

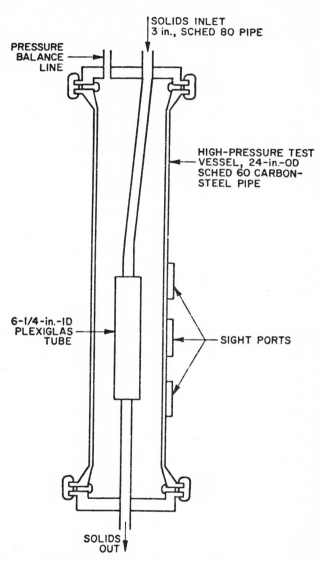

Fig. 3 High-pressure test vessel with 6.25 in. ID expanded plexiglas section.

Table 1 Physical characteristics of Lignite and Siderite

Screen Size Retained on, U.S.S.	Lignite, wt %	Siderite, wt %
10	0.4	0.0
14	5.8	0.0
20	20.6	0.0
30	13.5	0.1
40	15.6	0.3
60	21.5	28.2
80	9.1	24.0
100	4.3	13.1
200	6.5	28.2
325	1.8	5.7
Pan	0.9	0.4
Particle Density, lb/cu ft	78.6	244
Average Particle Size,* ft	0.00119	0.000515

$$* d_p = \frac{1.}{\Sigma x_i / d_i}$$

size range materials. The particle-size range and other solids characteristics of both materials are given in Table 1.

PROCEDURE

Two different types of runs were made in the study; one type to determine lift-line pressure drop as a function of solids mass velocity and a second type to determine choking velocities. Solids flow rates for both types of runs were determined by timing individual solid particles passing between two marks, 6 inches apart on a 6-1/4 inch ID Plexiglas tube (Figure 3).

To correlate the solids wall velocity with the actual solids velocity in the Plexiglas section, the tube was calibrated in a solids flow test at atmospheric pressure. In the test, the range of solids flow rates was determined by collecting the solids in a container over a certain time interval and then weighing the solids. During each solids rate measurement, the solids particle velocity at the wall was determined and correlated with the measured solids flow rate.

In a typical run to determine the effect of solids mass velocity on lift-line pressure drop, the solids to be studied were first charged to the system through a port on the top of the solids receiver. The system was then pressurized to the desired level utilizing compressors external to the system. Upon reaching this pressure, the recycle compressor was started and the lift gas flow rate was set. The solids flow rate into the lift line was varied from zero to its maximum value in several increments by adjusting the aeration rate to the L-valve. At each solids flow rate, the total lift-line pressure drop and the pressure drop over each 10-foot lift-line section were recorded.

In a typical choking velocity run, the solids were charged to the solids receiver and the system pressurized. The recycle compressor was started and the lift velocity was set at a relatively high value. The solids flow rate was then held constant while the lift gas velocity was decreased slowly. The solids flow rate was monitored at each reading and the total lift-line pressure drop and the pressure drop over each 10-foot section of the lift line were recorded. The choking velocity for a particular run was then determined graphically by plotting total lift-line pressure drop versus lift gas velocity. Near choking, the lift-line pressure drop would rise rapidly as the solids inventory in the lift line increased.

RESULTS AND DISCUSSION

Choking Velocity

If the gas velocity is decreased gradually in a lean-phase lift line and at the same time the solids mass velocity is kept constant, a pressure drop versus superficial velocity curve similar to that of Figure 4 results. A point A, the gas velocity is very high and the lift line is very dilute. As the lift velocity is decreased from point A to point B, the gas and solids both rise more slowly. The solids inventory in the lift line also rises, thus increasing the static head. However, the frictional resistance portion of the pressure drop predominates in this region; thus, as the velocity decreases, so does the pressure drop.

In the region B to C, the decreasing velocity causes a rapid rise in solids inventory. The static head now predominates over the frictional resistance and the pressure drop rises. Near point C, the bulk density of the mixture becomes too great for the gas velocity to support and the mixture starts to slug, or choke. The superficial gas velocity at point C then is termed the choking velocity. The choking velocity, V_{ch}, is the minimum velocity that can be used to transport solids in lift lines and is, obviously, an important design parameter.

Choking velocities for lignite and siderite using nitrogen as the conveying gas were determined as a function of gas density with solids mass velocity as a parameter. For lignite, three solids mass velocities (approximately 20, 30, and

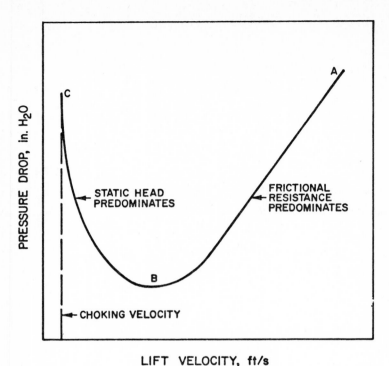

Fig. 4 Diagram of lean-phase gas-solid flow.

40 lb/sq ft-s) were studied at system pressures of 70, 155, 245, 490 and 700 psia. The siderite mass velocities and pressures were 50, 100, and 150 lb/sq ft-s and 156, 275, and 465 psia, respectively. Typical curves resulting from these runs are shown in Figures 5, 6, 7, and 8.

In each run, the solids flow rate was held constant as described above while the lift gas velocity was decreased slowly. The solids flow rate was monitored at each velocity and was held constant as the lift gas velocity was decreased. However, as the lift gas velocity was decreased, a point was reached where the solids flow rate suddenly decreased and could not be increased easily when more aeration was added to the controlling L-valve. This was not choking in the classical sense, however, because the lift-line and L-valve pressure drops did not increase and begin to cycle. A further decrease in the lift gas velocity would then cause the lift-line pressure drop to increase rapidly and cycle. However, because of the earlier decrease in solids flow, this point could not be taken as the choking velocity for the original mass velocity. Therefore, to estimate the choking velocity, the lift-line pressure drop-versus-gas velocity curves for a particular solids mass

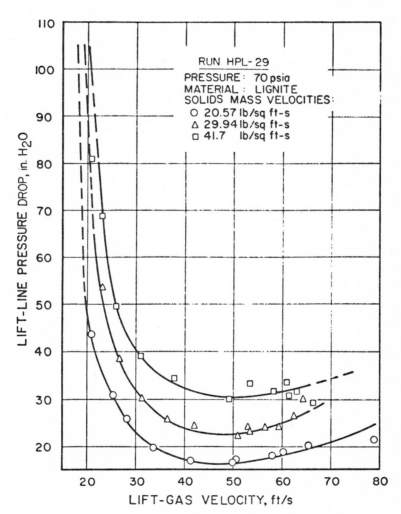

Fig. 5 Choking velocity as a function of solids mass velocity at 70 psia (run HPL-29).

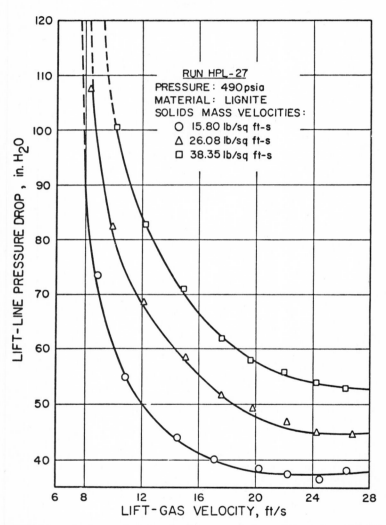

Fig. 6 Choking velocity as a function of solids mass velocity at 490 psia (run HPL-27).

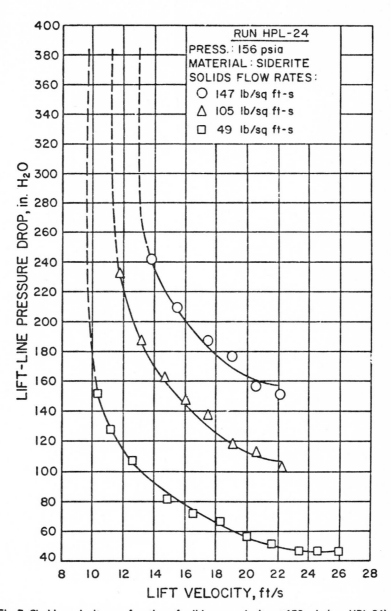

Fig. 7 Choking velocity as a function of solids mass velocity at 156 psia (run HPL-24).

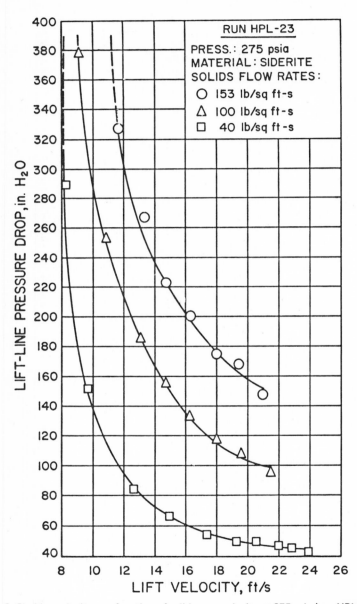

Fig. 8 Choking velocity as a function of solids mass velocity at 275 psia (run HPL-23).

velocity were extrapolated from the point immediately preceding the start of the decrease in solids flow rate. The extrapolated portions of the curves are shown as dashed lines in Figures 5, 6, 7, and 8.

Thus, rather than a "clear-cut phenomenon," there appears to be a gradual transition to choking over a range of gas velocities for wide-particle-size range materials. This is probably due to the refluxing of large particles near the wall at low gas velocities, whereas the smaller particles are still conveyed upward easily at these gas rates. This behavior has also been noticed by others.[1]

The pressure drop-versus-gas velocity curves shown in Figures 5 through 8 are almost totally in the region where the static head portion of the pressure drop predominates — region B to C in Figure 4. This region appears to be the "fast-fluid bed" region recently popularized by Squires.[12] The region A to B of Figure 4 was not extensively studied because solids carry-over was extremely heavy at the high lift velocities in this region and rapidly clogged the filters preceding the compressor.

The variation of choking velocity with gas density was also plotted for both siderite and lignite in Figures 9 and 10, respectively. The solid lines in these figures do not represent smoothed data curves, but are predicted choking velocity values obtained using a correlation developed from the siderite-lignite data. This correlation was developed using dimensional analysis using easily determined system parameters. A multiple linear regression technique was used to determine the exponents of the dimensionless variables. The correlation is presented below in Equation 1.

$$\frac{V_{ch}}{\sqrt{gd_p}} = 9.07 \left(\frac{\rho_p}{\rho_g}\right)^{0.347} \left(\frac{Wd_p}{\mu}\right)^{0.214} \left(\frac{d_p}{D_T}\right)^{0.246} \tag{1}$$

where

V_{ch} = choking velocity, ft/s
g = gravitational constant, 32.2 ft/s^2
d_p = average particle size, ft
ρ_p = particle density, lb/cu ft
ρ_g = gas density, lb/CF
W = solids mass velocity, lb/sq ft-s
μ = gas viscosity, lb/ft-s
D_T = tube diameter, ft.

This correlation predicts the experimental choking velocities obtained in this investigation to within 10% (Figure 11).

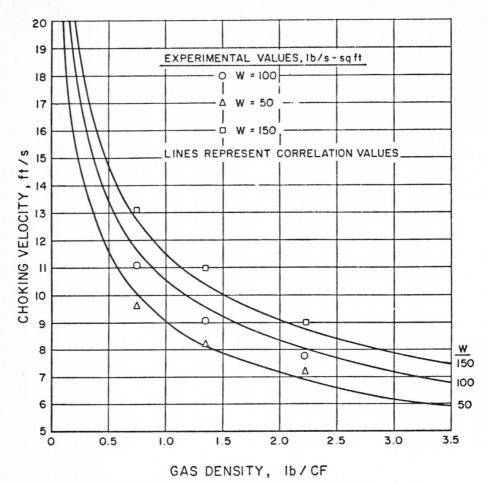

Fig. 9 Siderite choking velocity as a function of gas density.

Both the correlation and the data (Figures 9 and 10) show that lift-line velocities decrease nonlinearly as the gas density is increased. The greatest effect of density on choking velocity occurs at low gas densities (i.e., low pressures) where small changes in density cause relatively large changes in choking velocity. At nitrogen gas densities greater than 1 lb/cu ft, the effect of gas density on choking velocity was much less pronounced.

The pressure drop-versus-lift line velocity curves (Figures 5, 6, 7, and 8) for various mass velocities appeared to converge as they neared choking for a particular pressure. This has been observed by others.[12] This indicates that choking velcity is not a strong function of mass velocity and is reflected in the relatively low value of the exponent on the group containing mass velocity in the choking velocity correlation.

For the limited range of particle diameters studied, the correlation predicts an almost linear dependence of choking velocity on average particle size.

Minimum Lift-Line Pressure Drop

The lift-line pressure drop-versus-gas velocity curve (Figure 4) passes through a minimum in the transition region where the frictional and static head contributions to the lift-line pressure drop are equal. It is desirable to operate a lift line near or slightly to the right of the velocity where the lift-line pressure drop is a minimum. For instance, operating a lift line near the lowest possible pressure drop could mean shorter solids seal legs for solids transfer between

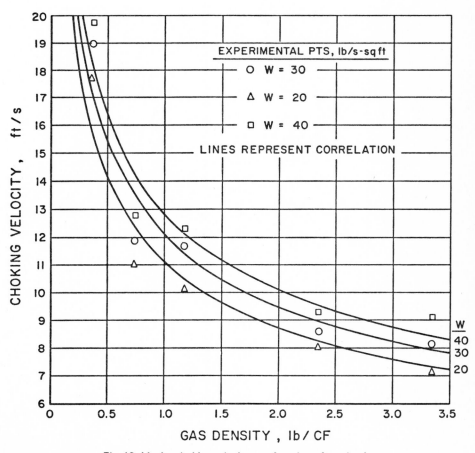

Fig. 10 Lignite choking velocity as a function of gas density.

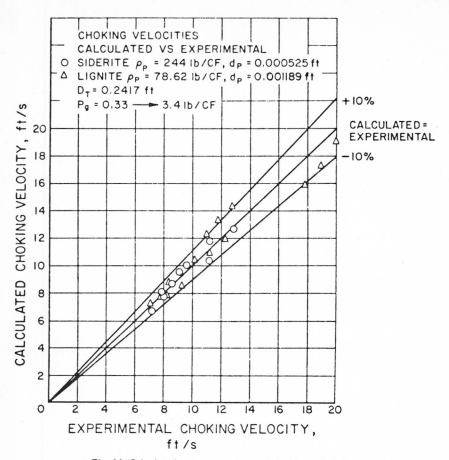

Fig. 11 Calculated versus experimental choking velocities.

stages of a fluidized bed gasifier. This would translate into shorter gasifier heights. Also, process upsets which could cause changes in the lift-line gas velocity would have a minimal effect on lift-line pressure drop. This is because the lift-line pressure drop is less sensitive to velocity upsets in this region than at any other.

A correlation to predict the velocity at which the lift-line pressure drop is a minimum was also developed. It is of the same form as the choking velocity correlation and is presented below.

$$\frac{V_{min}}{\sqrt{gd_p}} = 69.7 \left(\frac{\rho_p}{\rho_g}\right)^{0.273} \left(\frac{Wd_p}{\mu}\right)^{0.147} \left(\frac{d_p}{D_T}\right)^{0.272} \tag{2}$$

where

V_{min} = velocity at which pressure drop is a minimum, ft/s
d_p = average particle size, ft
ρ_p = particle density, lb/cu ft
ρ_g = gas density, lb/CF
W = solids mass velocity, lb/sq ft-s
μ = gas viscosity, lb/ft-s
D_T = tube diameter, ft.
g = gravitational constant, ft/s^2

This correlation predicts the experimental velocity at which the lift-line pressure drop is a minimum to within approximately 15% (Figure 12).

The ranges of all parameters varied in the above correlations and the ranges of the dimensionless groups are given below.

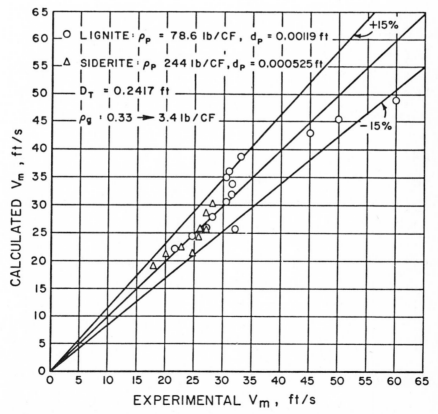

Fig. 12 Calculated versus experimental velocities for minimum lift-line pressure drop.

ρ_p : 78.6 and 244 lb/cu ft

ρ_g : 0.33 to 3.4 lb/cu ft

d_p : 0.000525 and 0.00119 ft

W   : 0 to 150 lb/sq ft-s

D_T : 0.2417 ft

μ : 0.0000124 lb/ft-s

$\left(\dfrac{\rho_p}{\rho_g}\right)$: 23.1 to 739.4

$\left(\dfrac{Wd_p}{\mu}\right)$: 0 to 6562.5

$\left(\dfrac{d_p}{D_T}\right)$: 0.000217 to 0.0049

Lift-Line Pressure Drop

Lift-line pressure drop was determined as a function of gas density (by varying system pressure) and solids mass velocity for both siderite and lignite. The siderite data obtained were used in the design of IGT's steam-iron hydrogen pilot plant in Chicago, and the lignite data were used for the evaluation of the operation of the hydrogasifier lift system in the HYGAS pilot plant at the same location. The lift-line pressure drops were measured for lignite over the pressure range of 70 to 700 psia, while the siderite lift-line pressure drops were studied over the pressure range of 30 to 465 psia.

One purpose of this study was to compare the measured lift-line pressure drops with those predicted by a low-pressure correlation. The Hinkle correlation,[11] a theoretical rather than an empirical relation, was selected for comparing the pressure drops. This correlation is given below in Equation 3.

$$\Delta P = \frac{V_g^2 \rho_g}{2g_c} + \frac{WV_p}{g_c} + \frac{2f\rho_g V_g^2 L}{g_c D_T}\left[1 + \left(\frac{f_p V_p}{fV_g}\right)\left(\frac{W}{V_g \rho_g}\right)\right] + \frac{WL}{V_p} + \rho_g L \qquad (3)$$

$$\quad\;\; 1 \qquad\quad 2 \qquad\qquad\quad 3 \qquad\qquad\qquad\qquad\quad 4 \qquad\quad 5 \qquad 6$$

where

ΔP = pressure drop, lb/sq ft

W   = solids mass velocity, lb/sq ft-s

g_c = gravitational conversion constant, 32.2 lb_m-ft/lb_f-s^2

V_g = gas velocity, ft/s

V_p = particle velocity, ft/s

f = fanning friction factor, dimensionless

f_p = particle friction factor, dimensionless

L = lift-line length, ft

ρ_g = gas density, lb/CF

D_T = tube diameter, ft

and where

term 1 is the pressure drop due to fluid acceleration

term 2 is the pressure drop due to particle acceleration

term 3 is the pressure drop due to fluid-to-pipe friction

term 4 is the pressure drop due to pipe-to-solid friction

term 5 is the pressure drop due to the static head of the solids

term 6 is the pressure drop due to the static head of the gas.

Because the lift-line gas was fully accelerated at the solids injection point for all runs in this investigation, term 1 was dropped in all analyses.

A typical comparison of the measured lift-line pressure drops and those predicted by the Hinkle correlation is shown in Figure 13. In this figure, the experimental lift-line pressure drop and the predicted lift-line pressure drop from the Hinkle correlation were plotted versus the lift gas velocity. The two curves intersected on the rising portion of the experimentally obtained curve in the region where the lift-line pressure drop rises as lift velocity is decreased. To the left of this intersection, the Hinkle correlation predicted low values of the lift-line pressure drop; and near choking, the prediction was extremely poor. To the right of the intersection, the Hinkle correlation predicted high values for the lift-line pressure drop. This result was similar for all pressures and mass velocities studied.

A literature search was made for other correlations. However, a comparative study by Khan and Pei[13] on vertical lift-line pressure drop correlations in the literature showed the unreliability of applying these correlations. In addition, all of these correlations were developed at low pressures, and therefore, did not cover the range of gas densities used in this investigation.

Because of the uncertainty in applying the correlations to even low-pressure data, it was decided to develop a correlation which would predict the lift-line pressure drops at high pressure. Instead of originating another dimensionless correla-

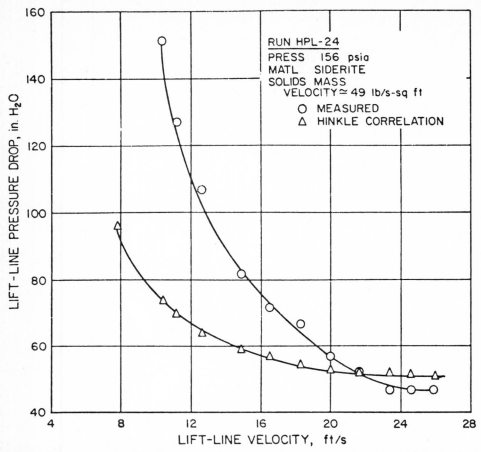

Fig. 13 Measured and calculated lift-line pressure drop versus velocity at 156 psia and 49 lb/sq ft-s (run HPL-24).

tion, the Hinkle correlation (Equation 3), which contains terms for all components of the pressure drop, was modified to fit the high-pressure data.

To use the Hinkle correlation, the particle velocity, V_p, and the particle friction factor, f_p, must be estimated. The usual procedure is to predict V_p by using the relationship

$$V_p = V_g - V_s \qquad (4)$$

where V_s is termed the slip velocity and is generally taken to be the terminal velocity of the average particle size as determined using the equation in Table 1.

The particle friction factor, f_p, was approximated using Hinkle's relationship

$$f_p = \frac{3}{2} \frac{\rho_g}{\rho_p} C_D \frac{D_T}{d_p} \left(\frac{V_g - V_p}{V_p} \right)^2 \tag{5}$$

Two approaches were taken to modify the Hinkle correlation. First, it was noted that term 5 in the correlation (the static head of solids) dominates the lift-line pressure drop near choking and the the Hinkle correlation predicted extremely low pressure drops in this region. Therefore, a method of calculating V_p was proposed to make the pressure drop-versus-velocity curve increase more sharply near choking. This was done by assuming

$$V_p = V_g - V_{ch} \tag{6}$$

where V_{ch} was calculated from Equation 1.

This assumption increased the predicted lift-line pressure drops too sharply. Therefore, the expression

$$V_p = \sqrt{(V_g - V_{ch})V_g} \tag{7}$$

was tried. This gave much better results. However, the best results were obtained using

$$V_p = \sqrt[3]{(V_g - V_{ch})V_g^2} . \tag{8}$$

The second modification made was to develop a correlation for f_p from the experimental data instead of using Equation 5. This correlation was developed using dimensional analysis and is given below.

$$f_p = \left[0.02515 \left(\frac{W}{\rho_g V_g} \right)^{0.0415} \left(\frac{V_p}{V_g} \right)^{-0.859} \right] - 0.03 . \tag{9}$$

The usual procedure in calculating solids-tube friction with the Hinkle correlation is to assume the ratio:

$$\frac{f_p V_p}{f V_g} = 1 \tag{10}$$

if the ratio exceeds unity. With the modified Hinkle correlation, this ratio was allowed to take on its calculated value.

The modified Hinkle correlation as described above predicted the experimentally obtained lift-line pressure drop very well over the entire pressure range studied. Figures 14, 15, 16, 17, 18, and 19 illustrate the agreement of both the Hinkle corre-

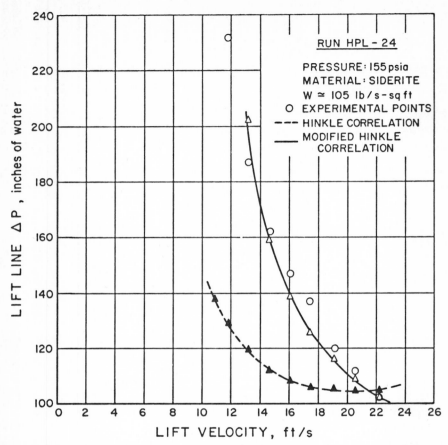

Fig. 14 Comparison of Hinkle Correlation and modified Hinkle Correlation with experimental lift-line pressure drop (run HPL-24).

lation and the modified Hinkle correlation with the experimentally obtained lift-line pressure drops. These plots cover the pressure range studied for both siderite and lignite and show the superiority of the modified Hinkle correlation.

The listing of all parameters varied in this investigation and their ranges, as well as the ranges of the dimensionless groups involved in the f_p correlation, are listed below:

$$V_g \quad : \quad \text{5 to 80 ft/s}$$
$$d_p \quad : \quad \text{0.000525 and 0.00119 ft}$$
$$W \quad : \quad \text{0 to 150 lb/sq ft-s}$$
$$\rho_g \quad : \quad \text{0.14 to 3.4 lb/CF}$$
$$\rho_p \quad : \quad \text{78.6 and 244 lb/cu ft}$$

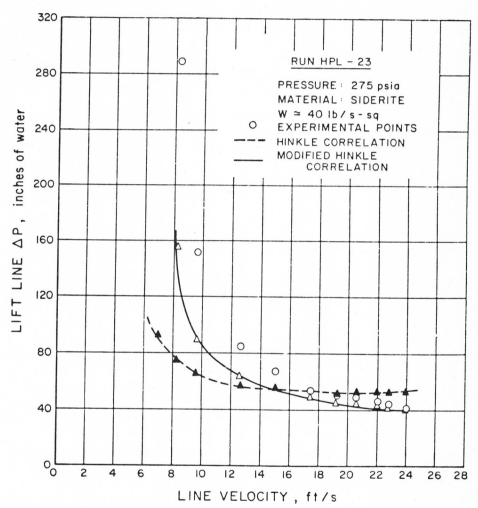

Fig. 15 Comparison of Hinkle Correlation and modified Hinkle Correlation with experimental lift-line pressure drop (run HPL-23).

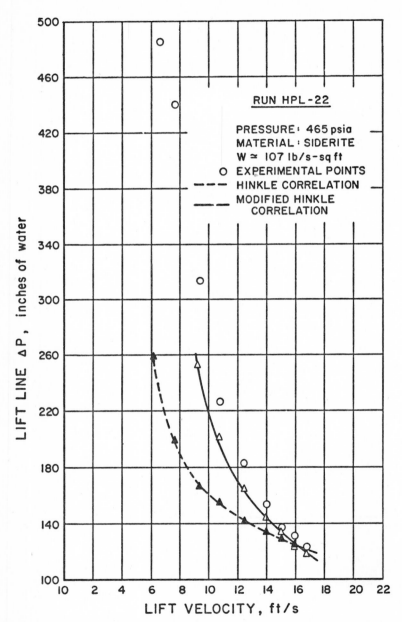

Fig. 16 Comparison of Hinkle Correlation and modified Hinkle Correlation with experimental lift-line pressure drop (run HPL-22).

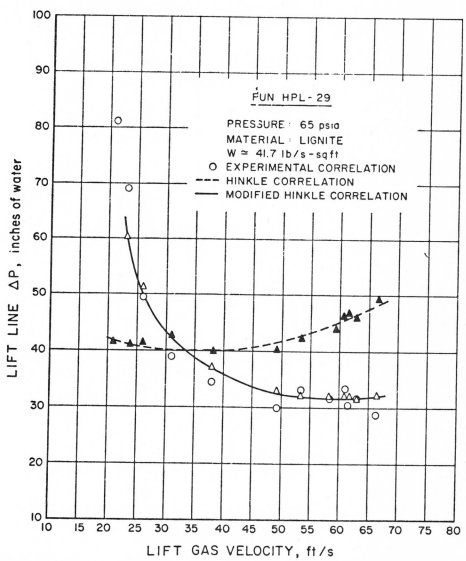

Fig. 17 Comparison of Hinkle Correlation and modified Hinkle Correlation with experimental lift-line pressure drop (run HPL-29).

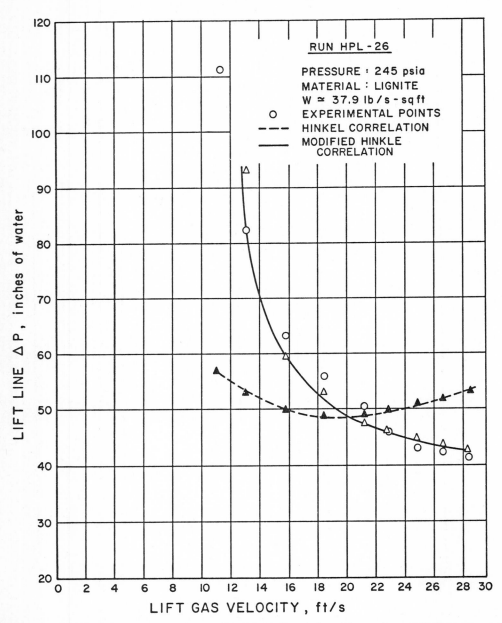

Fig. 18 Comparison of Hinkle Correlation and modified Hinkle Correlation with experimental lift-line pressure drop (run HPL-26).

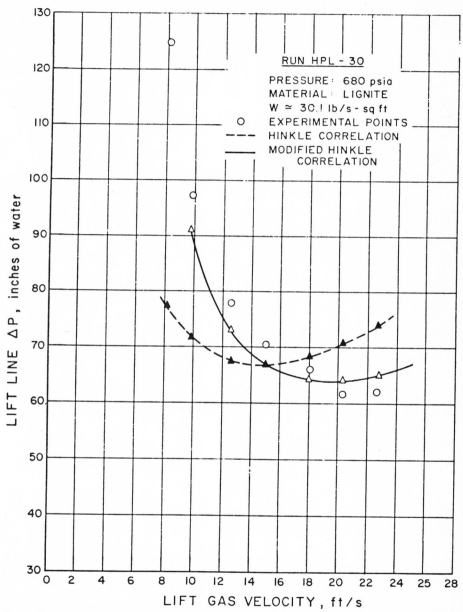

Fig. 19 Comparison of Hinkle Correlation and modified Hinkle Correlation with experimental lift-line pressure drop (run HPL-30).

$$D_T \quad : \quad 0.2415 \text{ ft}$$

$$\mu \quad : \quad 0.0000124 \text{ lb/ft-s}$$

$$\left(\frac{W}{\rho_g V_g}\right) \quad : \quad 0 \text{ to } 214$$

$$\left(\frac{V_p}{V_g}\right) \quad : \quad 0 \text{ to } 1$$

CONCLUSIONS

The Hinkle correlation did not predict lift-line pressure drops well at elevated pressures, especially at velocities near choking, for both siderite and lignite. After modification by 1) correlating the particle friction factor from the experimental data, and 2) developing a new expression for V_p, the Hinkle correlation predicted the measured lift-line pressure drops well.

Correlations were developed which predicted choking velocities to within 10% and V_{min} to within 15% (for both siderite and lignite) over their range of applicability.

NOMENCLATURE

A_T = tube cross-sectional area, sq ft

C_D = drag coefficient for a single particle

D_T = tube diameter, ft

d_i = average particle size of individual size fraction, ft

d_p = average particle size, ft

f = fanning friction factor for gas-to-wall friction

f_p = particle-to-wall friction factor

g = gravitational constant, 32.2 ft/s^2

g_c = gravitational conversion constant, 32.2 lb$_m$-ft/lb$_f$-s^2

K = constant of proportionality

L = lift-line length, ft

V_{ch} = choking velocity, ft/s

V_{min} = velocity at which lift-line pressure drop is a minimum, ft/s

V_p = particle velocity, ft/s

V_s = slip velocity, ft/s

V_w = velocity of particle at the wall, ft/s

W = solids mass velocity, lb/sq ft-s

x_i = weight fraction

μ = gas viscosity, lb/ft-s

ρ_B = solids bulk density, lb/cu ft

ρ_{eff} = solids effective bulk density, lb/cu ft

ρ_g = gas density, lb/cu ft

ρ_p = solids particle density, lb/cu ft

LITERATURE CITED

1. Capes, C. E. and Nakamura, K., "Vertical Pneumatic Conveying: An Experimental Study With Particles in the Intermediate and Turbulent Flow Regimes," Can. J. Chem. Eng. 51, 31-38 (1973) February.

2. Boothroyd, R. G., "Pressure Drop in Duct Flow of Gaseous Suspensions of Fine Particles," Trans. Instn. Chem. Engrs., Vol 44, 306-313 (1966).

3. Rose, H. E. and Duckworth, R. A., "Transport of Solid Particles in Liquids and Gases," The Engineer, 392-396, 429-433, 478-483 (1969) March.

4. Yousfi, Y. and Gau, G., "Aerodynamique de·L'Ecoulement Vertical de Suspensions Concentrees Gaz-Solides — I. Regimes D'Ecoulement et Stabilite Aerodynamique," Chem. Eng. Sci. 29, 1939-1946 (1974).

5. Leung, L. S., Wiles, R. J., and Nicklin, D. J., "Correlation for Predicting Flow Rates in Vertical Pneumatic Conveying," Ind. Eng. Chem. Process Design and Development, 10, 183 (1971).

6. Zenz, F. A., "Minimum Velocity for Catalyst Flow," Petroleum Refiner, 133-142 (1957) June.

7. Zenz, F. A., "Two-Phase Fluid-Solid Flow," Industrial and Engineering Chemistry 41, 2801-2806 (1949) November.

8. Zenz, F. A., "Conveyability of Materials of Mixed Particle Size," I. and E. C. Fundamentals 3, 65-75 (1964) February.

9. Hariu, O. H. and Molstad, M. C., "Pressure Drop in Vertical Tubes in Transport of Solids by Gases," Industrial and Engineering Chemistry 41, 1148-1160 (1949) June.

10. Belden, D. H. and Kassel, L. S., "Pressure Drops Encountered in Conveying Particles of Large Diameter in Vertical Transfer Lines," Industrial and Engineering Chemistry 41, 1174-1178 (1949) June.

11. Hinkle, B. L., Ph.D. Thesis, Atlanta, Georgia Institute of Technology (1953) June.

12. Squires, A. M., Yerushalmi, J., and McIver, A. E., "The Fast Fluidized Bed," Presented at AIChE Meeting in Washington, D. C., December 1-5, 1974.

13. Khan, J. and Pei, D., "Pressure Drop in Vertical Solid-Gas Suspension Flow," Ind. Eng. Chem. Process Design and Development, 12, 428-431 (1973).

ACKNOWLEDGEMENT

The author would like to express his appreciation to the American Gas Association, under whose auspices and funding the work was conducted.

PART III

Fossil Fuel Processing

COMBUSTION OF COAL IN FLUIDIZED BEDS

E. K. CAMPBELL and J. F. DAVIDSON

1.   Introduction. This paper summarises work carried out in a
laboratory scale fluidised bed combustion unit. A mathematical model
is proposed for the description of the combustion mechanism, and the
model is then tested by experiment. The model gives predictions for
(i) the burnout time of a batch charge of carbon
(ii) the size distribution of carbon particles in a bed which is con-
     tinuously fed with particles of uniform size.
The model predicts the parameters controlling bed combustion rates.
Values for these parameters were determined from batch experiments
and applied to the prediction of particle size distributions for
continuous combustion.

      Our objectives can be stated thus:
(i) The derivation of a model describing the combustion of carbon
     particles in fluidised beds.
(ii) A critical analysis of the combustion model by experiment.

2.   The Combustion Model. In a recent paper, AVEDESIAN & DAVIDSON[1]
have described a model for the combustion of char particles in fluidised beds. The model, described in detail by AVEDESIAN[2], assumed that
the particulate phase CO_2 concentration was always near zero. Extension of the model to cases where finite CO_2 concentrations exist yields
results not in agreement with experiment. Modification of the model
by assuming that all exothermic combustion reactions occur very near
the carbon particle surfaces, thus maintaining the particles at
temperatures several hundred degrees above the mean bed level, achieves
good agreement between theory and experiment. Evidence in support of
the relatively hot carbon particles has been given by CURR & WARREN[3]
and WHELLOCK[4].

      The model is based on the following principles:
(i)   At reactor temperatures (T = 900°C) all reactions are fast,

hence combustion rates are controlled by molecular diffusion of
reactants and products.

(ii) The two-phase theory of fluidisation, as outlined by DAVIDSON
& HARRISON[5], is assumed.

(iii) A reaction zone model, as described by FIELD[6] is assumed to
describe the reaction mechanism. We assume the reaction zone size to
be restricted to the maximum size R = b.d/2 where b is a constant near
unity.

(iv) The particulate phase, though perfectly mixed, is assumed to
approximate a stagnant gas for the description of the diffusion of
gases.

Analysis of the above assumptions yields an expression for the
burnout time of a batch charge of carbon particles. This expression
allows the computation of bed combustion rates and the Sherwood
number Sh for the diffusion of O_2 through the particulate phase, from
batch experiment data.

The model also allows the direct prediction of bed carbon
particle size analyses for continuous particle addition experiments,
in terms of the combustion rates and the value for Sh obtained from
batch experiments.

3.   Combustion Experiments

Combustion experiments were carried out in a 76 mm diameter
cylindrical quartz tube reactor. The bed was air fluidised coal
residue ash, at a temperature of $900^{\circ}C$. Results from experiments are
shown in figures 1 to 4. Conclusions drawn from results are listed
below.

(i)   Variation of the carbon particle size produces no noticeable
change in bed combustion rate.

(ii) About 90% of input oxygen is consumed regardless of bed ash
size or air fluidising velocity. This result is similar to that
obtained by ORCUTT[7] for temperatures at which reaction rates are
diffusion controlled.

(iii) Variation of bed height above shallow levels does not increase
the bed combustion rate. This indicates that most interphase gas
transfer occurs near the distributor where bubbles are small and
contact area between bubbles and the particulate phase is large.

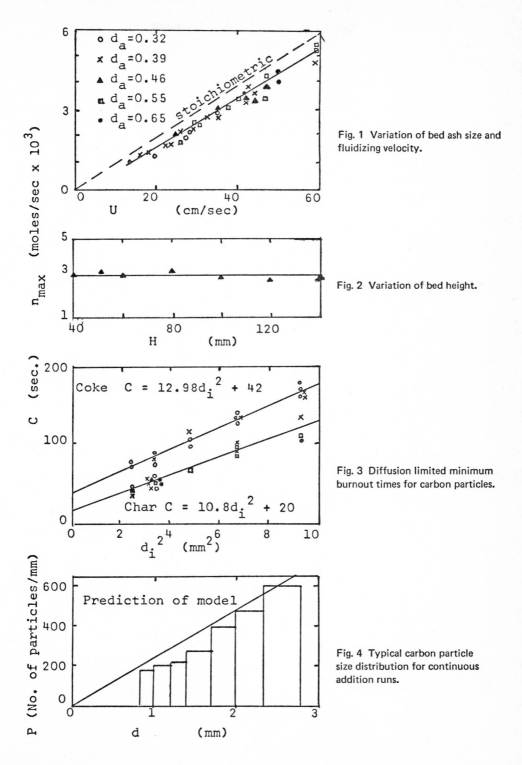

Fig. 1 Variation of bed ash size and fluidizing velocity.

Fig. 2 Variation of bed height.

Fig. 3 Diffusion limited minimum burnout times for carbon particles.

Fig. 4 Typical carbon particle size distribution for continuous addition runs.

(iv) Computation from the results of fig.3 gives b.Sh = 2.38.
Since we expect Sh = 2ϵ (see reference 2), we conclude that b = 2,
i.e. b is a small number as assumed in section 2.

(v) Experiments at lower temperatures indicate that the model
applies only at temperatures above about 820°C.

The results appear to be in good agreement with the theory,
hence implying the validity of the model. This work is continuing,
and more detailed continuous particle addition experiments are
planned.

NOTATION

b	constant defining reaction zone size	−
C	diffusion limited minimum particle burnout time	sec.
d	carbon particle diameter	mm.
d_a	ash particle diameter	mm.
d_i	initial carbon particle diameter	mm.
H	fluidised bed height	mm.
n_{max}	maximum bed carbon combustion rate	moles/sec x 10^3
P	carbon particle size distribution function	mm^{-1}
R	reaction zone radius	mm.
Sh	Sherwood number for particulate phase diffusion	−
T	bed temperature	$^\circ$C
u	superficial fluidising air velocity	cm/sec
ϵ	particulate phase voidage fraction	−

REFERENCES

1. AVEDESIAN, M.M. & DAVIDSON, J.F. Trans.Instn Chem.Engrs 1973, 51.

2. AVEDESIAN, M.M., Ph.D.Dissertation, University of Cambridge, 1972

3. CURR, T.N. & WARREN, R., Private communication, N.C.B. C.R.E.
            U.K. 1969

4. WHELLOCK, J.G., Private communication, University of Birmingham,
            1971

5. DAVIDSON, J.F. & HARRISON, D., "Fluidised Particles",
            Cambridge University Press, 1963.

6. FIELD, M.A. et al., "Combustion of Pulverised Coal",
            Leatherhead, 1967.

7. ORCUTT, J.C., Ph.D. Dissertation, University of Delaware, 1960.

ANALYSIS OF FLUIDIZED-BED COMBUSTION OF COAL WITH LIMESTONE INJECTION

M. HORIO and C. Y. WEN

ABSTRACT

A mathematical model of fluidized-bed combustor has been developed based on the Modified Bubble Assemblage Model. The distributions of coal-particle size and of limestone conversion are also considered in this model. The model is examined by the experimental data of SO_2 removal in literatures and is found to be satisfactory in the prediction of observed performance. The model may be helpful in scale-up and design of fluidized-bed combustor.

INTRODUCTION

Because of a number of inherent advantages of fluidization techniques considerable efforts have been focused during the past decade on the pilot-plant experiments of fluidized-bed combustion in the presence of limestone or dolomite. Considerable amount of data have been presented by Bishop et.al. (1968), Robinson, et.al. (1970), Coates et.al. (1970, 1973), Jonke et.al. (1968, 1970) and Davidson and Smale (1970). Recently Vogel et.al. (1974) reported an experiment using a pressurized fluidized-bed combustor containing dolomite particles.

However, so far, no theoretical models have been devised which

can consistently explain the results from these various pilot-plant
studies. It seems that the difficulty of the modeling arises mainly
from the following three reasons.

1) The complexity of the reaction kinetics for the limestone-
SO_2 reaction, which have been thoroughly studied recently by Coutant
(1970), Borgwardt (1970a, 1970b, 1972), and Wen and Ishida (1973).

2) The complicated fluidized bed reactions of simultaneous SO_2
generation and absorption occurring in series.

3) Unsteady-state analysis which is often necessary for the
analysis of some of the data; for example the semi-batch operation
data or Robinson et.al. (1970).

This paper develops a mathematical model for a fluidized bed
combustor emphasizing the incorporation of desulfurization phenomena.
The model derived must be able to simulate consistently the pilot-
plant operations listed above. The model is based on the Modified
Bubble Assemblage Model of Mori and Wen (1975b) and considers the
residence time distributions of coal and limestone particles within
the fluidized bed.

FORMULATION OF MODEL

The items listed in the following are considered to play an
important role in the operation of a fluidized-bed combustor.

1) Devolatilization and combustion of coal.

2) Absorption of SO_2 by limestone.

3) Residence time distributions of coal and limestone particles.

4) Jetting above the perforated plate, bubble formation and
   coalescence along the bed axis.

5) Gas interchange between bubbles and emulsion phase.

6) Heat transfer between gas and solid, and bed and heat exchange surfaces.

All items listed above except 6) have been considered in the model derived in this paper. Items 4) and 5) are expressed by Bubble Assemblage Model formulated by Kato and Wen (1969) and modified by Mori and Wen (1975b). With respect to item 3), the condition of the solid mixing in a fluidized-bed combustor has been assumed to be complete. This assumption is supported by the fact that the solid circulation rate estimated by the bubble rising velocity and wake volume using the same method as that of Wen et.al. (1974) is extremely large. The order of magnitude of the ratio of solid circulation rate to the solids feed rate is estimated to be about $10^{3} \sim 10^{4}$. The mixing experiments in a cold model of fluidized-combustor by Highley and Merrick (1971) also support rapid mixing of coal particles.

The following equation is derived based on the population balance in terms of the distribution function $\phi(x)$, where x denotes both the size of a coal particle, d_c, or the fractional conversion of limestone, f.

$$\frac{\partial N\phi}{\partial \theta} = n_f\phi_f - n_e\phi_e - n_w\phi_w + N\,\frac{\overline{\partial\phi(dx/d\theta)^*}}{\partial x} \qquad (1)$$

where $(dx/d\theta)^*$ is the rate of change in x corresponding to one particle and the bar on the last term means an average over the bed. $(dx/d\theta)^*$ is expressed as follows:

$$
(dx/d\theta)* =
\begin{cases}
\left(\dfrac{d(d_c)}{d\theta}\right)^* = -\dfrac{2M_C r_c^*}{\pi d_c^2 \rho_C} & \text{(for coal)} \quad (2\text{-}1) \\[20pt]
\left(\dfrac{df}{d\theta}\right)^* = \dfrac{6M_{Ca} r_1^*}{\pi d_p^3 \rho_{Ca}} & \text{(for lime-} \\
 & \text{stone)} \quad (2\text{-}2)
\end{cases}
$$

where r_c^* and r_1^* are combustion rate of a coal particle and the rate of limestone-SO_2 reaction per one limestone particle respectively and are estimated by the following equations:

$$
r_c^* =
\begin{cases}
\pi\, d_c^2\, k_c\, C_{O_2} & \text{(SRC)} \quad (3\text{-}1) \\[14pt]
2\pi\, d_c\, \mathcal{D}_{O_2}\, C_{O_2} & \text{(FDC)} \quad (3\text{-}2)
\end{cases}
$$

$$
r_1^* = (\pi/6)\, d_p^3\, k_v\, C_{SO_2}\, \lambda \qquad\qquad\qquad\qquad (4)
$$

where SRC and FDC denote the surface reaction control and gas film diffusion control respectively for a shrinking coal particle. In Equation (3-2) the Sherwood number is assumed to be 2. Parameter λ in Equation (4) represents the reactivity of partially reacted limestone and is defined as

$$
\lambda(f, d_p) = \frac{\text{actual } SO_2 \text{ absorption rate by calcined limestone at } f=f}{SO_2 \text{ absorption rate by calcined lime-stone at } f=0} \qquad (5)
$$

In the case of complete solid mixing, the distribution function ϕ is identical throughout the bed. $X(=d_c$ or $f)$ is an independent variable in Equation (1). Using Equations (2)-(4), the last term of Equation (1) becomes as follows.

For coal particles:

$$N_c \frac{\overline{\partial \phi_c (d(d_c)/d\theta)^*}}{\partial (d_c)} = \begin{cases} -\dfrac{2N_c M_C k_c \overline{C}_{O_2}}{\rho_C} \dfrac{\partial \phi_c}{\partial d_c} & \text{(SRC)} \quad (6\text{-}1) \\[2em] -\dfrac{4N_c M_C \mathcal{D}_{O_2} \overline{C}_{O_2}}{\rho_C} \dfrac{\partial (\phi_c/d_c)}{\partial d_c} & \text{(FDC)} \quad (6\text{-}2) \end{cases}$$

For limestone particles:

$$N_1 \frac{\overline{\partial \phi_1 (df/d\theta)^*}}{\partial f} = \frac{N_1 M_{Ca} k_v \overline{C}_{SO_2}}{\rho_{Ca}} \frac{\partial \lambda \phi_1}{\partial f} \qquad (7)$$

It can be seen from Equations (6) and (7) that only the average gas concentration is needed to evaluate the function ϕ. Therefore, the overall reaction rates for combustion and SO_2 absorption can be simply expressed as follows:

$$R_c^* = \begin{cases} \pi N_c \overline{d_c^2} k_c \overline{C}_{O_2} & \text{(SRC)} \quad (8\text{-}1) \\[1.5em] 2\pi N_2 \overline{d_c} \mathcal{D}_{O_2} \overline{C}_{O_2} & \text{(FDC)} \quad (8\text{-}2) \end{cases}$$

$$R_1^* = (\pi/6) N_1 d_p^3 k_v \overline{C}_{SO_2} \overline{\lambda} \qquad (9)$$

where $\overline{d_c^2}$, $\overline{d_c}$ and $\overline{\lambda}$ are defined as

$$\overline{d_c^2} = \int_o^{d_{cm}} \phi_c d_c^2 d(d_c)/d_{cm} \qquad (10\text{-}1)$$

$$\overline{d_c} = \int_o^{d_{cm}} \phi_c d_c d(d_c)/d_{cm} \qquad (10\text{-}2)$$

$$\overline{\lambda(f,d_p)} = \int_o^1 \phi_1 \lambda(f,d_p) df \qquad (11)$$

In order to develop a mathematical model convenient for computation the fluidized bed is divided into n compartments based on the concepts of the Bubble Assemblage Model. The material balance around the ith compartment gives the following equations in regard to the mole fraction of oxygen or SO_2 in both bubble and emulsion phases.

$$\delta(x_{ABi}-x_{ABi-1})=K_i(x_{AEi}-x_{ABi})-\alpha_{Ai}\gamma_i x_{ABi}+g_{ABi} \tag{12}$$

$$(1-\delta)(x_{AEi}-x_{AEi-1})=K_i(x_{ABi}-x_{AEi})-\alpha_{Ai}(1-\gamma_i)x_{AEi}+g_{AEi} \tag{13}$$

where subscript A denotes the species O_2 and/or SO_2. The dimensionless parameters in Equations (12) and (13) are defined as follows.

$$K_i = \frac{\text{total gas exchange rate in } i^{th} \text{ compartment}}{\text{total flow rate of gas}} \tag{14}$$

$$g_{ABi} \text{ or } g_{AEi} = \begin{cases} 0 \quad\cdots\cdots\cdots\cdots\cdots \text{ (if } A=O_2) & (15\text{-}1) \\[2ex] \dfrac{\begin{pmatrix} \text{the volumetric generation} \\ \text{rate of } SO_2 \text{ in bubble or} \\ \text{emulsion phase of } i^{th} \\ \text{compartment} \end{pmatrix}}{\text{total flow rate of gas}} \quad \text{(if } A=SO_2) & (15\text{-}2) \end{cases}$$

$$\alpha_{Ai} = \begin{cases} \alpha_{O_2,i} = \begin{cases} \dfrac{\pi N_{ci}\overline{d_c^2}k_c}{u_o A_t} & \text{(SRC)} \quad (16\text{-}1\text{-}1) \\[3ex] \dfrac{2\pi N_{ci}\overline{d_c}\mathcal{D}_{O_2}}{u_o A_t} & \text{(FDC)} \quad (16\text{-}1\text{-}2) \end{cases} \\[6ex] \alpha_{SO_2,i} = \dfrac{\pi N_{1i}\overline{d_p^3 k_v}\overline{\lambda}}{6u_o A_t} & \quad\quad\quad\quad (16\text{-}2) \end{cases}$$

$$\gamma_i = \frac{\begin{array}{c}\text{volume of solids dispersed in clouds and} \\ \text{wake of the bubble in } i^{th} \text{ compartment}\end{array}}{\text{volume of solids in } i^{th} \text{ compartment}} \tag{17}$$

$$\delta = \frac{\text{the rate of gas flow as bubbles}}{\text{total flow rate of gas}} \tag{18}$$

In the Bubble Assemblage Model, δ is assumed to be 0. In the analysis of Hydrane Process fluid-bed hydrogasifier Wen et.al. (1974) allocated that the cloud phase into emulsion phase. This treatment corresponds to the case of $\gamma_i = 0$ in Equations (12) and (13). The gas mixing in the first compartment is considered to be more vigorous compared to the remainder of the compartments, and therefore, is assumed to be complete mixing. This assumption and other features of the Bubble Assemblage Model are discussed in detail by Mori and Wen (1975b).

First, the population balance equations are solved to give the average diameter of coal particles and average value of limestone reactivity which appear in the parameter α_{Ai} of Equations (12) and (13). In the following two sections the approximate solutions for these variables under steady state are given.

Combustion of Coal and Generation of SO_2

It has been well documented that the combustion efficiency can be as high as 99% when the fines are returned to the bed (cf. Coates and Rice (1970) or Hoy (1970)). Therefore, although elutriation of fines and its effect on the the residence time of coal particles are important problems in the design of fluidized bed combustor, a simplifying assumption stating that the fines elutriated will be completely returned to the bed (n_{ce}=0) will bring about only little error.

Assuming n_{ce}= 0, ϕ_{cw} = ϕ_c and no size distribution of coal particles in the feed, the solution of Equation (1) becomes very simple as shown in the following:

$$\phi_c(d_c) = \begin{cases} (B_c a_c/d_{cm})\exp[B_c(d_c/d_{cm}-1)] & \text{(SRC)} \quad (19\text{-}1) \\ (B_c a_c d_c/d_{cm}^2)\exp[(B_c/2)(d_c^2/d_{cm}^2-1)] & \text{(FDC)} \quad (19\text{-}2) \end{cases}$$

where B_c and a_c are defined as

$$B_c = \begin{cases} \rho_C n_{cw} d_{cm}/2M_C N_c k_c \overline{C}_{O_2} & \text{(SRC)} \quad (20\text{-}1) \\ \rho_C n_{cw} d_{cm}^2/4M_C N_c \mathcal{D}_{O_2} \overline{C}_{O_2} & \text{(FDC)} \quad (20\text{-}2) \end{cases}$$

$$a_c = \begin{cases} 1/(1-e^{-B_c}) & \text{(SRC)} \quad (21\text{-}1) \\ 1/(1-e^{-B_c/2}) & \text{(FDC)} \quad (21\text{-}2) \end{cases}$$

Then overall combustion rate can be expressed as

$$R_c^* = \begin{cases} \pi N_c d_{cm}^2 \psi_a k_c \overline{C}_{O_2} & \text{(SRC)} \quad (22\text{-}1) \\ 2\pi N_c d_{cm} \psi_a \mathcal{D}_{O_2} \overline{C}_{O_2} & \text{(FDC)} \quad (22\text{-}2) \end{cases}$$

where the coefficient ψ_a is defined as

$$\psi_a \equiv \begin{cases} \overline{d_c^2}/d_{cm}^2 = a_c(1-2/B_c) + 2/a_c B_c^2 & \text{(SRC)} \quad (23\text{-}1) \\ \overline{d_c}/d_{cm} = 1-\exp(-B_c^2/2)\int_o^1 \exp(B_c x^2/2)\,dx & \text{(FDC)} \quad (23\text{-}2) \end{cases}$$

The total number of coal particles contained in the bed, N_c, is one of the unknown variables and is expressed by using a_c as follows:

$$N_c = 6W_{Cf}\,\Theta_1/\pi\rho_c d_{cm}^3 a_c \tag{24}$$

To derive this equation the following relationships have been applied.

$$N_c/N_1 = n_{cw}/n_{1w} \tag{25}$$

$$N_1/n_{1w} = N_1/n_{1f} = V_{Ca}/W_{Caf} = \Theta_1 \tag{26}$$

Since the hold up of coal particles in the bed is negligibly small, the hold up of calcium in the bed, V_{Ca}, in Equation (26) is approximately equal to $(1-\varepsilon_{mf})A_t L_{mf}\rho_{Ca}$. Therefore, the mean residence time of limestone, Θ_1, can be estimated only from the operating conditions and the number of carbon particles N_c can be calculated if $\overline{C}_{O_2}$ and the bed temperature are given.

However, the values of N_c can be calculated more easily because the value of a_c in Equation (24) usually is nearly equal to $1/B_c$ for SRC and $2/B_c$ for FDC. Then we have

$$N_c = \begin{cases} 3W_{Cf}/\pi d_{cm}^2 M_C k_c \overline{C}_{O_2} & \text{(SRC)} \quad (27\text{-}1) \\[2mm] 3W_{Cf}/4\pi d_{cm} M_C \mathcal{D}_{O_2} \overline{C}_{O_2} & \text{(FDC)} \quad (27\text{-}2) \end{cases}$$

The error of this approximation is less than 0.5% for the case when B_c is less than 0.01. In this situation the coefficient ψ_a is nearly equal to 1/3 for SRC and to 2/3 for FDC, and the size distribution function ψ_c becomes linear with respect to d_c.

These analyses shown above can be easily extended to the case when the size distribution of coal particles exists in the feed, but the results are not so different.

The oxygen concentration, C_{O_2}, along the bed is represented by the following equation when the homogeneous plug flow is assumed for the gas flow through the bed.

$$C_{O2} = C_{O2,0} \exp(-\alpha_c \zeta) \tag{28}$$

where ζ is dimensionless distance from the distributor and in the case of SRC,

$$\alpha_c \equiv \frac{\pi N_c d_c^2 k_c}{u_o A_t} \cdot \frac{\text{oxygen reacting with H and S in coal}}{\text{oxygen reacting with C in coal}} \tag{29}$$

By the assumption of complete combustion of coal,

$$\alpha_c = \frac{\text{theoretical oxygen rate}}{u_o A_t \overline{C}_{O2}} \tag{30}$$

From Equation (28) the average concentration is obtained as follows:

$$\overline{C}_{O2} = C_{O2,0} (1-e^{-\alpha_c})/\alpha_c \tag{31}$$

From Equations (28), (30) and (31) and from overall material balance the following expression is obtained for α_c.

$$\alpha_c = \ln\left[\frac{1 + \text{excess air ratio}}{\text{excess air ratio}} \right] \tag{32}$$

Since the excess air ratio is known from operating conditions, we can evaluate α_c by Equation (32), $\overline{C}_{O_2}$ by Equation (31) and N_c by equation (27).

The total amount of coal particles held in bed, V_c, is then

evaluated by using a coefficient ψ_v as follows:

$$V_c = \pi \rho_c N_c d_{cm}^3 \psi_v / 6 \qquad\qquad (33)$$

$$\psi_v \equiv \frac{\overline{d_c^3}}{d_{cm}^3} = \begin{cases} \alpha_c(1-\dfrac{3}{B_c} + \dfrac{6}{B_c^2}) - \dfrac{6}{B_c^3} & \text{(SRC)} \quad (34\text{-}1) \\[4mm] \alpha_c[1-\dfrac{3}{B_c} + \dfrac{3}{B_c}\exp(-\dfrac{B_c}{2})\displaystyle\int_0^1 \exp(\dfrac{B_c}{2}x^2)\,dx] & \text{(FDC)} \ (34\text{-}2) \end{cases}$$

When B_c approaches zero, ψ_v converges to 1/4 in the case of SRC and to

2/5 in the case of FDC.

It seems reasonable to assume that SO_2 is generated simultaneously

with the combustion reaction and thus, the specific generation rate

of SO_2 is equal to the specific combustion rate. The specific genera-

tion rate of SO_2 is defined by

$$R_{SO_2}^* \equiv \frac{\text{local generation rate}}{\text{average generation rate}} \simeq R_{O_2}^* \qquad\qquad (35)$$

and the specific combustion rate $R_{O_2}^*$ is derived from Equation (26) as

$$R_{O_2}^* = \frac{(dy_{O_2}/d\zeta)}{1-y_{O_2,\text{outlet}}} = \begin{cases} \alpha_c e^{-\alpha_c \zeta}/(1-e^{-\alpha_c}) & \text{(plug flow)} \qquad (36\text{-}1) \\[3mm] 1 & \text{(complete mixing)} \ (36\text{-}2) \end{cases}$$

As seen from Equation (36), the difference between the case with plug

flow and that with complete-mixing is reduced when α_c increases and

approaches to infinity. The effect of the excess air ratio on the

rate of SO_2 generation over the bed is shown in Figure 1. Equations

(35) and (36-1) are applied in the following simulation.

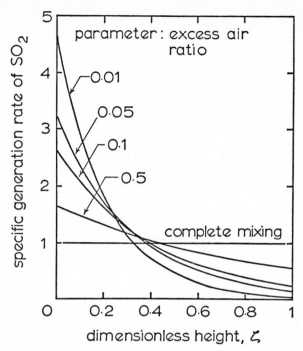

Fig. 1 Effect of excess air on the local rates of SO$_2$ generation in a fluidized bed combustor.

The Rate of SO$_2$ Absorption by Limestone

The effect of elutriation of limestone is also neglected when the fines are returned completely to the bed. Thus, $n_e = 0$ and for negligible size distribution of the limestone feed stream we obtained the following solution of Equation (1).

$$\phi_1(f) = \frac{B_1}{\lambda(f)} \exp \left[-\int_0^1 \frac{B}{\lambda(f)} \, df\right] \qquad (37)$$

where

$$B_1 \equiv \frac{(Ca/S)u_o}{(1-\epsilon_{mf}) \, L_{mf} \, k_v \, (C_{SO_2}/C_{SO_2,0})} \qquad (38)$$

Therefore, the average value of λ becomes,

$$\bar{\lambda} = B_1 \int_0^1 \exp \left[-\int_0^f \frac{B_1}{\lambda} \, df\right] df \qquad (39)$$

From the overall material balance of limestone particles, we have

$$\overline{\lambda} = B_1 \overline{f} \tag{40}$$

From Equations (38) and (39) the relation between $\overline{\lambda}$ vs. $\overline{f}$ is obtained by changing the value of the parameter B_1.

Figure 2 shows the reactivity of calcined limestone, $\overline{\lambda}(\overline{f})$, for the cases when shrinking core model is applied. In Figure 3 the values of $\overline{\lambda}(\overline{f}, d_p)$ calculated from Equation (38) based on the experimental data of Borgwardt (1970b) are plotted. It can be seen clearly that the actual reactivity of limestone is considerably different from that expected by shrinking core model. The compositions of limestone studied by Borgwardt (1970a, 1970b, 1972) are shown in Table 1 for comparison with

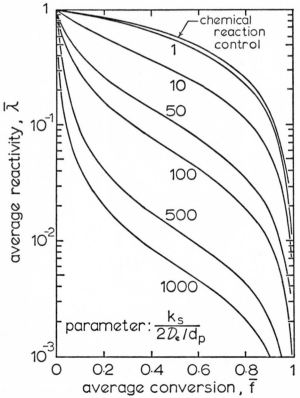

Fig. 2 relation between the average reactivity, $\overline{\lambda}$ and average limestone conversion, $\overline{f}$, based on the shrinking core model.

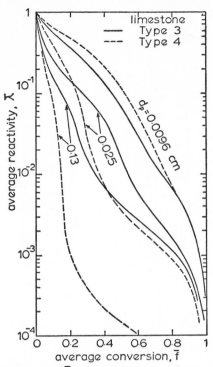

Fig. 3 Relation between the average reactivity, $\bar{\lambda}$, and average limestone conversion, $\bar{f}$, calculated using the data of Borgwardt (1970b).

those used in the pilot-plant studies.

The works of Borgwardt (1970b) are the only data covering the entire range of conversion and different limestone sizes. Therefore, in the following calculations, the curves shown in Figure 3 are interpolated to generate the reactivity of limestone, $\bar{\lambda}(\bar{f}, d_p)$.

Intrinsic rate constant of limestone-SO_2 reaction, k_v, is estimated by the following equation, Wen and Ishida (1973).

$$k_v = 3.72 \times 10^2 \exp(-7500/RT)S_g \quad 1/sec \qquad (41)$$

The specific surface area of calcined limestone, S_g, is affected by the temperature of calcination and can be approximated by

$$S_g = 13.2 \exp(9240/T) \quad cm^2/g \qquad (42)$$

Table 1 Composition of Limestones Used in Kinetic and in Pilot-Plant Studies[a]

Author	Borgwardt		Coates and Rice Davidson and Smale Jonke et.al. Robinson et.al.		Davidson and Smale
System	Kinetic Study		Fluidized-Bed Combustor		
Type of Limestone	Type 3[c]	Type 4[d]	BCR 1343	BCR 1359	British Limestone
$CaCO_3$	95.3	97.8	96.0	97.8	> 97.0
$MgCO_3$	3.89	0.00	0.96	1.3	—b
Fe_2O_3	0.20	0.31	0.38	0.12	—b
Al_2O_3	ND	0.01	—b	0.16	—b
SiO_2	ND	1.53	1.70	0.60	—b
Ignition loss	43.67	43.15	42.8	43.6	—b

a Only the data related to this paper are presented.

b Not reported.

c Type 3: calcite limestone with few scattered fine-grained dolomite rhombos.

d Type 4: calcitic limestone, very fine, equant granular, and dense.

In this study, the temperature of calcination is assumed to be equal to the bed temperature.

Method of Calculation

As can be seen from the above analysis, the fractional conversion or removal   efficiency of SO_2, η, is obtained by an iteration procedure since neither $\bar{\lambda}(\bar{f}, d_p)$ nor the fluidized bed model are linear. The method of calculation is shown schematically in Figure 4. In addition to the equations derived above, the following relationship between $\bar{f}$ and η from the overall sulfur balance is used.

$$\bar{f} = \eta/(Ca/S) \tag{43}$$

where (Ca/S) denotes the atomic ratio of Ca to S in the solid feed. Parameter α in Figure 4 is defined by

$$\alpha = \frac{(1-\varepsilon_{mf})L_{mf}k_v\bar{\lambda}}{u_o} \tag{44}$$

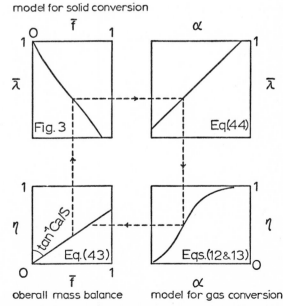

Fig. 4 Illustration of the algorithm used in calculation of SO_2 removal efficiency under steady-state operation.

This parameter corresponds to the parameter kL_{mf}/u_o in catalytic fluidized bed modeling (Mori and Wen (1975b)). However, in the noncatalytic fluidized bed the parameter α contains the average reactivity of solid reactant, $\overline{\lambda}$ which must be determined by the model.

By using Newton-Raphson method only few iterations are necessary to get successful convergence.

RESULTS

Calculations are performed for the data listed in Table 2. The data of Robinson et.al. (1970) are obtained from the experiments in which the bed is operated without withdrawal of spent solids. Thus, these data are analyzed based on the unsteady state condition.

The conditions of the experiments reported by Jonke et.al. (1970) and Davidson and Smale (1970) are such that the operation may be near the slugging region. The bubble diameters estimated by the equation of Mori and Wen (1975) for the operating conditions of these investigators equal to the tower diameter when bubbles reach near the top of the bed. The gas interchange coefficient for slugging zone is different from the value estimated by $K_{BE}=11/D$. Therefore, in the following calculation the gas exchange coefficient in the upper part of the bed is estimated on the basis of the slug-flow model of Hovmand and Davidson (1968).

Analysis of the Steady State Operations

The calculated SO_2 removal efficiency from the Bubble Assemblage Model are compared with the experimental data in Figure 5. The calculated efficiency has an average error of $\pm 6.5\%$ and a standard

Table 2 Summary of Experimental Conditions for Fluidized Combustor Studies Used in Simulation

Key*	Investigator	Coal rate lb/hr.	Ca/S	Fluidizing Velocity ft/sec	$\bar{d}_p$ cm	Temperature °K	Bed Height ft.	Bed Diameter ft.
△	Jonke et.al. (1970)	4.4	2.3~2.5	3.2~4.1	0.062	1143	2	0.5
□ ⊡ ⊟ ⊓	Coates and Rice (1973)	30.1 ~ 52.9	0.92~2.5	2.9~2.5	0.013~0.12**	1083~1133	2	1.5
o ⌀ ⊘ ⊗	Davidson and Smale (1970)	2.5 ~ 4.1	0.18~3	2.3	0.043~0.082	1073	2.3	0.5
—	Robinson et. al. (1970)	107 ~ 115	0.9~2.0	10.9~14.7	0.2	1088~1199	0.67 (L_{mf})	$A_t = 1 \times 1\ 1/3\ ft^2$

*) Each symbol represents one type of coal.

**) Some values are adjusted.

305

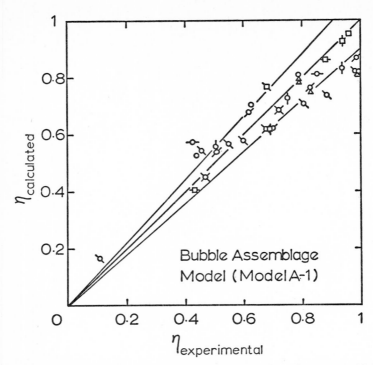

Fig. 5 Comparison of experimental SO₂ removal efficiency with that calculated from the model. (Symbols refer to Table 2).

deviation of ±4.8% relative to percentage of SO_2 removal.

The kinetic data (e.g. $\overline{\lambda}(\overline{f},d_p)$) of type 4 limestone shown in Figure 3 are used in the calculation. The reason the kinetic information of type 4 limestone has been selected for simulation in this study is that chemical composition of type 4 limestone is much closer to the limestones used by the investigators. In fact, type 3 limestone contains substantial amount of dolomite which resulted in much greater reactivity than any of the results of the investigators examined.

As shown in Figure 5, the agreement between the observed SO_2 absorption and the calculated results based on the proposed model must be considered satisfactory inspite of the lack of accurate limestone kinetic information and experimental errors associated with the measurement of SO_2 removal from pilot-plant unit (Cf. Bethell (1970)). The accuracy of limestone reaction kinetics seems to affect rather sensitively on the overall SO_2 removal in simulation of fluidized bed combustion and requires further investigation of the specific limestone employed.

In order to check the validity of the fluidized-bed model, several models defined schematically in Figure 6 are compared. The calculated results from Models A-2, A-3, C-1 and B-1 are compared with the experimental data in Figures 7-9, and 13 respectively. The complete mixing model generally under estimates the SO_2 removal efficiency as shown in Figure 9.

Among the three different modes of compartment-model, the Bubble Assemblage Model ($\delta=0$ and $\gamma_c=0$) gave the best results. Although the differences among these models are not always great, it is clear the complete mixing model fails to explain the magnitude of observed SO_2 conversion. To clarify this point further, SO_2 conversions are plotted against the parameter α in Figure 10. Using Equation (43) the average conversions of limestone are estimated based on the data of observed SO_2 removal efficiency. From Figure 3 $\bar{\lambda}$ is estimated which is then substituted into Equation (44) to obtain the parameter α corresponding to the experimental conditions. The relationship between η_{obs} and α is compared with that obtained from Bubble Assemblage Model. Here it must be recalled that the parameter α cannot be determined explicitly from the operating conditions as solid reactant reactivity varies with the extent of conversion. Therefore, in Figure 10 small arrows indicate the calculated points and the corresponding observed points. It can be seen from Figure 11, the SO_2 removal efficiency calculated from the Bubble Assemblage Model is higher than that

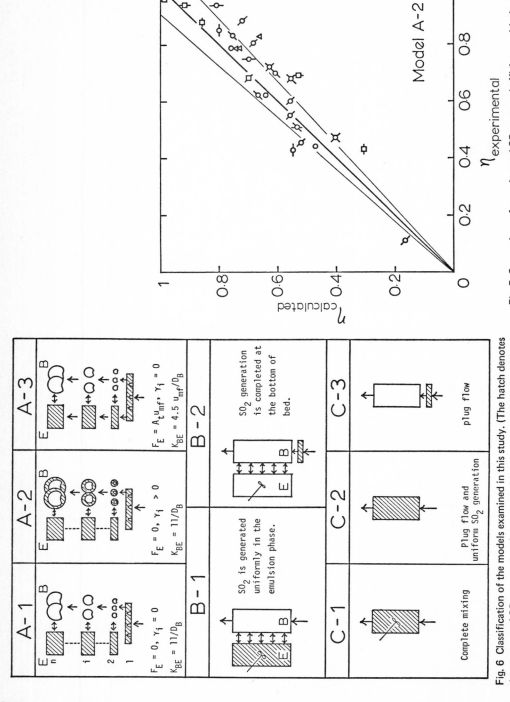

Fig. 7 Comparison of experimental SO_2 removal efficiency with that calculated from the model. (Symbols refer to Table 2).

Fig. 6 Classification of the models examined in this study. (The hatch denotes the presence of SO_2 generation. In every model the particles are assumed to be mixed completely).

A-1

$F_E = 0, \gamma_i = 0$

$K_{BE} = 11/D_B$

A-2

$F_E = 0, \gamma_i > 0$

$K_{BE} = 11/D_B$

A-3

$F_E = A_t u_{mf}, \gamma_i = 0$

$K_{BE} = 4.5\ u_{mf}/D_B$

B-1

SO_2 is generated uniformly in the emulsion phase.

B-2

SO_2 generation is completed at the bottom of bed.

C-1

Complete mixing

C-2

Plug flow and uniform SO_2 generation

C-3

plug flow

Model A-2

$\eta_{experimental}$

$\eta_{calculated}$

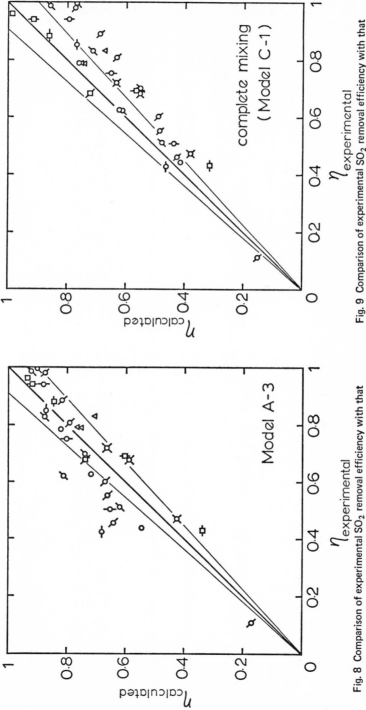

Fig. 8 Comparison of experimental SO_2 removal efficiency with that calculated from the model. (Symbols refer to Table 2).

Fig. 9 Comparison of experimental SO_2 removal efficiency with that calculated from the model. (Symbols refer to Table 2).

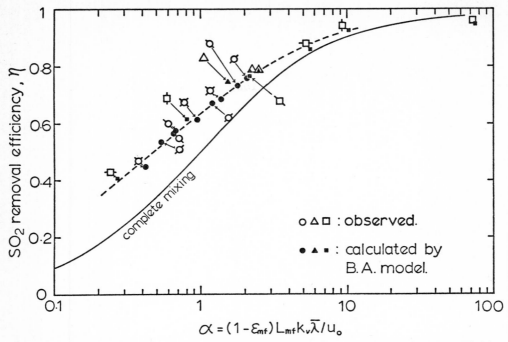

Fig. 10 Comparison of observed SO$_2$ removal efficiency as a function of a by Bubble Assemblage Model.

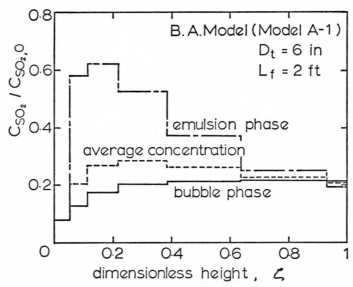

Fig. 11 Calculated SO$_2$ concentration profiles based on Jonke et al.'s experimental condition.

calculated from the complete mixing model. Such phenomena are
seldom observed in catalytic fluidized bed operation. But in the
fluidized bed combuster, SO_2 generation and absorption take place
simultaneously resulting the concentration of reactant (e.g. SO_2)
in the emulsion phase to be sometimes higher than that in the Bubble
phase. This is mainly due to the SO_2 generation in the emulsion
phase as shown in Figure 11.

In Figure 12, the characteristics of five models (models B-1,
B-2, C-1, C-2 and C-3 in Figure 6) are compared. Among these
models, Model B-1 gives the results nearest to that based on the
Bubble Assemblage model as shown in Figure 13. In Figure 13 the
average gas interchange coefficient is estimated by using the
equation for bubble diameter of Mori and Wen (1975a). The Model

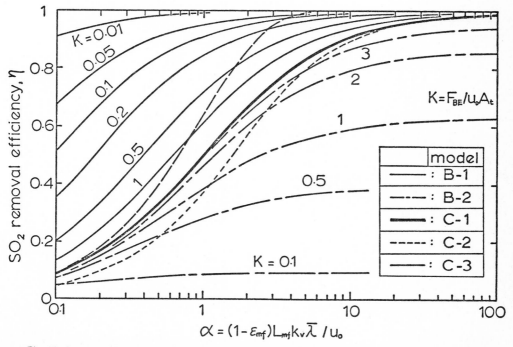

Fig. 12 Comparison of SO_2 removal efficiency calculated from homogeneous and two-phase models.

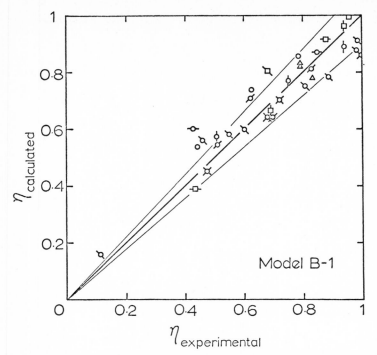

Fig. 13 Comparison of experimental SO_2 removal efficiency with that calculated from the model. (Symbols refer to Table 2).

B-1 gives almost similar results as that presented in Figure 6.

Analysis of Data from Semi-Batch Fluidized-Bed Operation

Here the analysis is restricted to the case of fluidized-bed combustor with no solid withdrawal and no elutriation. By solving Equation (1) SO_2 removal efficiency, η, distribution function, ϕ_1, and the average reactivity of limestone, $\overline{\lambda}$, are obtained as follows:

$$\eta = (Ca/S)\tau\overline{\lambda}(1-\overline{\eta})/B_1^{\circ *} \qquad (45)$$

$$\phi_1 = B_1^{\circ *}/\{1-\overline{\eta}(\tau_0)\}\tau\lambda(f) \qquad (46)$$

$$\overline{\lambda} = \frac{B_1^{\circ *}}{\tau} \int_0^{f_{max}(\tau)} \frac{df}{1-\overline{\eta}(\tau_0)} \qquad (47)$$

where $f_{max}(\tau)$ denotes the maximum conversion of limestone particles at time=τ.

The dimensionless time τ is defined by introducing the mean residence time of limestone at $L_{mf} = \overset{o}{L}_{mf}$ as follows:

$$\tau = \theta/\overset{o}{\theta}_1$$

where

$$\overset{o}{\theta}_1 = \rho_{Ca} A_t \overset{o}{L}_{mf}(1-\epsilon_{mf})/W_{Caf}$$

τ_o in Equation (47) is a function of f and τ and denotes the time of entry of a limestone particle into the fluidized bed. $\tau_o(f,\tau)$ can be obtained by solving the following differential equation with the initial condition.

$$df/d\tau = (1-\overline{\eta}(\tau))\lambda(f)/\overset{o}{B}_1^* \qquad (48)$$

$$\text{I.C. } f = 0 \text{ at } \tau = \tau_o \qquad (49)$$

The parameter $\overset{o}{B}_1^*$ is defined as

$$\overset{o}{B}_1^* = \frac{(Ca/S) u_o}{\overset{o}{L}_{mf}(1-\epsilon_{mf})k_v} \qquad (51)$$

The data by Robinson et.al. (1970) are analyzed by this transient state model. The SO_2 concentration in the outlet gas is seen to drop rapidly during the initial period for approximately one hour and gradually decrease thereafter. Since the data are not available for the initial period, in Table 3 the calculated removal efficiencies of SO_2 at θ=4-5 hrs. are compared with the experimental data. This period corresponds to the so called steady state period reported by Robinson et.al. (1970). Although good agreement is seen from Table 3, the results are still of preliminary nature and further study is needed to acertain conclusively the characteristics of the semi-batch operation.

Table 3 Comparison of calculated efficiency of SO_2 removal, η_{calc}, with the experimental results, η_{exp}. for semi-batch operation by Robinson et. al. (1970).

Operating Conditions			SO₂ Removal Efficiencies	
Coal rate lb/hr	Ca/S —	u_o ft/sec	η_{exp}	η_{calc}
110	1.15	14.1	0.154	0.169
110	1.93	13.1	0.270	0.269
107	1.15	14.6	0.118	0.184
112	1.47	14.5	0.266	0.234
110	2.0	14.3	0.342	0.287
115	0.9	13.2	0.132	0.139
112	1.57	13.9	0.231	0.228
112	2.0	13.7	0.236	0.300
112	1.0	14.7	0.154	0.158

CONCLUSIONS

A mathematical model of fluidized-bed combustor has been developed based on the Modified Bubble Assemblage Model by considering the effect of residence time distribution of particles in the bed. The distribution functions of coal-particle size and of limestone conversion within the fluidized-bed are obtained by solving the population balance equation. The results of simulation for the cases where coarse limestone is used are compared with the experimental data reported in the literature. The model is found to be able to predict the observed data within the accuracy of $\pm 6.5\%$ of SO_2 removal percentage. In order to obtain a simplified model, several two phase models and homogeneous models have been examined. A simple two phase model is shown to represent the performance of fluidized-bed combustor quite well provided proper exchange coefficient is calculated based on bubble diameter. However, the simple complete mixing model failed to explain the behavior of the observed data. It has been found that the generation of SO_2 in the emulsion phase caused the higher conversion than that estimated by the homogeneous contact model. Experimental data from semi-batch operations are also examined by using a transient state model developed in this paper.

LITERATURE CITED

Bethell, F. V., BCURA Report to NCB, "Retention of Sulphur in Fluid-Bed Combustion: Analysis of P.E.R. Data", Document No. CT/13/1, Nov. 1970.

Bishop, J. W., Robinson, E. B., Erlich, S., Jain, A. K., and Chen. P. M., Paper 68-WA/FU-4, Presented at Winter Annual Meeting, A.S.M.E., New York, NY, December, 1968.

Borgwardt, R. H., Environ. Sci. Technol., 4, (1), 59 (1970a).

Borgwardt, R. H., 4th Dry Limestone Process Symposium, Gilbertsville, KY, June 22-26, 1970b.

Borgwardt, R. H., Harvey, R. D., Environ. Sci. Technol., 6 (4), 350, 1972.

Coates, N. H. and Rice, R. L., Proceedings of 2nd International Conference of Fluidized-Bed Combustor, II-2-1, Hueston Woods, Ohio, October, 1970.

Coates, N. H. and Rice, R. L., 7th National Meeting of A.I.Ch.E., Detroit, Michigan, June, 1973.

Coutant, R. W., Barrett, R. E., Simon, R., Campbell, B. E., Longher, E. H., Summary Report on "Investigation of the Reactivity of Limestone and Dolomite for Capturing SO_2 from Flue Gas", Battelle Memorial Institute, November, 1970.

Davidson, D. C., and Smale, A. W., Proceedings of 2nd International Conference on Fluid-Bed Combustion, II-1-1, Hueston Woods, Ohio, October, 1970.

Highley J., and Merrick, D., A.I.Ch.E. Symposium Series, No. 116, 67, 219, 1971.

Hovmand, S., and Davidson, J. F., Trans. Instn. Chem. Engrs., 46, T190 (1968).

Hoy, H. R., Proceedings of 2nd International Conference on Fluidized-Bed Combustion, I-4-1, Hueston Woods, Ohio, October, 1970.

Jonke, A. A., Jarry, R. L., and Carls, E. L., Proceedings of the First International Conference on Fluidized Bed Combustion, Hueston Woods, Ohio, November, 1968.

Jonke, A. A., Carls, E. L., Jarry, R. L., Anastasia, L. J., Haas, M., Davlik, J. R., Murphy, W. A., Sthoffstoll, C. B., and Vargo, G. N., Annual Report of ANL, ANL/ES-CEN-1002, "Reduction of Atmospheric Pollution by the Application of Fluidized-Bed Combustion", 1970.

Kato, K., and Wen, C. Y., Chem. Eng. Sci., 24, 1351 1969.

Mori, S., and Wen, C. Y., A.I.Ch.E. Journal, 21, 109, 1975a.

Mori, S., and Wen, C. Y., Paper to be presented at the International Conference on Fluidization, Pacific Grove, California, June, 1975b.

Robinson, E. B., Bagnudo, A. H., Bishop, J. W., and Ehrlich, S., Interim Report to the National Air Pollution Control Administration "Characterization and Control of Gaseous Emissions from Coal-Fired Fluidized-Bed Boilers", October, 1970.

Vogel, G. J., Swift, W. M., Lenc, J. F., Cunningham, P. T., Wilson, W. I., Panek, A. F., Teats, F. G., and Jonke, A. A., Environmental Protection Technology Series, EPA-650/2-74-104, "Reduction of Atmospheric Pollution by the Application of Fluidized-Bed Combustion and Regeneration of Sulphur-Containing Additives", 1974.

Wen, C. Y., and Ishida, M., Environ. Sci. Technol., 7, (8), 703, 1973.

Wen, C. Y., Mori, S., Gray, J. A., and Yavorsky, P. M., Annual Meeting of A.I.Ch.E., Washington, DC, December, 1974.

NOMENCLATURE

a_c Dimensionless parameter defined by Equation (21).

A_t Cross sectional area of fluidized bed, cm^2.

B_c Dimensionless parameter defined by Equation (20).

C_A Gas concentration of species A, $gmol(A)/cm^3$.

$C_{O_2,0}$ Concentration of oxygen at the bottom of bed, $gmol(O_2)/cm^3$.

$C_{SO_2,0}$ Concentration of SO_2 when there is no absorption by limestone $gmol(SO_2)/cm^3$.

Ca/S   Ratio of the feed rate of Ca to the rate of S.

d_c Diameter of coal particle, cm.

d_{cm} Maximum diameter of coal particles, cm.

d_p Diameter of limestone particle, cm.

D_B Bubble diameter, cm.

D_{Bo} Initial bubble diameter, cm.

$\mathcal{D}_e$ Intraparticle diffusivity, cm^2/sec.

D_t Diameter of the fluidized bed, cm.

$\mathcal{D}_{O_2}$ Molecular diffusivity of oxygen, cm^2/sec.

f Fractional conversion of limestone.

$\overline{f}$ Average value of f in the bed.

F_E Flow rate of gas passing through the emulsion phase, cm^3/sec.

$g_{A,B,i}$ Generation rate of A in the bubble phase of ith compartment.

$g_{A,E,i}$ Generation rate of A in the emulsion phase of ith compartment.

k_c Surface reaction rate constant of coal combustion, cm/sec.

k_s Surface reaction rate constant of limestone-SO_2 reaction, cm/sec.

k_v Volumetric reaction rate constant of limestone-SO_2 reaction, sec^{-1}.

K Dimensionless parameter representing the effect of gas exchange.

K_{BE} Gas exchange coefficient, sec^{-1}.

K_i — Dimensionless parameter representing the effect of gas exchange in the $i\underline{th}$ compartment.

L_f — Fluidized bed height, cm.

L_{mf} — Bed height at the minimum fluidization velocity, cm.

M_C — Atomic weight of carbon, g/gatom.

M_{Ca} — Atomic weight of calcium, g/gatom.

N_{cw} — Number of coal particles withdrawn per unit time, sec^{-1}.

n_e — Number of particles elutriated per unit time, sec^{-1}.

n_f — Number of particles fed per unit time, sec^{-1}.

n_{1f} — Number of limestone particles fed per unit time, sec^{-1}.

n_{1w} — Number of limestone particles withdrawn per unit time, sec^{-1}.

n_w — Number of particles withdrawn per unit time, sec^{-1}.

N_c — Total number of coal particles held up in the bed.

N_{ci} — Total number of coal particles held up in $i\underline{th}$ compartment.

N_1 — Total number of limestone particles held up in the bed.

N_{1i} — Number of limestone particles held up in $i\underline{th}$ compartment.

r_c^* — Combustion rate of one particle, gatom(C)/sec.

r_1^* — Rate of limestone-SO_2 reaction of one particle, gmol(SO_2)/sec.

R_c^* — Overall combustion rate of the bed, gatom(C)/sec.

R_1^* — Overall absorption rate of SO_2 in the bed, gmol(SO_2)/sec.

R_{O_2} — Specific combustion rate.

R_{SO_2} — Specific generation rate of SO_2.

S_g — Internal surface area of calcined limestone, cm^2/g.

T — Temperature, °K.

u_o — Superficial gas velocity, cm/sec.

u_{mf} — Minimum fluidization velocity, cm/sec.

V_{Ca}	Amount of Ca contained in the bed, g(Ca).
$W_{C,f}$	Feed rate of carbon in coal, g(C)/sec.
$W_{Ca,f}$	Feed rate of calcium in limestone, g(Ca)/sec.
x_A	Mole fraction of species A.
X	Independent variable of the distribution function ϕ.
z	Height from the distributor, cm.
α	Dimensionless parameter defined by Equation (44).
α_c	Dimensionless parameter defined by Equation (29).
α_{Ai}	Dimensionless parameter defined by Equation (16).
γ_i	Dimensionless parameter defined by Equation (17).
δ	Dimensionless parameter defined by Equation (18).
ϵ_i	Void fraction of ith compartment.
ϵ_{mf}	Void fraction in the bed at u_{mf}.
η	Removal efficiency of SO_2.
θ	Time, sec.
Θ_1	Mean residence time of limestone, sec.
λ	Reactivity of limestone defined by Equation (5).
$\bar{\lambda}$	Average reactivity of limestone in the bed.
ρ_C	Density of carbon in coal particle, g(C)/cm^3(coal).
ρ_{Ca}	Density of carbon in limestone particle, g(Ca)/cm^3(limestone).
τ	Dimensionless time.
ϕ	Distribution function of X.
ϕ_c	Distribution function of dc, cm^{-1}.
ϕ_e	Distribution function with respect to the particles elutriated.
ϕ_1	Distribution function of f.
ϕ_w	Distribution function with respect to the particle withdrawn.

ψ_a Coefficient defined by Equation (23).

ψ_v Coefficient defined by Equation (34).

ACKNOWLEDGEMENTS

We wish to thank Dr. S. Mori for helpful advice and discussions regarding the development of the mathematical model and the Energy Research Development Administration, Washington, DC, for financial support.

COMBUSTION AND DESULFURIZATION OF COAL IN A FLUIDIZED BED OF LIMESTONE

LAWRENCE A. RUTH

Exxon Research and Engineering Company has been studying fluidized bed coal combustion for several years, under contract to the U.S. Environmental Protection Agency. This paper summarizes our recent work, carried out under contracts CPA 70-19 and 68-02-1451.[1]

Our objectives are to: (1) develop the equipment required; (2) determine the effect of operating parameters (temperature, pressure, excess air level, superficial velocity, bed depth) on SO_2 and NO_x emissions and combustion efficiency; (3) determine the effect of sorbent conversion on desulfurization; (4) investigate trace element emissions; (5) measure heat transfer coefficients; and (6) determine effect of sorbent particle size.

EQUIPMENT AND OPERATING CONDITIONS

The combustor vessel was constructed from 10 inch steel pipe, refractory lined to an inside diameter of 4-1/2 inches (11.4 cm). Heat could be removed from the fluidized bed by cooling coils made of 1/4 in (6 mm) diameter 316 stainless steel tubing. Water was pumped into the coils to produce a steam-water mixture. Coal was added continuously to the combustor using a system which consisted of a conical bottom hopper that held coal to be fed and an orifice and mixing tee assembly that mixed coal with carrier gas (air). Coal

flowed through the orifice at the bottom of the hopper, was picked up
by air in the mixing tee, and conveyed through an injection tube and
a specially designed probe, which was inserted into the combustor.
Flue gas leaving the combustor was cleaned of solids with two cyclones
and a filter. A small portion of the off-gas was sent to analytical
equipment, which included infrared analyzers for SO_2, CO, and CO_2,
a chemi-luminescence analyzer for NO and NO_x, and a polarographic
analyzer for O_2.

The range of operating conditions was 1400-1800°F (760-980°C)
bed temperature, 5-9 atm (510-910 kPa) pressure, 3-6 ft/sec (0.9-1.8
m/sec) superficial velocity, and 1-5 ft (0.3-1.5 m) expanded bed height.
The coal used for most of the work was a W. Virginia bituminous coal
with 2.6% sulfur, sized to -16 mesh. The sorbents tested were Grove
limestone (BCR No. 1359), Tymochtee dolomite, and Gibsonburg dolomite
(BCR No. 1337). The stone particle size was -8 +25 mesh.

EXPERIMENTAL RESULTS

Temperature Profile

A serious problem which occurred during the development of
the combustor was a poor temperature profile in the bed, characterized
by severe hot spots at the coal feed point. We deduced the cause of
this problem to be that combustion of the powdered coal occurred at
a much faster rate than solids could mix and distribute heat throughout
the bed. Top-to-bottom mixing of solids was poor because of the high
L/D ratio of the bed and the tightly wound horizontal serpentine
coils. In order to improve mixing, the coils were changed from the
horizontal serpentine design to vertical loops. Figure 1 is a comparison

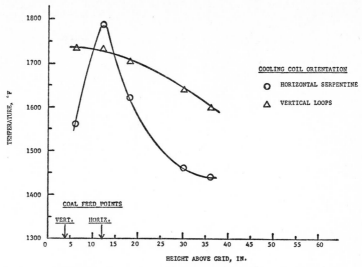

Fig. 1 Comparison of temperature profiles for horizontal and vertical coils.

of temperature profiles obtained with horizontal and vertical coils. It is clear that vertical coils produced a dramatic improvement; hot spots were nearly eliminated and ΔT's across the bed were reduced from about 400°F to 100°F.

Emission of SO_2, NO_x, and Trace Elements

SO_2 emissions have been low, with desulfurization of 85-95% routinely obtained. This is well below the EPA standard for new coal fired plants of 1.2 lb SO_2/MBtu of fuel burned, which corresponds to about 70 percent desulfurization for our 2.6% sulfur coal.

Figure 2 shows that NO_x emissions increase with excess air, but that they are still below the EPA standard for excess air levels up to about 100% (2X stoichiometric air). NO_x emissions are higher for runs made with horizontal coils than with vertical coils because with vertical coils the coal burning zone is spread out over more of the bed, which results in a lower combustion intensity. Larger com-

bustors can be expected to emit even less NO_x because of more uniform
temperatures and lower combustion intensities. For example, Figure 2
shows that NO_x levels are lower from the British NRDC's combustor,
which has a 6 ft^2 cross-sectional area.

A potential advantage for the fluidized bed over conventional
coal combustors is that the fluidized bed system may retain, in a
solid constituent, hazardous trace elements released during the combustion of coal. To investigate this possibility, samples of W. Virginia
coal, Tymochtee dolomite, and the bed and overhead (material collected
in cyclone diplegs and filter) were analyzed by neutron activation.
Material balances were made for trace elements. Table 1 gives concentrations for some elements and percentage recoveries. Because recoveries
are fairly high, especially when solids' losses in handling are taken
into account, it appears that only small amounts of trace elements are
emitted with the flue gas.

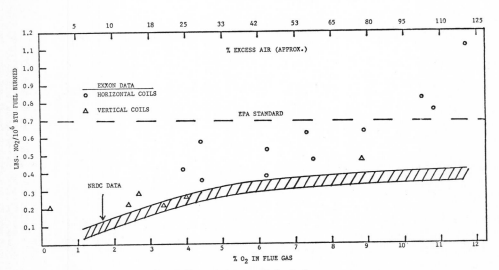

Fig. 2 No emissions vs. % O_2 in flue gas.

Table 1 Concentrations and Recoveries of Trace Elements

Element	W. Virginia Coal	CONCENTRATIONS (PPM) Tymochtee Dolomite	Bed	Overhead[1]	Recovery(%)[2]
Al	27000	10900	27000	71300	65.3
Sb	0.253	0.0527	0.501	0.606	78.2
As	3.7	0.566	8.02	9.4	85.9
Cs	0.349	0.439	0.403	2.04	100.0
Cl	1050	447	787	627	21.2
Fe	13800	3240	19600	39900	80.5
Mn	20	42	102	105	95.5
K	599	2180	1280	4040	74.8
Rb	3.74	12.2	9.17	31.1	97.2
Na	784	303	364	3200	88.1
Sr	205	130	278	690	77.6
U	0.652	2.23	3.13	4.41	93.9

(1) Overhead consists of flyash + entrained dolomite.
(2) Recovery = percentage of element present in coal and dolomite that can be
       accounted for in solids removed from combustor (bed + overhead).

Combustion Efficiency

Combustion efficiencies (without carbon recycle) ranged from about 93-98 percent; raising the temperature or excess air level caused efficiency to increase. Most of the losses were the result of unburned carbon being entrained from the bed; CO losses were small.

Heat Transfer Coefficients

Overall coefficients (U) were measured for heat transfer from the bed to vertical cooling coils. Results were in the range of 70-80 Btu/hr ft^2 °F (400-450 W/m^2 °C). Coefficients were also measured for coils in the freeboard and were about 20 Btu/hr ft^2 °F (110 W/m^2 °C).

Effect of Coal Type

Substituting a non-caking, low-sulfur, Wyoming coal for highly caking Arkwright coal caused no noticeable difference in operation. Neither did the use of a moderately caking Illinois coal with 4.1 percent sulfur.

CONCLUSIONS

This work has shown that fluidized bed coal combustion can reduce SO_2 and NO_x emissions to levels that are well below the EPA standards. It appears that emissions of hazardous trace elements are also reduced. Early commercialization of this process would be an important step toward the burning of coal without severe environmental penalties.

(1) Hoke, R. C., Nutkis, M. S., Ruth, L. A., and Shaw, H., "A Regenerative Limestone Process for Fluidized Bed Coal Combustion and Desulfurization," (Exxon Research and Engineering Company), EPA report EPA-650/2-74-001, January, 1974.

PRESSURIZED FLUIDIZED BED COAL COMBUSTION

MELVYN S. NUTKIS

INTRODUCTION

Fluidized bed combustion provides a new boiler technique where coal is combusted in a bed of particles maintained in a state of fluidization by the air required for combustion. The use of limestone or another suitable sorbent as the bed material permits the capture and removal of sulfur dioxide simultaneously with the combustion process.

Within the fluidized bed combustor, limestone is calcined to calcium oxide which reacts with sulfur dioxide and oxygen in the flue gas to form calcium sulfate. Fluidized bed combustion offers the potential of an efficient and compact boiler technique also capable of providing acceptable pollution control.

Some advantages of the fluidized bed combustion process are improved combustion efficiency, ability to combust a wide variety of fuels, improved heat transfer rates, longer boiler tube life, "in-situ" sulfur removal, compact boiler because of high volumetric heat release, higher power cycle efficiencies, and low NO_x emissions because of lower combustion temperatures. Pressurized fluidized bed combustion offers even greater benefits in size reduction, efficiency and load control.

PROGRAM OBJECTIVES

Under contracts sponsored by the Environmental Protection Agency, Exxon Research & Engineering has designed, constructed, and is

329

operating a pressurized fluidized bed combustion facility, ("Mini-plant"),
to demonstrate the feasibility of this coal combustion technique for
energy production. The program objectives are the investigation of the
effect of operating variables such as combustor temperature and pressure,
fluidization velocity and bed depth upon the efficiency, control and
emissions reduction of the fluidized bed combustor.

DESIGN AND DESCRIPTION

A 12.5 inch diameter combustor size was selected as a basis
for the design because it provided a system capable of rapid construction
at reasonable cost and could supply essential data needed for future
development of pressurized fluidized bed combustion technology.[1] The
overall system flow plan for the Fluidized Bed Combustion Miniplant is
presented in Figure 1.

Main fluidizing air for the combustor is supplied at operating
pressures to 150 psig and measured by an orifice flow meter and regulated
by a control valve. The air passes through a distributing grid, up
through the fluidized bed of stone, and is discharged through two stages
of cyclones (for solids removal) before entering the discharge line
where a control valve automatically regulates the combustor pressure.

Coal and make-up limestone are continuously injected into the
combustor from a system designed for controlled solids injection under
pressure. Heat extraction in the fluidized bed is achieved by 6 cooling
water loops located in discrete vertical zones of the combustor. The

(1) Design of the Fluidized Bed Miniplant
    M. S. Nutkis and A. Skopp
    Third International Fluidized Bed Combustion Symposium, Hueston Wood,
    Ohio, Oct. 29 - Nov. 1, 1972

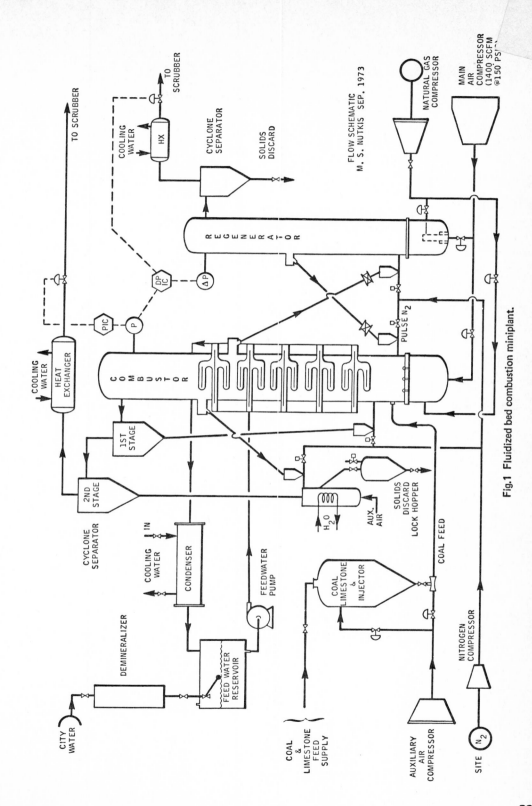

Fig.1 Fluidized bed combustion miniplant.

FLOW SCHEMATIC   M. S. NUTKIS   SEP. 1973

cooling coils are constructed of 3/4-inch OD type 316 stainless steel
tubing and arranged in a serpentine vertical orientation.

START-UP PROGRESS AND OPERATING CONDITIONS

The fluidized bed coal combustor is designed for pressures to
10 atmospheres, superficial bed velocities of 10 ft/sec and coal feed
rates of 480 #/hr. Operating levels achieved have satisfied most design
conditions as listed in Table 1.

Process control of combustor pressure and fluidization air flow
rate has been excellent, and mechanical operation has been good. Continuous runs have exceeded 24 hours with total operating time in excess of
300 hours.

Table 1 Design and Operating Conditions

	Design Basis (Maximum)	Operating Level Achieved
Pressure (atm)	10	10.2
Temperature – Power Zone (°F)	1750	1750
Temperature – Cyclone Discharge (°F)	1500	1350
Superficial Bed Velocity (ft/sec)	10	10.5
Bed# Depth – Static (in)	48	71
Bed Depth – Expanded (in)	180	140
Coal* Feed Rate (#/hr)	480	340
Combustion Heat Release (BTU/hr)	6,200,000	4,420,000
Heat Release/Ft Expanded Bed (BTU/hr/ft)	413,000	490,000

* Arkwright Mine Coal (2.6% Sulfur), –#8 mesh size,

Bed Material – Grove Limestone, #7 – #25 mesh

At a fixed combustor pressure and bed velocity, combustor bed temperature control is achieved primarily by regulation of the coal feed rate. Initial operating problems with the pressurized solids feed system were resolved by modification and re-design of components and operating techniques.

For optimum combustor temperature uniformity and maximum flue gas discharge temperature, it is necessary that all operating cooling coils be immersed in the fluidized bed. Stable control and steady operation has been achieved in the FBC and combustor conditions are reproducible. Two typical run summaries for the Miniplant are presented in Table 2.

RESULTS AND DISCUSSION

Substantial improvements in fluidization and heat dispersion throughout the Fluidized Bed Combustor have been accomplished, as evinced by the typically uniform bed temperature profiles (Figure 2). FBC Miniplant bed temperature variations are currently less than 120°F within the first 10 feet of expanded bed with the combustor operating at 1700°F. Improved bed fluidization, circulation and mixing patterns allowing the uniform temperature profile are attributed to modifications to the coil configuration, orientation and elevation.

The overall heat transfer coefficient for the cooling coils in the FBC Miniplant have been measured to be 50-55 BTU/hr-ft^2-°F with the fluidized bed at 1700°F, 6 ft/sec velocity and a pressure of 9 atm. This is in agreement with the published ranges of heat transfer coefficients for fluidized beds[2] of 35-90 BTU/hr-ft^2-°F reported by various fluidized combustion research groups.

(2) The Fluidized Combustion of Coal, D. G. Skinner, Mills & Boone Ltd. (1971)

Table 2 Fluidized Bed Combustion Miniplant Run Summary Data

Run #	12.1	13.2
Date	1/8/75	1/30/75
Time	18:20	19:10
Run Duration (hrs)	10.5	11.0
Combustor Pressure (atm)	9.1	9.0
Superficial Bed Velocity (ft/sec)	5.9	6.0
Temperature - Lower Zone Avg. (^{O}F)	1684	1667
Combustor Exit Temperature (^{O}F)	1090	1307
Bed Depth - Settled (inches)	48	59
Bed Depth - Expanded (inches)	92	112
Coal Feed Rate (#/hr)	215	264
Limestone Feed Rate (#/hr)	24	29
Calcium/Sulfur Molal Ratio	1.2	1.2
Excess Air (%)	24	6
Overall Run Ranges:		
SO_2 Emissions (ppm)	0-120	0-300
NO_x Emissions (ppm)	150-270	75-140
CO Emission (ppm)	~0	~0
SO_2 Removal (%)	94-100	86-100
Combustion Efficiency (%)	97	96.7

Sulfur dioxide emissions have typically been 0-250 ppm, which at coal feed rates of 250-300 #/hr yield approximately 0.40 $^{\#SO}2/_{10}6$ BTU, well below the EPA standard of 1.2 $^{\#SO}2/_{10}6$ BTU). Sulfur removal efficiencies have been about 90 per cent, and combustion efficiencies are generally 96-98%. Oxides of nitrogen emission levels have varied from 0.34 to 0.20 $^{\#NO}2/_{10}6$ BTU, well below the EPA standard of 0.7 $^{\#NO}2/_{10}6$ BTU, and in agreement with results from other fluidized bed combustors (Fig. 3).

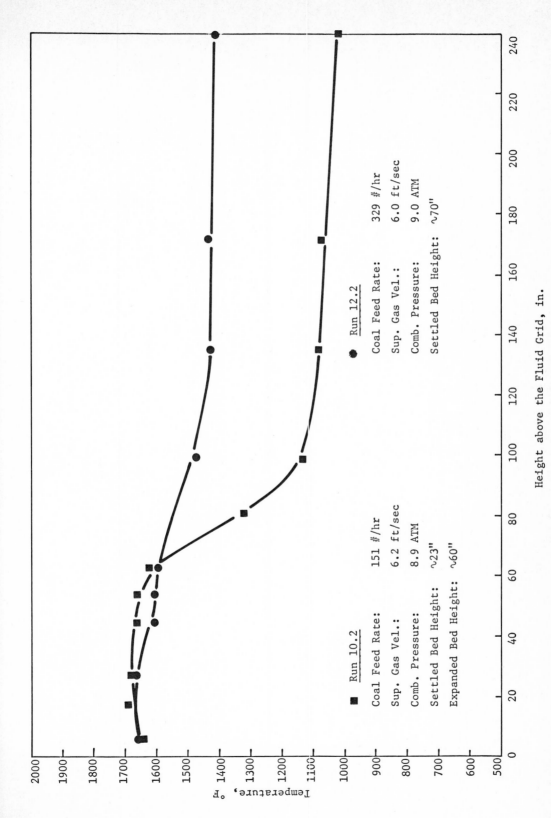

Height above the Fluid Grid, in.

Fig. 2 Miniplant bed temperature profiles.

■ Run 10.2

Coal Feed Rate:	151 #/hr
Sup. Gas Vel.:	6.2 ft/sec
Comb. Pressure:	8.9 ATM
Settled Bed Height:	~23"
Expanded Bed Height:	~60"

● Run 12.2

Coal Feed Rate:	329 #/hr
Sup. Gas Vel.:	6.0 ft/sec
Comb. Pressure:	9.0 ATM
Settled Bed Height:	~70"

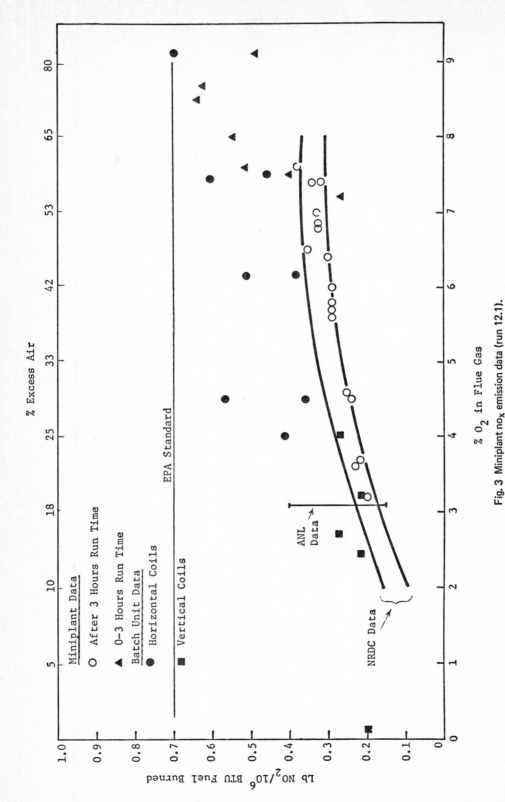

Fig. 3 Miniplant no$_x$ emission data (run 12.1).

FUTURE ACTIVITIES

Investigations intended in the Fluidized Bed Combustion Miniplant will be to study the effect and interaction of variables such as bed velocity, temperature, pressure, bed depth and coal rate and excess air on such parameters as emissions, heat transfer coefficients and combustion efficiency.

Start-up of a combined system for the continuous fluidized bed combustion and desulfurization with limestone regeneration is scheduled for late 1975, and incorporation of a system capable of reliable particulate clean-up suitable for application to a gas turbine assembly is currently under design.

FLUIDIZED-BED COMBUSTION REVIEW

H. NACK, K. D. KIANG, K. T. LIU, K. S. MURTHY,
G. R. SMITHSON, Jr. and J. H. OXLEY

ABSTRACT

Presented in the paper is a state-of-the-art review on the fluidized-bed combustion of coal, in light of currently available data. Several aspects of this subject are discussed, which include history of development, engineering data, and cost estimation.

INTRODUCTION

Current estimates of U.S. coal reserves by the U.S. Geological Survey are 3.2 trillion tons. About 2 trillion tons of this amount are potentially recoverable. This represents the largest fossil fuel reserve in the country. The development of non-polluting, higher economy, power-generating systems using coal as the fuel has received increasing importance as a national objective in the area of energy development during the past few years.

Conventional pulverized coal-fired boilers suffer the problems of low thermal efficiency and critical antipollution control. To circumvent these problems, several fluidized-bed combustion (FBC) systems are being developed. This paper reviews the current state-of-the-art of this technology including the following subjects:

- History of development
- FBC systems currently under development
- Engineering data
- Cost analyses

HISTORY OF DEVELOPMENT

Fluidized-bed combustion of coal dates back to 1928 when Stratton developed a spouting fluidized-bed boiler to combust crushed coal at gas velocities of about 10-40 ft/sec in the furnace[1]. As shown in Figure 1, the Stratton furnace operated near 2000 F to agglomerate the ash which was removed after falling through the grate. There were no tubes in the bed to recover heat. A unit of 5000 lb coal/hr capacity was installed at a U.S. Gypsum Company paper mill.

In the early 1950's a number of fluidized bed combustion processes were proposed in Europe and the United States, which includes patents filed by Badishe Anulin[2,3], Union Carbide[4], Combustion Engineering[5,6], and Standard Oil of Delaware[7]. These proposed processes employ fluidized-beds with cooling surfaces which are immersed in or surrounding the beds. Although the advantages of fluidized-bed combustion were recognized, these processes were not developed to the extent of full commercial applications. However, these processes resemble closely those evolved in the later U.K. development work on fluidized-bed combustion. Early work on sulfur reactions during the fluidized-bed combustion of coal was also carried out in the Coal Research Laboratory of Carnegie Institute of Technology.[8]

The development of the Ignifluid boiler[9] was started about 1953 in France by Soceite Anonyme Activit and Babcock-Atlantique to burn low sulfur coal and anthracite fines. A conventional stoker is modified to create a fluidized-bed on the grate as shown in Figure 2. The fluidized-bed is operated between 1000 and 1200 C to agglomerate coal ash which is carried out of the bed on the moving grate. Partial combustion occurs in the bed and secondary air (~50%) is supplied above the bed to complete the combustion. A 25-MW plant has been in operation since 1968. The system offers low particulate emission as more than 50 percent of coal ash is removed from the grate. However, the high combustion temperature in the Ignifluid and Stratton boilers renders sulfur removal rather difficult when burning high sulfur coal.

Research work on the fluidized-bed combustion of low grade fuel has been conducted in the Fuel Research Institute[10], Czechoslovakia, since 1952. It involves a two-stage fluidized-bed combustion concept as shown in Figure 3. Combustion of solid fuel with ash content as high as 75 percent

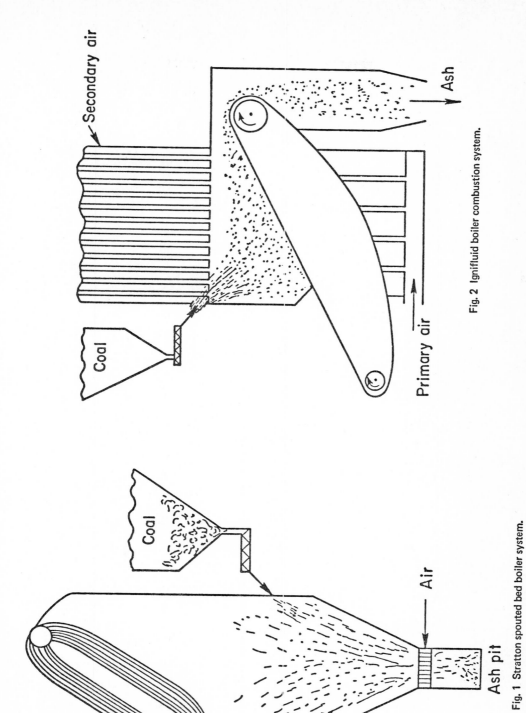

Fig. 2 Ignifluid boiler combustion system.

Fig. 1 Stratton spouted bed boiler system.

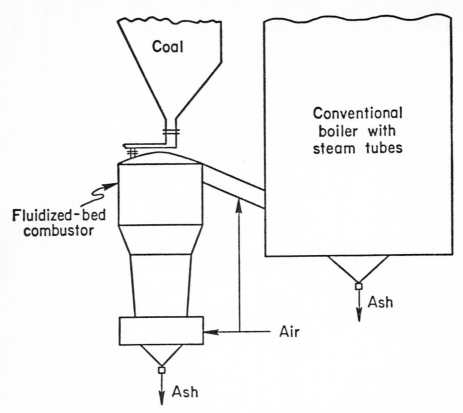

Fig. 3 Two-stage combustion system of institute of fuel research (Czechoslovakia).

without clinker formation is carried out in the first stage fluid bed combustor at temperatures below 1000 C. There is no need for immersing tubes
in the fluid bed to recover heat for burning low grade fuels. Unburned
combustion gas and particles from the first stage are burned in a second
combustion space at temperatures from 1000 to 1200 C (such as cyclone
furnace). It was also demonstrated that the fluidized-bed furnace can burn
solid and liquid fuels with 100 percent interchangeability. Semi-full scale
and full-scale boilers and retrofit of old boilers up to 25-MW were demonstrated using the above concepts.

        Research and development work on fluidized-bed combustion of
coal has been in progress in the U.K. since about 1963[11], first under
the C.E.G.B. (Central Electricity Generating Board) and later under
B.C.U.R.A. (British Coal Utilization Research Association) and the N.C.B.
(National Coal Board). The British recognized the superior heat transfer

properties of the fluidized bed and located steam tubes within it. Experimental efforts and conceptual design studies were aimed at the development of four types of fluidized-bed combustion systems: (1) atmospheric pressure utility-sized system (660-MW and 120-MW), (2) pressurized fluidized-bed combustion system for combined cycle power generation (120-MW and 20-MW), (3) industrial shell boiler (<50,000 lb steam/hour), and (4) packaged water-tube boiler (50,000~200,000 lb steam/hour and 120-MW). The nominal gas velocity was designed at 2 fps for utility boilers and 10-14 fps for industrial boilers. Conceptual design studies were undertaken by Babcock and Wilcox Ltd., John Thompson Ltd., and Preece, Cardew and Rider Ltd. Experimental support efforts were provided by BCURA in a 4 ft x 4 ft pressurized combustor (5 atm) and a 27-inch diameter atmospheric combustor, and by CRE (Coal Research Establishment) in atmospheric pressure combustors of 6-inch and 12-inch diameter and 36-inch x 18-inch in size. Experimental data at 1470 F indicated an overall combustion efficiency of 95 percent for industrial type boilers (14 fps) and over 99 percent for utility boilers (2 fps) when elutriated fines were recycled. Heat transfer coefficients to tubes immersed in bed were found to vary from 40 to 100 Btu/hr-ft^2-F. Preliminary economic study indicated a capital cost saving of 9 percent for fluidized-bed boiler plant as compared to pulverized coal boiler plant.

Following the promising results obtained in the early U.K. work, experimental projects were formulated in the U.S. by the Office of Coal Research (OCR), the Bureau of Mines, the Consolidation Coal Company, and in Australia by the Commonwealth Scientific and Industrial Research Organization (CSIRO). These have been in progress since about 1965.

OCR has contracted with Pope, Evans, and Robbins [PER[12]] to develop a fluidized-bed industrial packaged boiler. A modular cell concept with water wall was selected by PER and experimental efforts were initiated on a 16-inch x 12-inch fluidized-bed column. This eventually led to the development of the full-scale boiler module, a 72-inch x 20-inch combustor, which constitutes a half cell of a full-scale multi-cell boiler concept with a heat input rate of 10 million Btu/hr. In order to keep the boiler compact, high gas velocity (6~14 fps) is used, which resulted in high carbon carried over (~15 percent) from the bed. To achieve 99 percent carbon burnup, a carbon burnup cell concept was developed to complete combustion of elutriated carbon in a separate cell at higher temperatures.

The U.S. Bureau of Mines[13] studied the fluidized-bed combustion of various American coals in an 18-inch atmospheric combustor. A variety of coals, including highly caking types, were burned successfully under a nominal operating condition of 3 fps and 1600 F with mullite, zirconia, or alumina as bed material. Overall heat transfer coefficients from the bed to a water-cooled tube averaged about 53 Btu/hr-ft^2-F. For most coals, the coal ash was entrained with combustion gas. In the case of two coals containing ash in excess of 20 percent, approximately half of the ash remained in the bed, indicating that for these coals, the ash could be bed material.

The Consolidation Coal Company[14] investigated the fluidized-bed combustion of coal and char in a 4-inch combustor at temperatures of 1700-1900 F and gas velocities 1.5 to 3 ft/second in the late 1960's. Desulfurization efficiency with a Ca/S ratio of 1 was 78 percent and with Ca/S of 2 or higher desulfurization efficiency exceeded 90 percent. Regeneration of the sulfated dolomite using partial combustion of carbon monoxide gas at 1950 F was demonstrated. Effective sorbent activity was maintained up to 4 regeneration cycles.

The CSIRO in Australia[15] investigated combustion in fluidized-beds with considerable emphasis on heat transfer. Early tests were made in a 9-inch combustor of fluidized sand at temperatures in the range 750 to 1850 F. A heat transfer coefficient of 50 Btu/hr-ft^2-F was reported. Investigations were continued in a 3-stage combustor one foot square and 12 feet high burning Australian coals.

In 1967, what is now the Control Systems Laboratory of the Environmental Protection Agency[16] initiated a program to develop low-pollution fluidized-bed combustion systems. While pollution control was not a prime consideration in various earlier investigations, evidence of anti-pollution potential for fluidized-bed combustion systems was beginning to accumulate. Past work by EPA had shown that limestone was most reactive with sulfur dioxide at temperatures typically used in fluidized-bed combustors (1400-1800 F) and it seemed reasonable to expect that NO_x emission would be low at these temperatures. Recognizing the possibility that fluidized-bed systems might have the combined attributes of low-pollution emissions and lower cost than conventional systems, EPA thus directed immediate efforts toward development of a system which burned coal in a chemically active fluidized-bed of limestone.

As a first step, ongoing work at PER which has been supported by OCR was expanded to look at the air pollution control possibilities on their pilot-scale equipment. Early in 1968, studies were funded by EPA at Argonne National Laboratory (ANL) so that existing competence with fluidized-bed systems could be used to generate bench-scale information. Late in 1968, contact with researchers in the U.K. was established and negotiations for corporative efforts were begun. In 1969, a U.S. team[17] visited British organizations concerned with the development of fluidized-bed combustion systems for fossil fuels and an information exchange agreement was signed by the NCB and EPA. In late 1969 and early 1970, the EPA program was expanded to include work at the Bureau of Mines Morgantown Laboratory, at the laboratories of Esso (now Exxon) Research and Engineering (ER&E), and at Westinghouse Research Laboratories (WRL). Also in 1970, EPA began joint support of the work being funded by the NCB. A brief description of the work at the various locations is given below.

The earlier program at PER concentrated on the once-through sulfur removal system[18]. Injection of fine limestone into a fluidized-bed boiler was found effective for over 90 percent sulfur removal. The high limestone requirement (Ca/S ratio in excess of 3 at high gas velocity) prompted effort on the development of a regeneration system[19] with a coarse (8-20 mesh) limestone bed. High concentrations of SO_2 (3 to 10 percent) could be produced by operating a regeneration cell at high temperature (>1850 F) and low oxygen level (<0.5 percent). On the other hand, the carbon burnup cell[20] needed to be operated at high temperature (1950 to 2050 F) and high oxygen level (>3 percent) to achieve over 99 percent overall combustion efficiency. Addition of salt into the fluidized boiler[21] enhanced greatly the SO_2 capture effectiveness of the system, however, corrosion would certainly be expected to be more of a problem with salt additions.

The experimental efforts at ANL in a 6-inch diameter atmospheric combustor and 6-inch and 3-inch pressurized combustors were aimed at providing fundamental information on combustion efficiency, SO_2 and NO_x emission, particulate emission, and sorbent regeneration and attrition in both atmospheric and pressurized fluidized-bed combustion. The atmospheric data[22] suggested an optimum temperature (1400 to 1600 F) for sulfur retention depending on coal and sorbent types. For 90 percent sulfur retention, the Ca/S ratio must be maintained over 3. In contrast, the pressurized data (10 atm)[23] indicated little temperature dependence for sulfur capture

(1450 to 1650 F) and a Ca/S ratio of only two for over 90 percent sulfur removal. NO_x emission in pressurized combustion (120-270 ppm) was also much lower than in the atmospheric system (215-350 ppm). Similar combustion and desulfurization behavior was reported on lignite and subbituminous coal combustion.

Research activities at Esso R&E[24,25] included bench-scale investigation using 3-inch diameter atmospheric and pressurized combustors and regenerators, and a 0.65-MW pressurized continuous miniplant. The miniplant consists of a 12.5-inch diameter combustor and a 5-inch diameter regenerator capable of operating at temperatures up to 1700 F (combustor) and 2000 F (regenerator) and velocities up to 10 fps. Both ER&E and ANL data indicated NO_x emission in fluidized-bed combustion is derived from nitrogen in coal rather than from nitrogen fixation. Regeneration studies at both ANL and ER&E revealed technical difficulties with the pressurized system. The Esso miniplant has been in operation recently without the regeneration unit. The present goal of Esso's investigation is to improve fluidization quality with deep-bed operation and demonstrate the feasibility of deep-bed combustion with immersed cooling coils. WRL has been designing a turbine blade cascade to be appended to the miniplant.

The EPA-NCB joint study[26], which began June 1970 and ended July 1971, was very comprehensive and included pilot-plant tests of American coals in five British test rigs at BCURA and CRE for atmospheric and pressurized fluidized-bed combustion. For the atmospheric system, the gas velocity was varied from 2 to 11 fps, temperature from 1420 to 1680 F, and bed depth from 2 to 7 feet. For pressurized combustion at BCURA (5 atm), operating conditions were 2 fps, 1470 F, and bed depth of 4 feet. In general, these data using much larger combustors (48-inch x 24-inch pressurized combustor at BCURA, 27-inch diameter at BCURA, and 36-inch, 12-inch, and 6-inch at CRE for atmospheric combustors) verify the findings at ANL and EXXON R&E. The BCURA data at low temperature (1470 F) and low gas velocity revealed little sintered deposits or erosion on the turbine blade cascade at the outlet of the pressure combustor. BCURA[26] later (August 1972 to September 1973) extended investigation into the high temperature ranges (1650 to 1750 F) under a joint contract between National Research Development Corporation (NRDC) and the Office of Coal Research of the U.S. Department of Interior. These later studies[27] revealed that deposition on turbine blades cascade was not significant at bed temperatures less than 1600 F but was sufficiently extensive

at a bed temperature of 1750 F that blade cleaning by injection of a proprietary fruit stone material was required.

Based on data accumulated from experimental work at BCURA, NCB-
CRE, PER, ANL, and ER&E, Westinghouse Research Laboratories (WRL) under an
EPA contract evaluated the technical and economic feasibility of FBC
systems[28]. With subcontracts to United Engineers and Constructors, and
Foster-Wheeler, WRL prepared conceptual designs of a 250,000 lb/hr industrial
boiler plant, and 635-MW atmospheric and pressurized fluidized-bed boiler
utility plants in 1971. The industrial boiler design was based on a gas
velocity of 12.5 fps and bed depth of 2.5 feet; the atmospheric utility
boiler on 10-15 fps and bed depth of 2.5 feet; and pressurized utility boiler
on 6 to 9 ft/sec and bed depth of 10 to 15 feet. The capital cost for the
fluidized-bed industrial boiler was $7.40/lb/hr steam as compared to
$7.60/lb/hr for a conventional coal-fired plant and $2.44/lb/hr for a conventional gas/oil-fired plant. The marginal economic advantage of the
fluidized-bed industrial boiler and the availability of clean fuel at the
time of the Westinghouse study prompted WRL to recommend that development of
the industrial fluidized-bed boiler be suspended until future availability
of clean fuel can be assessed. For utility application, the pressurized
fluidized-bed combustion boiler plant (with sorbent regeneration) represented
a capital cost saving of 18 percent and power cost saving of 7 percent over
the conventional plant with a stack gas clean-up system. The potential advantages of the pressurized system as revealed by WRL's evaluation thus
shifted the EPA's contract activity at ANL and ER&E to emphasize the
pressurized systems.

Continued effort of WRL for EPA culminated in a three-volume
report on the evaluation of the fluidized-bed combustion process in 1973[29].
In this later report, WRL advocated the once-through pressurized fluidized
boiler plant for the first generation utility application and presented a
preliminary design of a 30-MW pressurized fluidized-bed boiler development
plant. Alternate pressurized fluidized-bed combustion systems such as the
adiabatic combustion concept and recirculating fluidized-bed combustion concept were also analyzed and recommended for further development.

Development work for OCR and EPA by PER has led to an OCR contract[30] for the design, construction, and operation of a 30-MW coal-fired
atmospheric pressure, fluidized-bed boiler plant at the Rivesville Station
of the Monogahela Power Company at Fairmont, West Virginia. The design and

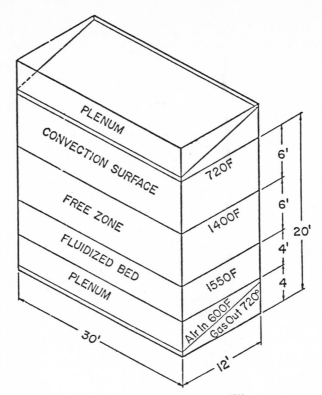

Fig. 4 FWC-PER FBC boiler cell[31].

construction of the Rivesville plant has been in progress since October,
1972.   Foster Wheeler Corporation has been subcontracted to provide final
design guidance for the 30-MW Rivesville demonstration plant and to develop a
design concept of a 800-MW plant[31]. To provide maximum operating availability, load turn-down, and flexibility of control, the 800-MW plant has
been divided into four 200-MW capacity modules that may be completely isolated
from one another for inspection or maintenance of any module while the
remaining modules stay in operation. Each of the four modules has been
divided into seven cells stacked vertically one above the other which serve
the functions of evaporator (path 2), evaporator (path 1), finishing reheater,
finishing superheater, intermediate superheater, primary superheater, and
FBC cell. The physical dimensions of each cell are based on process and
mechanical criteria and a typical cell size is shown in Figure 4. Following
the development of the concept for an 800-MW fluidized-bed boiler plant,
Foster Wheeler evaluated the PER concept of the Rivesville unit and revised

the design based on mechanical and heat transfer considerations and the requirement for this unit to produce design information required to scale up to the 800-MW utility unit size. The resulting arrangement of the Rivesville unit (Figure 5) incorporates multiple cells and horizontal boiling surface which requires the use of forced circulation pumps to insure proper fluid mass-flow through the boiler tubes. To support the Rivesville plant, laboratory work at PER has been continued to investigate coal feeding system, distributor design, and sorbent regeneration system. An 18-inch combustor, and 6' x 6' and 5" x 21" cold models at Foster Wheeler have also been constructed to investigate boiler tube bank design and fluidization characteristics.

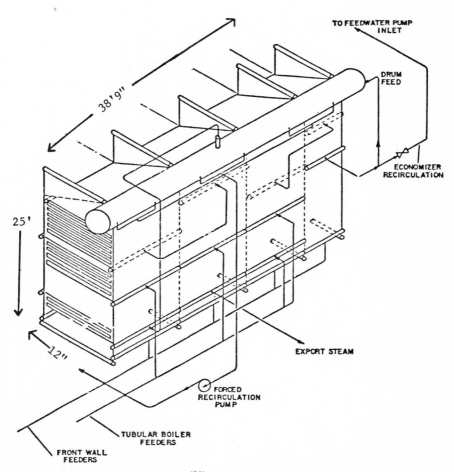

Fig. 5 FWC-PER final design[31] of Rivesville FBC boiler plant, circuit schematic.

In 1973, the National Coal Board, British Petroleum, and the National Research Development Corporation united to form Combustion Systems. BCURA has accumulated several hundred hours operation experience in their 48" atmospheric fluidized-bed shell boiler with a steam capacity of 8000 lb/hr and has been building a 44,000 lb/hr shell boiler in coporation with a boiler maker[32]. The Combustion Systems has also been building a 10,000 lb/hr shell boiler in South Africa in cooperation with Overint Trust.

Under a contract with EPA, the Combustion Power Company has been studying the fluidized-bed combustion of municipal solid waste in a pressurized (4 atm) 7.1 foot ID combustor (CPU-400). The pressurized combustor does not contain steam tubes and delivers hot gas (about 1500 F) after cleaning to a gas turbine to drive an electric generator and the fluidizing air compressor. Recently, the CPC has been contracted by OCR to investigate the pressurized adiabatic combustion of coal in their CPU-400 pilot facility.[33] The adiabatic combusion of coal in an atmospheric 20-inch ID combustor was carried out in the temperature of 1400-1800 F, velocity from 4-10 ft/sec, and the exit oxygen concentrations of 11-15 percent. The study reveals that dolomite is a better sorbent for SO_2 absorption than limestone at the same Ca/S ratio. At a Ca/S ratio of 1.5 and gas velocity of 10 ft/sec, a sulfur retention of 80 percent was achieved. The high sulfur retention was attained by reinjection of the primary cyclone fines in the CPC unit.

The Bergbau-Forschung[34] (a research organization for German coal industry, Steinkohlenbergbauverein) in Essen, Germany, has been working on fluidized-bed combustion of coal in their 16-inch x 31-inch atmospheric combustor for 2 years. A coal feed rate of about 100 lb/hr has been used. The combustion conditions investigated include temperature from 1470-1560 F and superficial gas velocity from 1.3 to 2.5 m/sec. For a 2 percent sulfur coal, a Ca/S ratio of 4 is normally required for over 90 percent sulfur removal, although sometimes ratios as low as 1.3 to 2 have been observed. A special feature of this test rig is the unique grid design which utilizes metal bars for directing combustion air downward first into inverted cone-shaped plenums to avoid sifting of coal ash. Bergbau-Forschung is planning an 8-atm pressure fluidized-bed combustion unit with a capacity of 10-15 MW using an experimental gas turbine.

The turbulent layer furnace (fluidized-bed furnace)[35] developed by Lurgi Gesellschaft für Chemie und Hüttenwesen mbH has been used for

combustion of low-grade fuels such as carboniferous tailings resulting from
coal preparation, coal-containing clays, low-grade lignites and lignite-
containing strippings, oil sands, oil shales, and bituminous marls which
have heating values as low as 1,800 Btu/lb. For fuels with higher heating
value, the extraction of heat from the fluidized combustor is necessary.
The power output for a 1,000 ton/day plant (on the order of 10-MW) is in
commercial practice. The combustion residues could be used for the production
of building aggregates. Lurgi has also been developing the fast-bed combustion technique for the chemical processing of fine materials. For possible
application to coal combustion, it appears that considerable research needs
to be done in regard to particle agglomeration, elutriation, particle
strength, and erosion problems.

Experimental work on FBC at Battelle dates back to 1949 with a
bench-scale fluidized-bed combustion study for bloating shale to produce
expanded aggregates. Later work involved roasting of ores[36], oxidation
of waste pulping liquors[37,38] in 10-inch and 24-inch fluidized-bed
combustion units, and more recently on combustion of coal in 6-inch unit
under support of the Battelle Energy Progarm[39]. A 24-inch pressurized
(8 atm) ash agglomerating fluidized-bed combustor is being built employing the Battelle-Union Carbide gasification process.[40] A 4-inch fast
fluidized bed cold model and 6-inch combustor are being used to explore
the ramifications of Dr. Reh's[41] and Professor Squire's[42] work in this
area.

A summary of these FBC experimental devleopment efforts is given
in Table 1.

FBC SYSTEMS CURRENTLY UNDER DEVELOPMENT

Combustion of coal in a fluidized-bed in the presence of limestone
or dolomite may be advantageously conducted at temperatures of 1400-1800 F
to maximize sulfur retention. To maintain this optimal combustion temperature
for desulfurization, the combustion heat must be extracted by steam or air
tubes immersed in the bed or carried away by the sensible heat of the excess
air (300-500 percent stoichiometric air). When the FBC is operated under
pressure, it is coupled with a gas turbine to improve the thermal efficiency
of the overall system. The following systems are currently under active

Table 1 Fluidized-Bed Combustion Facilities

Organization	Combustion and Regenerator Size	Pressure, atm	Gas Velocity, fps	Temperature, F	Status/Purpose
National Coal Board (NCB) and British Coal Utiliza-tion Research Association (BCURA)	6" diam, 12" x 12"	1	2-3	1240-1680	62-73/Combustion
	36" x 18"	1	2-8	1420-1680	
	48" x 24" (36" x 24")	5	2	1470-1745	
	27" diam	1	6-14	1420-1680	
Pope, Evans & Robbins (PER)	12" x 16" (FBC)	1	8-14	1450-2000	66-85/Combustion and regeneration, development plant
	20" x 6' (FBM)	1	12-15	1600-2050	
	20" x 20" (CBC)	1	12-15	1450-2000	
	28' 8.2" x 12' (Rivesville)	1	9-12	1550	
U.S. Bureau of Mines	18" diam	1	2-3	1450-1650	66-72/Combustion heat transfer
Consolidation Coal	4" diam	1.5	2-3	1700-1900	68-70/Combustion and regeneration
CSIRO	9" diam	1	2-7	750-1850	67-68/Combustion
Argonne National Lab.	6" diam	1	3-7	1400-1700	68-75/Combustion, regeneration
	3" and 6" diam	8	2-5	1450-1650	
EXXON R&D	3" diam	1	2-4	1500-2000	68-75/Combustion, regeneration
	3" diam	10	5	1500-2000	
	12.5" and 5" diam	10	5-10	1700-2000	
Foster-Wheeler	18" diam	1	--	--	Operating/Combustion cold flow model
	6' x 6'; 5" x 21"	1	4-12	70	
Combustion Power	20" diam	1	4-10	1400-1800	73-75/Combustion (adiabatic)
	7' diam	4	5-10	1400-1800	
Bergbau-Forschung	16" x 31"	1	4-8	1470-1560	74-75/Combustion
Lurgi	5" diam	1	>10	2000-2200	62-70/Fast bed combustion of oil for Al_2O_3 production
	40" diam	1	>10	2000-2200	
	12' 5" diam	1	>10	2000-2200	
Battelle-Columbus Laboratories	6" diam	1	1-5	1400-2500	49/Bloating
	24" diam	1	2-10	1400-1700	60-75/Combustion
	6" diam	1	3-30	1400-1700	74-75/Combustion
	24" diam	8	10	2100	75/Combustion (gasification)

development:

(1) Pressurized FBC boiler for utility application

(2) Atmospheric FBC boiler for utility or industrial application

(3) Pressurized adiabatic FBC for utility and industrial application

(4) Pressurized air tube FBC for industrial and commercial application.

A brief description of each system and of typical development programs are given below.

Pressurized FBC Boiler for Utility Application

The FBC is operated under pressure with boiler tubes immersed in the bed to control the bed temperature. The cleaned pressurized combustion gases are discharged through a gas turbine for additional power generation. The Westinghouse conceptual design study[29] has indicated that such a FBC power plant has the potential of a capit cost reduction of 12 and 23 percent and an operating cost reduction of 1 and 16 percent as compared to a conventional coal-fired plant with and without air pollution control, respectively. EXXON R&D[25] has been undertaking a pressurized FBC development program which is funded by EPA. This EXXON miniplant is designed with an equivalent capacity of 0.65-MW. The combustor has a 12.5-inch ID and is capable of operating at a pressure of 10 atm and gas velocities up to 10 fps. The miniplant has been in operation recently without the regeneration unit. Westinghouse is designing a turbine blade cascade to be appended to the miniplant.

Pressurized FBC has also been studied at ANL and at BCURA in England.

Atmospheric FBC Boiler for Utility or Industrial Application

Steam may be generated in an atmospheric fluidized-bed with heat transfer tubes immersed in and above the bed. Pope, Evans & Robbins (PER)[31] has been contracted by OCR to design and operate a 30-MW atmospheric FBC

power plant at Rivesville, West Virginia. Foster Wheeler Corporation has
been subcontracted for final design and to install the boiler. It consists
of three fluidized-bed primary boiler cells and one carbon burn-up cell (CBC).
The unit incorporates horizontal boiling surface which requires the use of
forced circulation pumps to insure proper fluid mass-flows through the
boiler tubes. The design also incorporates the use of manufacturing
techniques that will be used for the fabrication of future fluidized-
bed steam generators. Construction of the Rivesville plant is in progress.

        Besides the PER work, the atmospheric FBC system has been studied
in smaller units at ANL, at NCB in England, and at Bergbau Forschung in
Germany.

Pressurized Adiabatic FBC for Utility and Industrial Application

        In this system, heat transfer tubes are not used to extract heat
from the fluidized bed or combustion gases above the bed. The combustion
temperature is controlled by the admission of 200 to 500 percent excess air
and the energy of the hot combustion gas is then recovered in a gas turbine.
        The Combustion Power Company has been under contract with OCR
to investigate such a FBC system in their CPU-400 facility[33]. The combustor is 9.5 feet diameter and 22 feet high and capable of operating up to
4 atm pressure. The CPU-400 unit is being modified to burn coal. Parametric
evaluation of coal combustion in a 20-inch atmospheric combustor at excess
air (10-15 percent O_2 in combustion gas) condition provided information on
SO_2 and NO_x emission. Desulfurization and combustion were improved by reinjection of primary cyclone fines into the combustor.

Pressurized Air Tube FBC for Industrial and Commercial Application

        In this system the FBC temperature is controlled by heat transfer
tubes in and above the beds but the working medium is air instead of water.
The system is thus operated under pressure and incorporated with a gas turbine.
Various configurations of closed or open cycles are possible. This air
heater concept was first proposed by H. Harboe in the EPA 3rd International

Fluidized–Bed Combustion Conference, Hueston Woods, Ohio, October 1972.
The greater variations in the difference between bed temperature and tube
wall temperature allow higher degrees of load change. In the United States,
the Modular Integrated Utility System (MIUS) has been studied by Oak Ridge
National Laboratory[43] under the sponsorship of the Office of Policy
Development and Research, the Department of Housing and Urban Development,
and the Office of Coal Research. It makes use of a small on-site fluidized-
bed boiler utility plant located adjacent to a new housing development to
provide electricity, hot water for building heating, domestic hot water, and
chilled water for air conditioning.

ENGINEERING DATA

        The available engineering data a number of critical components
of FBC will be briefly reviewed and assessed.

Coal and Sorbent Feeding and Ash Reinjection

        The primary objective of a coal feeding or ash reinjection system is
to evenly distribute the fuel into the fluidized bed combustor for maximizing
combustion and desulfurization efficiencies. This can be achieved quite
easily in fluidized-bed combustors of small bed areas (less than 10 ft^2).
However, as the bed area increases to the 300 – 400 ft^2 of commercial size
range, the question arises as to how much bed area can be fed by one feed
injection point. Other important questions are: what is the most effective
device for metering and providing solid fuel to the injection probes; is
water cooling necessary to prevent the coal from caking; do heat transfer
tube banks above the coal feeding point have any effect on the mixing; and
how well can cold flow data be used to predict the feeding characteristics
in a hot combustor? The British experimental data indicated that for com-
combustors with immersed water tubes, each coal injection probe will feed
4.5 ft^2 of atmospheric combustor and 2 ft^2 of pressurized combustor (5 atm).
Some units have required water cooling on the probe to prevent coal from
coking. Without heat transfer tube banks in the bed, PER had demonstrated
coal injection using a long uninsulated tube at a 45 degree angle from the

verticle to feed a bed area of 9.1 ft^2. The CPC pressurized combustor
(without tubes in bed) has a bed area of 39.4 ft^2. Preliminary data seem to
indicate one coal feeding probe would be sufficient. Further long duration
combustion tests are needed to confirm this preliminary result.

Based on the Foster Wheeler cold model study, the vibrating coal
distributor feeder appears to have a better operating record than the screw
feeder. Fuel injection up through the air distributor appears desirable;
however, it faces the potential danger of fuel line plugging and erosion.
The straight down flow arrangement for fuel injection has demonstrated good
operation, but it requires a long length of fuel line.

The PER data appear to indicate a necessity for a different feeding
device for fly ash reinjection into the carbon burn-up cell (mushroom feeder).
More development data are needed to provide the basis for sound commercial
design.

Air Distributor Design

A good air distributor design is necessary to provide uniform air
distribution and maximize combustion efficiency. Normally the pressure
drop across the grate drops as the gas velocity is reduced to meet turn
down requirements. The experimental data on the PER nozzle button grate
design indicate that if the pressure drop across grate drops below 4 inches
of water, the coal feeder will plug and poor air distribution will occur.
Five different types of air distributor design have been tested in a small
cold model unit by Foster Wheeler:

(a) Perforated plate
(b) Johnson screen grate (closely spaced
     wires)
(c) Procedyne®plate nozzle (no machined elements)
(d) Foster-Wheeler punched plate with fabricated
     covers
(e) PER nozzle button plate.

No significant difference was observed between the types in bubble formation
or in bed mixing. However, the PER pressure drop measurements and hot combustion tests reveal quite different results for some types of air distributors.
The perforated plate gave low pressure drop but performed poorly at reduced
velocity during the hot test. The Johnson screen had high pressure drop and

openings were bridged during the hot test. The Procedyne plate nozzle had
high pressure drop but appeared to give good performance; it did not sift,
operated well at low load, and the minimum fluidization velocty was independent
of bed depth.

It is evident that further development work on air distributors is
needed. Desirable features for the air distributor include low cost, low
and constant pressure drop, non-sifting, and high resistance to corrosion and
thermal stress.

Attrition and Elutriation of Bed Materials

The attrition and subsequent elutriation of bed materials from
fluidized-bed combustors depends on the properties of the materials and the
gas velocity. In EXXON's 4-inch pressurized (6-9 atm) batch reactor at
superficial gas velocities ranging from 4 to 8 ft/sec, the observed attrition
rates of three batch charged 8 x 25 mesh stones were

Grove limestone	15%/hr
Tymochtee Dolomite	20-25%/hr
Pfizer Dolomite (1337)	50%/hr

The dolomite 1337 showed excessive attrition rates during calcination.
These data appear to indicate high attrition rates during combustion. In
ANL's 6-inch combustor (8 atm), a fluidized bed of 14 x 45 mesh partially
sulfated dolomite (Tymochtee) was fed with 14 x 80 mesh (80% in the
14 x 45 mesh range) dolomite additive. Elutriation increased from a low
of 5% of the additive being fed to the combustor at 2 ft/sec to as high as
80% at 5 ft/sec. ANL, estimated the terminal velocity of 16 mesh additive
at 5 ft/sec for a sphericity factor of 0.6. The high elutriation rate is
thus related to the predominance of fine particles in the feed. Measurements
of the + 45 mesh additive entering and leaving the combustor were made and
an average recovery for all experiments was 96%. This appears to indicate
very little attrition of the dolomite occurred over the range of ANL's experimental conditions.

An interesting observation has been reported by PER that if partially
sulfated limestone is fluidized with humid air for an extended period,
the particle will hydrate and turn to powder.

Few attrition data have been reported. Such information is needed

because the attrition and elutriation of bed materials affect the control of
bed level and can reduce the amount of sorbent in the bed. Systematic
investigation and reporting of attrition and elutriation of bed materials
is needed to support further development work.

Combustion Efficiency

Atmospheric FBC. The ANL data indicate that the combustion efficiency of the FBC system is determined primarily by the elutriation of unburned
carbon from the bed as the carbon loss in the bed and in the unburned gases is
quite negligible. In the ESSO data, the carbon loss was shown to decrease
with the temperature and increase with gas velocity. At 3 ft/sec, the combustion efficiency of ANL's experiment ranged from85 percent to 97 percent
(1600 F) and in ESSO's experiment ranged from 89 to 97 percent (1500-1800 F).
The average combustion efficiency in NCB-CRE's combustor at 3-4 ft/sec and
1380-1560 F was 92 percent. However, the combustion efficiency ranged from
85 percent to 91 percent when the gas velocity was 8 ft/sec at 1560 F. At
the operating conditions of 12-14 ft/sec and temperature 1700-1800 F, the
combustion efficiency of PER's unit was about 85 percent. These data appear
quite consistent.

The carbon burn-up cell (CBC) was developed by PER to raise the
combustion efficiency of FB combustors operated at high velocity. To achieve
over 99 percent overall combustion efficiency, the carbon rich fly ash is
burned in a CBC at a temperature of 1950-2050 F with high oxygen concentration (>3%). A mushroom fly ash injector was developed by PER to feed the
fine carbon rich ash. This concept has been tested on a 1.1 ft^2 CBC with
ample free board area and some screen baffles to reduce the carry-over of
fine ash before combustion is complete.

Further improvement of CBC's to reduce the high combustion temperature requirement may be advantageous to increase the desulfurization efficiency.
Improvement in the main fluidized bed combustor may eventually lead to the
elimination of CBC's.

Pressurized FBC. The ANL data indicate that carbon loss as unburned hydrocarbon was very low (<.1 percent of the feed carbon) and unburned
carbon retained in the bed was also low (<1 percent of feed C). The primary
mode of carbon loss was by elutriation of the unburned carbon particles with

ash from the bed. ANL has investigated pressurized fluidized-bed combustion
(8 atm) of three different coals (-14 mesh); Bituminous coal, subbituminous
coal, and lignite with a bed height of 3 ft. These coals had the following
properties:

	Arkwight Bituminous Coal	San Juan Subbituminous Coal	Alenharold Lignite
Moisture, wt%	2.89	9.28	30.90
Volatile matter, wt%	38.51	33.28	30.00
Fixed carbon, wt%	50.92	40.48	32.99
Ash, wt%	7.68	16.96	6.11
Sulfur, wt%	2.82	.78	.53
Heating value, Btu/lb	13,706	9,621	7,625

The combustion efficiency of Akwright coal at the experimental conditions
of 8 atm, Ca/S ratios of 1 to 3, and gas velocities of 2 to 5 ft/sec varied
from 89% at 1450 F to 97% at 1650 F. The experimental conditions for the
subbituminous coal and lignite were: bed temperature of 1550 F, gas velocity
of 3.5 ft/sec, and Ca/S ratio of 1. The combustion efficiencies for
the San Juan and Alenharold coals were 94 and 97% respectively. These values
agree well with the combustion efficiency of 94% reported for the Akwright
coal at a combustion temperature of 1550 F.

In the BCURA-EPA study, coal combustion was conducted at 1470 F,
2 ft/sec, bed depth of 3.75 ft, and pressures of 3.5 and 5 atm. The combustion efficiency at low pressure (3.5 atm) was 97 percent, which was lower
than that (99 percent) at high pressure (5 atm). In the BCURA-OCR study,
coal combustion was conducted at higher temperatures (1650 and 1750 F), 2 ft/
sec, bed depth of 4.5 ft, and pressure of 4.8 atm. The combustion efficiency
was 99 percent at 1650 F and 99.5 percent at 1750 F.

In the CPC 20" fluidized bed of 6 x 16 mesh sand, Ill. No. 6 coal was
burned at combustion conditions of 1500 F, 4 ft/sec., and exit O_2 concentrations
of 11-15%. A temperature difference between the fluid bed and the free
board of 300 F was observed for a coal feed of 7/8" x 6 mesh, and a temperature
difference of 100 F for a coal feed of 1/4" x 30 mesh. The coal particle size
thus appears to have some effect on the combustion characteristic in the
fluidized-bed combustor. More in-depth investigations of the fundamentals of
coal combustion in fluidized beds are needed. Information is needed to
permit prediction of the effect of coal types and particle sizes, and of the bed
operating variables on the combustion efficiency and combustion intensity.

SO$_2$ Retention

Atmospheric FBC. The most important operating parameters which affect
the sulfur retention in fluidized-bed combustion are temperature, Ca/S ratio
and superficial gas velocity. Experimental evidence suggests an optimum temperature range of 1400-1600 F for sulfur retention. The mechanism causing
lower retention at temperatures above 1600 F is not well understood. The
experimental data also indicate that sulfur retention decreases as the gas
velocity increases and the Ca/S ratio decreases. At low gas velocity (3-4
ft/sec) range, above 80 percent sulfur retention may be achieved with a Ca/S
ratio of 2. To achieve the same level of sulfur retention at a high velocity
range (8-14 ft/sec), a Ca/S ratio in excess of 4 is required. The other
parameters such as bed height, particle size, and excess air level had only
a slight effect on the sulfur retention.

The above discussion refers to the method of operation whereby sorbent
is injected into an inert bed. The latest results of PER runs with limestone
as the bed material (8 x 20 mesh) revealed that sulfur retention in excess
of 90 percent could be maintained by the simultaneous operation of the combustor
and sorbent regenerator, whereby the spent sorbent bed material was continously
withdrawn into the regenerator and the regenerated sorbent recharged into the
combustor. The limestone makeup requirements were low, about 5 percent of
coal input by weight (equivalent to a Ca/S ratio of 0.4).

PER also reported recently that a sulfur retention of 93% was achieved
burning Indiana No. 6 coal (4.17% S) and Sewickly coal (3.88% S) by the injection of limestone with a Ca/S ratio of 1 and common salt in amounts equivalent to 1-0.7% of the coal feed. The role of salt in increasing stone activity
is not understood. Furthermore, addition of salt to a boiler may be undesirable.
However, an understanding of the reaction mechanism may help to develop methds of
maximizing stone utilization. Fundamental studies in this area appear worthwhile.

Recent experimental data from the CPC 20" adiabatic combustor (0$_2$ concentration 11-15%) reveals that dolomite is a better sorbent than limestone
at the same Ca/S ratio. An optimum temperature effect was observed at low
Ca/S ratios (1.5); little temperature effect was observed at high Ca/S ratios
(3-5). For the high Ca/S ratios (3-5), the sulfur retention is little affected
by gas velocity up to 10 ft/sec. However, for low Ca/S ratios (1.5), the
sulfur retention decreases with increase in gas velocity. At a Ca/S ratio
of 1.5 and gas velocity of 10 ft/sec, a sulfur retention of 80% was achieved

in the CPC unit. This high sulfur retention efficiency appears to be related to the reinjection of the primary cyclone fines in the CPC unit.

Pressurized FBC. The ANL data indicate that the most pronounced factors affecting SO_2 emission at normal operating temperatures are gas velocity and Ca/S ratio. On increasing the combustion temperatures from 1450 to 1650 F, the SO_2 emission increased very slightly. For Ca/S ratios above 2, the SO_2 removal is generally greater than 90 percent and only decreased slightly with increasing gas velocity (from 2 ft/sec to 5 ft/sec). However, at a Ca/S ratio of 1, the SO_2 emission increases rapidly with increasing gas velocity. At low temperature (1450) and low velocity (2 ft/sec), the sulfur retention is 83 percent for Ca/S of 1, which corresponds to SO_2 emission (370 ppm) well below the equivalent EPA standard.

A somewhat better retention was reported for the combustion experiment with lignite, which suggests that the calcium in the coal may be an active agent in helping to retain SO_2 during combustion.

The BCURA data at low gas velocity (2 ft/sec) indicate a slight temperature effect on SO_2 absorption and a much greater effect of the Ca/S ratio. To achieve over 90 percent sulfur retention, a Ca/S ratio in excess of 1.6 was required at low gas velocity (2 ft/sec).

The sulfur retention data recently reported by EXXON R&D[44] on a fluidized bed of Tymochtee dolomite at experimental conditions of 6.5 atm and fluidizing velocity of 5-8 ft/sec showed a decrease in sulfur retention as the combustion temperature increased from 1700 F to 2000 F.

NO_x EMISSION

Atmospheric FBC. The measured NO_x emission from atmospheric FBC with limestone addition or in limestone beds ranges from 250 to 600 ppm. For FBC in an inert bed, NO_x emission ranges from 450 to 630 ppm at similar conditions. These NO_x emission levels are consistently higher than the equilibrium NO_x concentration (200 ppm) based on the gas phase nitrogen-oxygen reaction. Additional experiments in gas phase combustion by Shaw and Thomas[45] and in coal combustion by ANL,[46] in which argon was substituted for nitrogen in the air suggested little change in NO_x emission. A typical coal contains 1.4 percent nitrogen, equivalent to an NO_x emission of 2500 ppm. All this evidence points to the fact that NO_x emission from FBC systems is probably

derived from fuel nitrogen and the presence of limestone helps to reduce
the NO_x formation.

The ESSO[24] data indicate that reduced NO emissions may be effected
by burning under reducing conditions and in the presence of CaO and $CaSO_4$.
This is supported by the evidence of low NO_x emissions (70-100 ppm) in two-
stage combustion experiments conducted at ANL and ESSO.

Pressurized FBC. The NO_x emissions in ANL experiments varied from
120 to 270 ppm and was shown to correlate with the Ca/S ratio. The ANL data
also indicate that for pressures less than 4 atm, the NO_x emission increases
with reduced pressure for a lime bed, and increased even more for an inert
bed. In the BCURA experiments, the NO_x emission ranged from 70 to 250 ppm.

Recent data reported by EXXON R&D[44] indicate that NO_x emissions
increased with excess air for experimental conditions of 6-9 atm pressure,
1460 to 1700 F, and 4 to 7 ft/sec. To meet the EPA standard of 0.7 lb/10^6 Btu,
the excess air level must be less than 80% in EXXON's experimental unit
(4" combustor).

In contrast to EXXON's data, an NO_x emission of 0.683 lb/10^6 Btu was
reported by CPC in their 20" atmospheric fluid-bed combustor operated at 135%
excess air. The other experimental conditions of this 120 hour test run were
1600 F, 1.28 Ca/S and 7 ft/sec.

Stone Utilization

Except for the research at ANL, most investigations have paid
little attention to stone utilization. PER has reported stone utilization
of the fine limestone (-325 mesh) to vary from 30-40 percent at 90 percent
sulfur removal to 60 percent at 20 percent sulfur removal. For coarse limestone (8 x 18 mesh), the normal stone utilization was 10-15 percent in the
PER experiments.

The experimental data of ANL indicate that dolomite stone utilization
varied inversely with the Ca/S ratio and was relatively unaffected by fluid-
izing-gas velocity or bed temperature. The dolomite utilization in various
ANL experiments ranged from 80% at a Ca/S ratio of 1 to 35% at a Ca/S ratio
of 3 for the material remaining in the bed at the conclusion of the experiment.

The dolomite stone utilization was consistently higher for material retained in the bed than for the elutriated solids collected in the primary cyclone. In most runs the dolomite stone utilization in the overflow bed material was lower than that of the dolomite in the final bed. Dolomite utilization for elutriated solids generally increased the farther downstream the particulate matter in the flue gas was collected. For example, dolomite utilization up to 88-93% was observed for particulate collected in the secondary filter. The latter result is attributed to increasing reactivity with decreasing particle size.

Improvement in stone utilization offers great potential for reduced operating cost. Further development work to maximize stone utilization is needed.

Sorbent Regeneration

Sorbent for desulfurization in fluidized-bed combustion systems can be regenerated in one step by decomposition of sulfate in reducing gases at high temperature:

$$CaSO_4 + \begin{Bmatrix} CO \\ H_2 \end{Bmatrix} \xrightarrow{\ 1900-2000\ F\ } CaO + \begin{Bmatrix} CO_2 \\ H_2O \end{Bmatrix} + SO_2 \ .$$

The reducing gases may be produced by combustion of coal with air-steam in the sorbent regenerator. This one step regeneration of sorbent at reducing condition at high temperature (2000 F) was demonstrated in various research centers. Reasonable sorbent activity was maintained after regeneration up to 4 cycles in Consol experiments and 7 cycles in EXXON R&D experiments. The SO_2 concentration in the off-gas ranged from 3 to 9 percent, which is in the range suitable for economic sulfur recovery.

In the PER regenerator, the sulfated limestone circulated from the primary combustor was regenerated by burning with the carbon-bearing fly ash or coal at high temperature (2000 F) with low excess air. The SO_2 concentration can be maintained at levels of 3-4 percent if oxygen concentration is 0.5 percent or less. An SO_2 concentration of up to 8.5 percent was recorded.

The SO_2 concentration was low ($\sim$ 2%) when the one-step regeneration was operated under pressure as reported by EXXON R&D. EXXON's data also showed that the approach to equilibrium increased from 20% to 50% as the pressure increased from 1 to 10 atm apparently due to more favorable reaction rates. This effect presumably compensates for the lower SO_2 concentrations to be expected at higher pressures due to a less favorable chemical equilibrium.

The longer the contact time between gaseous and solid phases, the higher the SO_2 concentration in the regenerator off-gas. It was also found that injection of secondary air above the fluidized-bed regenerator reduced the CaS formation in the regenerated product.

The regeneration of $CaSO_4$ to CaO and SO_2 appears to be limited by reaction rates. Additional work will be necessary to determine if conditions can be found to allow a closer approach to equilibrium or raise the reaction rates.

The sorbent may also be regenerated by using a two-step reaction scheme at elevated pressure (10 atm) producing hydrogen sulfide:

$$CaSO_4 + \begin{Bmatrix} 4H_2 \\ 4CO \end{Bmatrix} \xrightarrow{\text{1500 F}} CaS + \begin{Bmatrix} 4H_2O \\ 4CO_2 \end{Bmatrix}$$

$$CaS + H_2O + CO_2 \xrightarrow{\text{1100 F}} CaCO_3 + H_2S$$

The reducing gas for the first reaction step may be provided using a gas producer by combustion of coal with air and steam. The first reaction is thermodynamically favored by low temperature and unaffected by pressure. Pressurized operation, however, helps to suppress undesirable side reactions. The second step reaction is thermodynamically favored by the combination of low temperature and high pressure, which also suppresses competing side reactions. The feed gas ($CO_2 + H_2O$) for the second reaction may be obtained by stripping CO_2 from boiler flue gas with a regenerable hot-carbonate or amine scrubber solution. It also may be provided from an air-coal-steam combustor; however, in this case, H_2S concentration in the off-gas is diluted by N_2 and an enrichment process for H_2S is needed.

The two-stage scheme appears very suitable for pressurized operation. However, it was reported by O'Neill[47] that conversion of calcium sulfate to calcium sulfide could be carried almost to completion at temperatures between 1382-1562 F and 10 atm, but regeneration of carbonate from the sulfide was difficult.

Results of ANL studies on the two-step scheme conducted in a 2-inch diameter batch reactor indicate that:

(1) The reaction producing H_2S was initially rapid, but the rate decreased to zero after a short time. Typically, the reaction rate dropped to zero after 12 minutes.

(2) The peak concentration of H_2S in the outlet gas was high, near the expected equilibrium value.

(3) Typically half or less of the CaS reacted.

These results suggest that considerable additional work is needed to determine or improve the technical feasibility of using the pressurized one-step or two-step regeneration scheme and research work on other novel approaches are definitely worth while.

Boiler Tube Bundle Design

In the BCURA pressurized test rig (5 atm, 2' x 3'), the bed cooling coils consist of 1 inch outside diameter x 16 g stainless steel tubing, of specification AISI 310. The tubes are made up into looped rows which cross and recross the bed casing. The horizontal rows are arranged on a staggered pitch. The horizontal distance between tube centers is 6 inches, and the vertical distance between tube centers is 1-1/2 inches. The pattern of this staggered pitching repeats itself every four rows. The surface area of tubing in contact with the bed material per unit volume of the bed is 3.97 ft^2/ft^3 vol. The excellent combustion and desulfurization efficiency of BCURA experiments indicates this configuration offers good fluidization quality.

In the Foster-Wheeler design of the Rivesville plant,[31] three primary fluidized beds (cells) are used to heat the primary superheater, the finishing superheater, and the evaporating tubes. These cells have bed cross section areas of 12' x 11.25', 12' x 11.25', and 12' x 10' respectively. The horizontal tube bundle is made of 2" OD tubes arranged on 6" centers with the horizontal tubes parallel to the 12' dimension of each cell. The vertical spacing is 2-5/8". The superheater material selections for different superheater locations are shown in Figure 6. The average heat transfer surface area is 2.5 ft^2/ft^3 of fluidized bed volume.

Foster-Wheeler has studied the tube bundle design in a 5" x 20" x 12' Plexi glass cold model. Four variables were studied: longitudinal pitch,

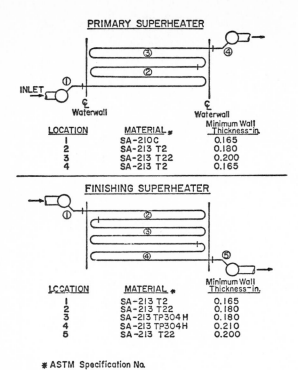

Fig. 6   Foster-Wheeler superheater material selection[31] of Rivesville plant.

transverse pitch, tube diameter and tube arrangement (either staggered or in
line).   The bed material for all tests was limestone sized to 6 x 16 mesh.
The air velocities in each test varied from 2 to 8 ft/sec.   High speed motion
pictures were taken of each test for visual comparison of the fluidizing
characteristics.   Analysis of the motion pictures by Foster-Wheeler indicated
that the configuration of the tube bundle had an affect on the pressure loss
through the bed and on the quality of fluidization.   The staggered tube
bundle arrangement in general gave better fluidization characteristics than
the inline arrangements which showed problems of channeling.   The tests indicate that the violent bubbling of the fluidized bed can be dampened by
locating the tube bundle close to the air distributor, increasing the mean
particle size of the bed material, and decreasing the vertical pitch of the
tube bundle.

Tube Corrosion, Erosion, and Deposition

The National Coal Board of England conducted a corrosion study of tubes in three different fluidized-bed combustors (4' x 2' pressurized, 27' dia. and 12" x 12" atmospheric) for EPA in 1970-1971. The weight loss data for 100, 500, and 1000 hr duration tests were reported[26]. The maximum tolerable penetration rate on the fire-side of conventional boiler tubing is generally assumed to be about 1.5×10^{-6} in/hr. This is equivalent to a specimen weight loss rate of 30 $\mu g/cm^2$ hr, if there is no preferential loss around the circumference of the tube nor any intergranular penetration of the metal surface.

Visual observation of the corrosion specimens by NCB revealed no direct erosion of the specimens by the ash particles at the experimental condition encountered. This is not surprising as the gas velocity (<11 ft/sec) was quite low compared to the flue gas velocity of around 100 ft/sec in conventional coal fired boilers where steel just begins to be eroded by particle impaction.

Visual examination of the test specimens also suggested that the scouring action by the ash particles was not sufficiently erosive to remove the protective oxide cover and expose clean metal surface to the environment. This was confirmed by the specimen weight loss results, as the errosion rate decreased with time indicating the formation of a stable protective oxide layer.

However, the presence of the fluidized bed had some influence on the formation of deposits on the tubes. The tubes immersed in the bed were mostly covered with a thin, even, hard and adherent deposit (<.1 mm thick). A thicker (up to 1.5 mm) but more loosely adherent ash-type deposit was sometimes formed on the underside of the tubes cooled to 750-900 F which were situated close to a coal feed nozzle. The underside of the tubes situated above the bed in the 12-inch combustor was covered with a layer of fine ash, and coarse ash particles formed a layer on the top of these tubes. Both types of deposit were loosely adherent and were not found on the tubes situated in the bed.

The deposits of material on the tubes were no thicker after 500 or 1000 hour operation than they were after 100 hours. They were insignificant compared with those occuring in conventional coal-fired boilers, and were predominantly ash.

Very little surface attack was observed on the alloys tested at their usual operating temperatures. These include the low alloys, namely, medium carbon and 2-1/4% Cr steels, which were only slightly blistered. These two alloys had somewhat thicker scales when tested above their normal maximum operating temperatures of 900 F and 1100 F respectively.

The tubes in the free board area showed consistently lower weight loss rates than their counterparts immersed in the bed at the same temperatures.

Metallographic examination of the tested specimens revealed that materials maintained at or below their accepted maximum working temperature in conventional plants did not suffer significant intergranular penetration by sulfur.

The surface of the austenitic steels from the 100 hour tests were mostly smooth (surface irregularities 4 µm). The low chromium steels exhibited somewhat rougher surfaces. The surfaces of the specimens from the 500 hour and 1000 hour tests were on the whole slightly rougher. The oxide layer for all steels was up to 10 µm thick with no significant effect of test duration or metal temperature.

The austenitic steels showed little evidence of sulfide penetration at temperatures below 1100 F. In the temperature range of 1290 to 1500 F, sulfide penetration of 10 to 20 µm was general in 100 hour tests and up to 70 µm in 1000 hour tests.

For the low chrome steels, sulfide penetration was generally absent in 100 hour tests at temperatures below 930 F. It was about 50 µm in 500 hour tests at temperatures of 1100 F.

Specimens of 2-1/4% Cr steel held at 1110 F for 500-hour tests had a sulfide penetration of about 50 µm.

An additional corrosion study was undertaken by BCURA for OCR during 1972-1973 in the 3' x 2' pressurized combustor at higher combustion temperatures up to 1740 F. The weight loss data for the 100 hour duration test was reported. These data are compared with previous NCB-EPA data and appear to fall into the existing pattern.

Metallographic examination of the 1%, 2-1/4%, and 12% chrome steel specimens showed that at temperatures up to their respective normal maximum limits for use in steam boilers, attack was slight, with no evidence of intergranular penetration. Attack of the AISI 321 steel at 1500 F was severe with pitting and intergranular penetration. The Incolloy 800 showed negligble

attack at 1510 F and could be a potential tube material for use in an air
heating combined cycle plant (closed cycle gas turbine).

Hot Gas Cleaning and Gas Turbine Blade Corrosion and Erosion

The Westinghouse Research Laboratory studied previous turbine blade
erosion work in coal-fired gas turbines and proposed a dust loading distribution as shown in Table 2 for acceptable turbine blade life. This distribution
indicates that dust particles greater than 6 µm should not be present in the
feed gas; that particles larger than 2 µm should be kept below 0.0115 gr/SCF;
and that dust particles less than 2 µ could have a dust loading of 0.1 - 0.15
gr/SCF with particle size predominantly less than 1 µm. However, safe loadings
are probably dependent on composition and structure of the particles as well.

The dust loadings of various gas streams leaving the ANL 6" combustor (8 atm) varied from 4.1 to 23.4 gr/SCF and of gas streams after the
secondary cyclone varied from .3 to 2.1 gr/SCF. These data show that the
dust loading of flue gas leaving the combustor, before cleaning, increased
with the Ca/S ratio and gas velocity. The dust loadings obtained with a
Brink impactor after two cyclone stages and a single filter are high, ranging
from 0.051 to 0.089 gr/SCF. Only the results (~.001 gr/SCF) with two cyclone
stages and two filter stages in the clean-up system, exhibited concentration
levels compatible with the Westinghouse recommendation.

Test data on gas cleaning by conventional cyclones in series (up to
3), conventional cyclones plus an Aerodyne cyclone, and conventional cyclones

Table 2   Proposed Dust Loading Distribution in Gas Turbine Feed[15]

Ave. particle dia. µm	Wt. % in Size range	Concentrations gr/SCF
< 1	76.8	.08–.12
1–2	16.6	.02–.03
2–3	4.7	.005–.008
3–4	1.6	.002–.003
4–5	.2	.0002–.0003
5–6	.1	.0001–.0002
6–7	0	0

plus a panel bed filter are needed on a much larger scale, and in conjunction
with turbine blade erosion testing.

       In the BCURA study, an assembly of the turbine blade cascade and
target rod was appended to the 2' x 3' pressurized combustor after two internal cyclones. The cascade was a nozzle guide vane segment from the Rolls
Royce "Protens" marine gas turbine. Two 3/32 inch diameter target rods
(Nimonic 75 & EPK 55) were arranged at 90° to the major axis of the blade
passage ways at a point where the Mach No. was circa 0.6. The flow conditions at the mid-point at these rods would be similar to those at the mid-
point of the leading edges of the first row of turbine blades. About 400
hours of operation were accumulated at a temperature of 1470 F (BCURA-EPA
series) and 200 hours at temperatures of 1630 to 1740 F (BCURA-OCR series).

       Deposition of material on the cascade did not occur to a significant
extent at bed and cascade temperatures of 1650 F and 1600 F respectively with
the pressurized combustor (66-73 psig) burning Illinois and Pittsburgh coals
at 2.2 to 2.5 ft/sec. The deposits at a bed temperature of 1750 F
and cascade temperatures of 1650 to 1700 F were sufficiently extensive to be
a potential operating problem. It was found that most of the deposit could
be removed by the injection of a "soft blast" (a proprietary fruit stone
material used for compressor/turbine cleaning) in a stream of cold air.

       The bulk of the material on the blades was derived from the dust passing
over the cascade -- there was no enrichment of alkali· The measured
alkali content of the exhaust gases was typically 5 ppm Na and 2 ppm K
at bed temperatures of 1750 F. These values are higher than those measured
at a bed temperature of 1470 F (2 ppm Na and 0.5 ppm K), but still well below
the 10-140 ppm values measured in conventional combustion systems.

       The quantity of dust passing over the cascade at 1750 F bed tem-
perature was about 250 ppm which is higher than the 150 ppm (0.1 gr/SCF)
measured at 1470 F bed temperature. The size of the dust, as determined by
a Coulter Counter, was typically 95% less than 10 micron and 85% less than
5 micron.

       The leading edge of the target rods usually had slight accumulations
of dust, similar to those which adhered to the leading edges of the cascade.
These accumulations were extremely fragile and broke away easily. The target
rod of Nimonic 75* after 400 hours of exposure to low temperature (1490 F)
combustion, showed no evidence of corrosion or erosion under microcopic
examinations. However, after an additional 200 hour exposure to high

temperature (1630 - 1740 F) operation, some pitting, oxidation and sulfidation
was noted. Sulfide penetration was detectable to a depth of 120 - 150 μm.

The second rod of EPK 55** was exposed only to 200 hours of high
temperature (1630 - 1740 F) operation and was not significantly attacked.
There were no localized pits; oxide penetration was visible around the periphery to a depth of about 10 μm.

START-UP AND TURN-DOWN

Design for start-up of a fluidized-bed boiler presents a difficult
problem. On the one hand it is desired to minimize the cost of auxiliary
start-up burners and the amount of costly gas or oil required for start-ups.
On the other hand, the time required for a start-up must be minimized so
that the fluidized-bed boiler does not suffer a competitive disadvantage
when compared with conventional boilers. The start-up problem is
basically one of raising the bed temperature, by external means, to the
point where coal is autogeneous while protecting the boiler tubes which are
in contact with the bed. The following alternatives have been considered
or practiced at various times.

Preheating the Fluidized Air Via Duct Burners

This technique suffers from the major flaw that the duct work and
the air distributor must be designed for high-temperature service for the
relatively infrequent start-ups. This technique was practiced by PER in the
period 1966 through 1968.

Premixing Gas and Air Beneath the Grate and Burning Above and in the Bed

This technique was used by BCURA in the 3-ft x 2-ft pressurized
combustor. Lurgi has also applied the technique safely in commercial apparatus.

---

* Nimonic 75 has the composition 64% Nu, 20% Cr, 0.2% ti, 1% Si, 5% Fe max.
** EPK 55 has the composition 43.6% Nu, 28.5% Cr, 20% Co, 2.3% Ti, 1.2% Al.

Firing Above the Bed or Back Firing

This method is used to start roasters, etc., in which a thick refractory lining can be used to store heat. It is used by CPC in the CPU-400 combustor. It would be useful in a fluidized-bed boiler in bringing the boiler to pressure and hence the tubes to the saturation temperatures.

Firing into the Bed

This technique is currently used by PER in the Alexandria facility. By premixing a certain amount of coal with the bed material in the fluidized-bed boiler before light-off by the propane burner, the time required to bring the unit to operating condition was reduced from more than 2 hours to about 30 minutes.

Certainly the start-up procedure needs to be carefully studied and optimized for time, cost, and considerations of safety. The techniques of horizontal firing also need to be developed as it appears best suited to the stacked-cell concept.

PER has reported their operating experience with FBM (Fluidized Boiler Modular) that when the bed was slumped it took 5 hours for the bed temperature to drop from 1550 F to 1000 F, but it took 5 minutes for the bed temperature to drop from 1480 F to 1000 F after a fuel failure. Further studies need to be directed to the problems surrounding the start-up procedure, such as the lowest temperature for autogeneous ignition of bituminous coal, reliable control signals for fuel failure, and tube damage during banking, etc. The light-off procedure for a multi-cell fluidized-bed boiler also needs to be developed.

A comprehensive analysis of applicable turn-down techniques for an atmospheric fluidized-bed boiler was reported by Westinghouse.[28] The various techniques considered include changing air- and coal-feed rate, adjusting excess air, control of bed level, flue gas recirculation, and defluidization. Experimental development work needs to be established to demonstrate and optimize the turn-down techniques.

To ensure the safe operation of the fluidized-bed boiler, the dynamic behavior of fluidized-bed boilers needs to be studied further. Investigation into the safety procedure in the commercial fluidized-bed practice appears to be helpful in establishing a practical safety procedure for operating the fluidized-bed boilers.

HEAT TRANSFER AND HEAT RELEASE RATES

Conventional water-tube boilers have convective heat transfer coefficients of 10-15 $Btu/hr-ft^2-F$.[49]   FBC data from PER indicate values ranging from 45 to 50 $Btu/hr-ft^2-F$.[49]   The heat transfer coefficients in Bureau of Mines[50] experiments ranged from 38 to 63 $Btu/hr-ft^2-F$. The Australian[15] data varied from 35 to 52 $Btu/hr-ft^2-F$. Also, CRE (NCB) data confirm that 80 to 90 $Btu/hr-ft^2-F$ can be obtained in FBC systems. BCURA's data in pressurized FBC are about 5 times the values obtainable in conventional boilers. Since the temperature driving force for both convective and radiant heat transfer in the fluidized-bed will be lower than for conventional boilers, the net heat transfer rate for the fluid-bed will be only about 2 to 3 greater than that in a conventional boiler. There are data indicating higher heat transfer rates[51] may be feasible, further experimentation in this area is necessary.

Skinner[11] reports that volumetric heat release rates in fluidized-bed combustion can be at least twice as great as those for a conventional chain grate-fired shell boiler. Although this type of boiler is too small to compare with a 1000 MW utility boiler, the general trend seems clear. The heat release rate in FBC will be about 100,000 $Btu/hr-ft^3$ if the volume of the free board is included. BCURA's[27] data confirm that bed heat release rates of 200,000 $Btu/hr-ft^3$ can be obtained. Further work to verify applicability of these high release rates to an operating boiler is necessary.

COST ANALYSIS

The above review has provided conclusive evidence regarding the technical feasibility of the FBC systems. However, the applicability and acceptance by the market will very definitely depend, among many other factors, on the economic advantages and viability of the systems. This section summarizes the cost data from various available sources.

The engineering cost of pressurized FBC boiler plant was studied by Westinghouse[29] and Combustion Systems, Ltd.[48] Shown in Table 3 are

Table 3 Cost Saving of Fluidized-Bed Boiler Plant as Compared to Conventional Practice[a]

System	Capital Cost Saving (%) WRL[b]	CS[c]	Operation Cost Saving (%) WRL[b]	CS[c]
Pressurized, with SO_2 control	12	21	2	10-14
Pressurized, without SO_2 control	23	21	16	12-16
Atmospheric, with SO_2 control	13	--	7	--

(a) Based on plant size of 600-650 MW.
(b) Estimates of Westinghouse Research Laboratory[29].
(c) Estimates of Combustion System, Ltd.[48].

the capital and operating cost saving of several FBC systems over the conventional power plant. These estimates indicate that FBC systems offer significant cost savings over the conventional plants.

As for the comparison of atmospheric FBC and pressurized FBC system, the Westinghouse study[29] favored the pressurized system; for a 600 MW plant, the pressurized system would give 11 percent in capital cost savings and 5 percent in operating cost savings over the atmospheric system.

For an atmospheric FBC industrial steam plant (250,000 lb/hr)[28], the capital cost for a once-through system was $7.40/lr/hr steam as compared to $7.60/lb/hr steam for a conventional coal-fired boiler with limestone scrubbing system. A capital cost estimate by PER[19] for a 300,000 lb/hr steam boiler plant was $4.50/lb/hr. However, Combustion System, Ltd.[48] reported a capital cost saving of 13-16 percent for 120-140 MW industrial FBC boiler plant as compared to conventional plant with or without SO_2 control.

The capital cost for a 600 MW pressurized FBC boiler plant by Westinghouse[29] in 1973 cost was 269 $/KW as compared to 350 $/KW for a conventional plant with Wellman Lord Scrubbing System. For a 50 MW adiabatic pressurized FBC combined cycle power plant, a preliminary cost study by CPC[33] indicates that the capital cost is 275 $/KW. For the MIUS, a capital cost estimated by Oak Ridge National Laboratory[43] was about 350-400 $/KW.

REFERENCES

(1)   Stratton, J.F.O., Power 68, 486 (September 1928).

(2)   Badische Anilin-u-Soda Fabrik Aktiengesellschaft. Brit. Pat. 768,656, 20th February, 1965 (Application date in Germany 15th September, 1953).

(3)   Badische Anilin-u-Soda Fabrik Aktiengesellschaft. Brit. Pat. 776,791, 12th June, 1957 (Application date in Germany 15th September, 1953).

(4)   Union Carbide Corp. Brit. Pat. 935,658, 4th September, 1963 (Application date in U.S.A., 30th December, 1959).

(5)   Combustion Engineering Inc. Brit. Pat. 784,595, Application date 22nd July, 1955 (Application date in U.S.A., 5th August, 1954).

(6)   Combustion Engineering Inc. Brit. Pat. 785,398, 30th October, 1957 (Application date in U.S.A., 16th November, 1954).

(7)   Standard Oil Development Corp., Delaware. U.S. Pat. 2,591,595, 1st April, 1952 (Application 29th September, 1949).

(8)   Oxley, J. H., Ph.D. Thesis, Carnegie Institute of Technology (1956).

(9)   Godel, A. and Cosar, P., AIChE Symposium Series 116, Vol 67, 210 (1971).

(10) Novotny, P., Sb. Prednasek 50 (Padesatemu) Vyroci Ustavu Vyzk. Vyuziti, Paliv, 104-11 (1972).

(11) Skinner, D. G., "The Fluidized Combustion of Coal", Mills & Boon Limited, London (1971).

(12) Robison, E. B., et al. "Development of Coal Fired Fluidized-Bed Boilers", OCR R&D Report No. 36, Vol. 1, II (1972).

(13) Coates, N. H., and Rice, R. L., "Proceedings of Second International Conference on Fluidized-Bed Combustion", sponsored by NAPCA, Houston Woods, Ohio (1970).

(14) Zielke, C. W., et al., "Sulfur Removal During Combustion of Solid Fuels in a Fluidized-Bed of Dolomite", Journal of Air Pollution Control Association, 20, 3 (1970).

(15) CSIRO, Journal of Fuel and Heat Technology, 15 (5), 11-13 (1968).

(16) Henschel, D. B., "Status of the Development of Fluidized-Bed Boilers", 1971 Industrial Coal Conference (October 1971).

(17) Carls, E. L., "Review of British Program on Fluidized-Bed Combustion", Report of the U.S. Team Visit to England, ANL/ES-CEN 1000 (1969).

(18) Robison, E. B., et al., "Characterization and Control of Gaseous
      Emissions from Coal-Fired Fluidized-Bed Boilers", PER Report to
      EPA (1970).

(19) Gordon, J. S., et al., "Study of the Characterization and Control of
      Air Pollutants from a Fluidized-Bed Boiler--The SO_2 Acceptor Process",
      PER Report to EPA (1972).

(20) Robison, E. B., et al., "Study of Characterization and Control of
      Air Pollutants from a Fluidized-Bed Combustion Unit--The Carbon-
      Burnup Cell", PER Report to EPA (1972).

(21) Pope, Evans & Robbins Incorporated, "Optimization of Limestone Utiliza-
      tion and Sulfur Capture in a Single Combustion Zone Fluidized-Bed
      Boiler," OCR R&D Report No. 88 (1974).

(22) Jonke, A. E., et al., "Reduction of Atmospheric Pollution by the
      Application of Fluidized-Bed Combustion", Annual Report ANL/ES-CEN
      1004 (1971).

(23) Jonke, A. E., et al., "Reduction of Atmospheric Pollution by the
      Application of Fluidized-Bed Combustion and Regeneration of Sulfur-
      Containing Additives", ANL to EPA Report, EPA 650/2-74-104 (1974).

(24) Skopp, W., et al., "A Regenerative Limestone Process for Fluidized-
      Bed Coal Combustion and Desulfurization", Final Report to EPA by
      Esso R&E (1971).

(25) Hoke, R. C., et al., "A Regenerative Limestone Process for Fluidized-
      Bed Coal Combustion and Desulfurization", EPA-650/2-74-001, Esso R&E
      Report to EPA (1974).

(26) National Coal Board, "Reduction of Atmospheric Pollution", Vol. I,
      II, and III to EPA-OAP, (1971).

(27) National Research Development Corporation, "Pressurized Fluidized-
      Bed Combustion", OCR R&D Report No. 85 (1974).

(28) Archer, D. H., et al., "Evaluation of the Fluidized-Bed Combustion
      Process", Vol. I, II, and III. A Report to EPA by Westinghouse
      Research Laboratories (1971).

(29) Keairns, D. L., et al., "Evaluation of the Fluidized-Bed Combustion
      Process", Vol. I, II, and III, Westinghouse Report to EPA (1973).

(30) Office of Coal Research, Press Release (July 3, 1973).

(31) Mesko, J. E., et al., "Multicell Fluidized-Bed Boiler Design, Con-
      struction, and Test Program", Pope, Evans, and Robbins Report to OCR,
      PB 236-254 (August 1974).

(32) Hoy, H. R., BCURA, private communication, (1974).

(33) Combustion Power Company, "Energy Conversion from Coal Utilizing
      CPU-400 Technology", OCR R&D Report No. 94 (September 1974).

(34) Janssen, K., Bergbau-forschung, private communication (1974).

(35) Reh, L., Lurgi Gesellschaft fur Chemie und Huttenwesen, private communication.

(36) Stephens, F. M., "The Fluidized-Bed Sulfate Roasting of Nonferrous Metals", Chem. Engr. Prog., 49 (9), 455 (1953).

(37) Copeland, G. G., and Hanway, J. E., Jr., "Fluidized-Bed Oxidation of Waste Liquors Resulting from the Digestion of Cellulosic Materials for Paper Making", U.S. Patent No. 3,309,262 (March 14, 1967).

(38) Smithson, G. R., Jr., and Hanway, J. E., Jr., "Process of Converting Sodium Sulfate to Sodium Sulfite, Particularly for Pulping Processes", U.S. Patent No. 3,397,957 (August 20, 1968).

(39) Locklin, D. W., Hazard, H. R., Bloom, S. G., and Nack, H., "Power Plant Utilization of Coal", A Battelle Energy Program Report, Columbus, Ohio (1974). 96 p.

(40) "Status of the Battelle/Union Carbide Coal Gasification Process Development Unit Installation", Corder, W. C., and Goldberger, W. M., Sixth Synthetic Pipeline Gas Symposium Proceedings, Chicago, Illinois, October 28-30, 1974, 21 pp, American Gas Association.

(41) Reh, L., "Fluidized-Bed Processing", Chem. Eng. Prog., 67 (2), 58-63 (1971).

(42) Yerashalmi, J., McIver, A. E., and Squires, A. M., "The Fast Fluidized Bed", GVC/AIChE Joint Meeting, Munich, Germany (September 17-20, 1974).

(43) Fraas, A. P., "Concept Preliminary Evaluation Small Coal Burning Gas Turbine for Modular Integrated Utility System", OCR R&D Report No. 96 (September 1974).

(44) Hoke, R. C., et al., "Combustion and Desulfurization of Coal in a Fluidized-Bed of Limestone", Combustion (January 1975).

(45) Shaw, J. T., and Thomas, A. C., 7th International Conference on Coal Science, Progue (June 10-14, 1968).

(46) Jonke, A. A., et al., "Reduction of Atmospheric Pollution by the Application of Fluidized-Bed Combustion", Argonne National Laboratory Report, ANL/ES-CEN-1001 (1969).

(47) O'Neill, E. P., et al., "Kinetic Studies Related to the Use of Limestone and Dolomite as Sulfur Removal Agents in the Fuel Processing", 3rd International Conference on FBC, EPA (1972).

(48) Locke, B., "Fluidized Bed Combustion for Advanced Power Generation with Minimal Atmospheric Pollution", paper presented at Achema 1973, June 24, Frankfurt, Germany.

(49) Bishop, J. W., Robinson, E. B., Ehrlich, S., Jain, L. K., and Chen,
      P. M., "Status of the Direct Contact Heat Transferring Fluidized Bed
      Boiler", Paper 68-WA/FU-4, presented at Winter Annual Meeting, ASME,
      New York, NY (December 1-5, 1968).

(50) Rice, R. L., and Coates, N. H., "Fluidized Bed Combustion--Suitability
      of Coals and Bed Materials", Power Engineering (December 1971)

(51) Botterill, J.S.M., et al., Proceedings of International Symposium on
      Fluidization, Drinkenburg A.A.H., Eindhoven (1967).

THE CHEMICALLY ACTIVE FLUIDIZED BED GASIFIER

G. MOSS

INTRODUCTION

The Chemically Active Fluidised Bed Gasifier is a simple piece of process equipment which partially oxidises and desulphurises heavy fuel oil. As shown in figure 1 it comprises two fluidised bed reactors, a relatively large gasifier and a smaller regenerator. Means are provided for the continuous exchange of bed material between the two reactors at controlled rates. Sulphur is fixed by the bed material in the gasifier as calcium sulphide, at temperatures in the region of 900°C and is released within the regenerator as SO_2 at about 1050°C.

Under the sponsorship of the Environmental Protection Agency of the United States Government a pilot scale unit has been installed at the Esso Research Centre at Abingdon and has been operated under desulphurising conditions for about 2700 hours. The purpose of this paper is to summarise the very considerable amount of technical information which the work has produced and to indicate how this information may influence the design of a larger scale plant.

THE PILOT SCALE GASIFIER

A detailed description of the original pilot scale gasifier has been published elsewhere(1). As can be seen in figure 2 it was of somewhat unusual construction being in the form of an encased monolithic block of

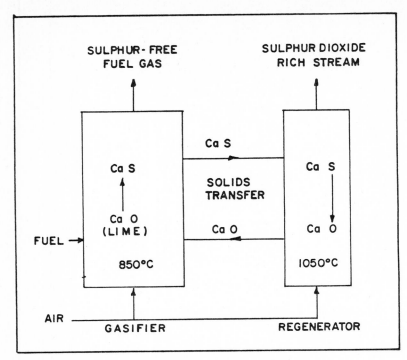

Fig. 1 Chemically active fluid bed gasifier.

refractory concrete. It was capable of handling a throughput of about 50

gallons an hour of 2.5% by weight sulphur residual fuel oil and could reduce

the sulphur content of this fuel by about 80-90% over a range of operating

conditions. The hot gas was used to fire a standard fire tube pressurised

hot water boiler and a gas fired pilot light was used for ignition.

The SO_2 content of the boiler flue gas was generally in the range 150-250 ppm

at an O_2 concentration of approximately 2% by volume. The tail gas from the

regenerator normally contained 6-8% by volume of SO_2 and, after some initial

difficulties, sulphur balances within $\pm 5\%$ were obtained over the course of

test runs of some hundreds of hours duration.

A flow plan for the unit in its original configuration is shown in

figure 3. It will be seen that flue gas recycle to the gasifier air blowers

was provided for temperature control, whilst the gasifier cyclones drained

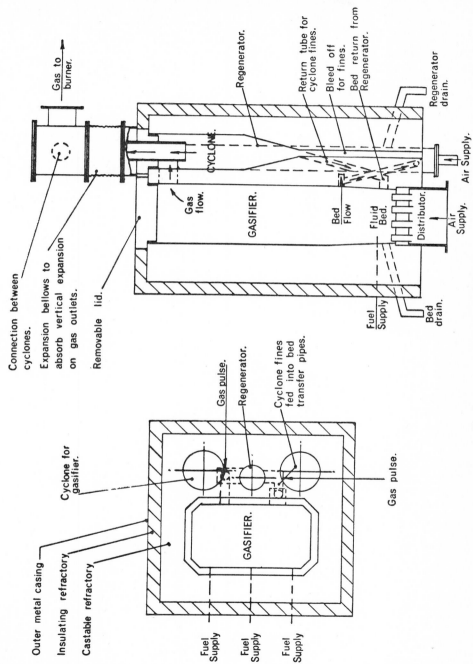

Gas to burner.

Connection between cyclones.

Expansion bellows to absorb vertical expansion on gas outlets.

Removable lid.

Regenerator.

Return tube for cyclone fines.

Bleed off for fines.

Bed return from Regenerator.

Regenerator drain.

CYCLONE.

Gas flow.

GASIFIER.

Bed Flow.

Fluid Bed.

Distributor.

Air Supply.

Air Supply.

Fuel Supply.

Bed drain.

Cyclone for gasifier.

Outer metal casing

Insulating refractory

Castable refractory

Gas pulse.

Regenerator.

Cyclone fines fed into bed transfer pipes.

GASIFIER.

Gas pulse.

Fuel Supply

Fuel Supply

Fuel Supply

Fig. 2 Layout of continuous gasifier unit.

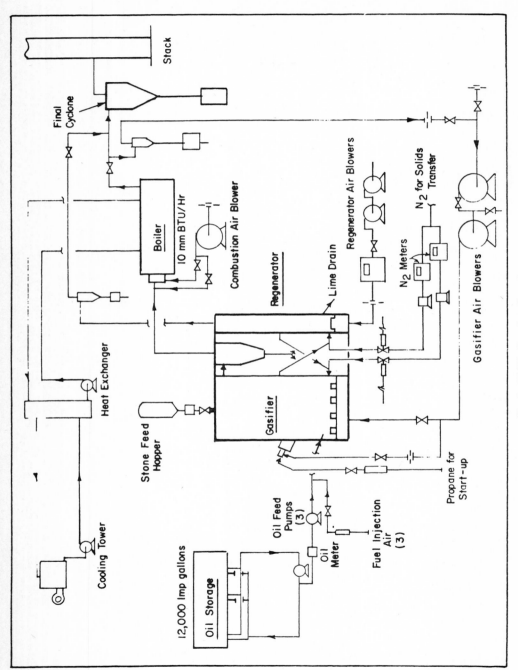

Fig. 3 CAFB pilot plant flow plan.

into the bed transfer ducts. Provision was made for metering stone into
the unit on a continuous basis via a lock hopper and a weighed dispensing
hopper which supplied stone to a vibrating feeder. Since the regenerator
operates at a temperature about 200°C higher than that of the gasifier it
was possible to control its temperature by exchanging bed material between
the two reactors at a rate higher than that required to react with the
fluidising air in the regenerator. The flow of bed material was controlled
by intermittent pulses of inert gas in the transfer ducts, the pulse rate
being varied according to the regenerator temperature.

     When after the initial runs it was decided to deepen the gasifier bed
it was found that this flooded the cyclones and rendered them inoperative.
In order to overcome this problem the cyclones were subsequently drained
externally and the fines were re-injected into the unit with the aid of a
commercial powder transport system. A number of modifications were subsequently
made to the fines return system during the course of the programme, but whilst
these had a significant effect on the operability of the pilot scale unit
they were not particularly relevant to the design of a full scale gasifier.

PROGRAMME OBJECTIVES

     The information produced during the pilot plant programme was copious,
complex and diverse. Three detailed reports have been published (2)(3)(4)
and a fourth is in preparation. Only a brief summary can be presented in
this paper and in order that this may be done in a logical fashion it is
necessary to relate it to the objectives of the programme. These were as
follows:

1.   To identify and solve any operational problems and to prove   the
     operability of the process.

2.   To determine the optimum operating conditions.

3.   To test materials of construction.

INFORMATION CONCERNING OPERABILITY

The operability of a process is determined by the constraints which it imposes on its operators in terms of supervision, maintenance and availability. In order to be considered operable, any process applied to a power station boiler must not reduce the rate at which the boiler can respond to changes in load and must not increase outage time. It must be inherently stable in operation, requiring minimal supervision, and it must not introduce any new safety hazards.

The gasifier is effectively the first stage of a two-stage combustion system. In practice its response to changes in fuel rate was virtually instantaneous, the only appreciable lag, a matter of a few seconds, was due to the residence time of the hot gas in the reactor and in the piping connecting the reactor to the burner. The unit was also found to be inherently safe. If the power was cut off then the fuel supply was immediately interrupted but the blower ran on, due to its inertia, long enough to purge the system. It was found that in normal operation the bed material contained about 4% by weight of sulphur as calcium sulphide and around 0.4% by weight of carbon. Once the fuel was cut off, the gas leaving the gasifier reactor was virtually inert and if the blower was then stopped the bed could be slumped in the sulphided condition. Provided that steps were then taken to prevent air from entering the slumped bed, the gasifier could be left in the slumped state for several hours but could be put back on stream virtually instantaneously simply by re-starting the air/blower and the fuel pumps. The factor limiting the duration of a shut-down period in the sulphided condition was the rate of

heat loss from the bed. Although the issue was not put to the test, it was
considered to be unwise to let the bed temperature fall below about 750°C
before re-starting. The shut-down period could be extended by re-fluidising
the bed in the absence of fuel and bringing it up to temperature periodically
by oxidising the sulphur. If this was taken to the limit then intermittent
oil firing with excess air could be used in the sulphated condition.
When however a sulphated bed was switched to gasifying conditions the efficiency
of desulphurisation was impaired for the first few minutes of operation until
the sulphate was reduced to sulphide.

Three operational problems became apparent during the first attempt
to operate the unit for an extended period. These were:

1.   The formation of lime/coke deposits   (approx. 80% C) in the hot gas
     ducting.

2.   Poor containment of fines.

3.   A tendency to form deposits of cemented bed material within the
     regenerator.

The deposits in the gas ducting were most severe in areas of high
turbulence. In the first instance deposits tended to choke the inlet ports
to the cyclones which were originally square edged. This problem was
greatly alleviated by shaping the inlet ports to give a smoother flow
transition and similar action was taken elsewhere in the hot gas ducting
where there were sharp bends. The deposits were detected by the increased
pressure drop downstream of the gasifier bed and could be removed by a
controlled burn-out procedure carried out above the slumped gasifier bed.
When this was done the pressure drop could be restored to its initial value.
The rate of deposit formation was found to depend to some extent on the rate
at which bed material   entered the cyclones, less coke being laid down
as the fines circulation rate was increased. The longest period of

continuous operation between burn-outs was of 211 hours duration and the burn-out procedure occupied about 4 hours.

The poor containment of fines was due in part to the coking of the cyclones and in part to malfunctions of the cyclone drainage system. When the cyclones were clean and the fines return system was operating correctly the fines were returned quite effectively. During one period of 50 hours duration the stack losses amounted to approximately 33 mg/nm^3. A more normal stone loss rate however was in the region of 0.5-1.0% by weight of fuel and efforts to improve the efficiency and reliability of the stone containment system are still in progress.

The problem of deposits in the regenerator was solved by dropping the level of the regenerator distributor so that the sulphided stone entering the regenerator from the gasifier did not meet the fluidising air until it had dispersed into the rest of the bed material. The effect of this was to raise the temperature at the bottom of the regenerator bed and to improve the selectivity of the system for the regeneration reaction producing lime and SO_2 rather than for the sulphation reaction. It had been observed previously (5) that a transient liquid phase occurs during the conversion of CaS to $CaSO_4$ and that this can cement particles together. If air was allowed to enter the gasifier bed in the slumped condition   agglomerates were formed, but once inert gas was employed for all bleeds the problem disappeared.

It may reasonably be concluded that provided that the gasifier cyclones and hot gas ducting can be kept clean then a large scale unit will present no major operational problems. A normal mid-merit plant is unlikely to operate continuously for more than a few days at a time and the ducts can be burnt out during each shut-down. In the case of high load factor
      burn-back burner
applications a system has been devised which allows individual ducts and

cyclones to be burnt out whilst the gasifier is on load. This system has
been tested on a small scale.

OPTIMUM OPERATING CONDITIONS

It was an apparently simple objective of the test programme to determine
the effects of variations in process conditions on the desulphurising
performance of the gasifier.

Variables of interest were:

1. Gasifier air/fuel ratio

2. Gasifier temperature.

3. Gasifier superficial gas velocity

4. Sulphur content of the bed material

5. Bed depth

6. Particle size of bed material

7. Type of bed material

8. Stone replacement rate

9. Regenerator air rate

10. Regenerator temperature

In practice it was much easier to establish which variables did not
have an appreciable effect on desulphurising performance than to demonstrate
the precise effects of those variables which did appear to be relevant.
The reason for this was the very flat desulphurising performance of the
unit, which remained within the 70-85% bracket over a wide range of
operating conditions. It was in fact not easy to obtain either a very
good result or a very bad one, though desulphurisation efficiencies
higher than 90% were obtained on a few occasions.

A number of the variables listed above are interrelated, the first

three for example,because a change in air rate at a given fuel throughput
will alter the air/fuel ratio, the bed temperature and the superficial gas
velocity. It was possible to maintain a constant superficial gas velocity
whilst varying both the air/fuel ratio and the bed temperature by using flue
gas recycle, but it was difficult to detect an appreciable effect for any
of these variables over quite wide ranges. Desulphurisation results better
than 80% were observed with bed temperatures in the range 870-990°C,
though there were indications that a combination of high bed temperature
and low air/fuel ratio is advantageous and desulphurisation efficiencies
exceeding 90% were recorded under these conditions. Superficial gas velocity
seemed to have very little effect within the range of 1.1 - 2.0 metres/sec.
Too low an air/fuel ratio caused carbon to accumulate on the bed material to
an extent which would ultimately de-activate the regenerator because all
of the oxygen in its fluidising air was used to oxidise carbon. Oddly
enough, in these circumstances the gasifier would desulphurise quite adequately
until the stone contained 8% by weight of sulphur, whereas under leaner
conditions, air/fuel ratios in the range 20-25% of stoichiometric, there was
evidence of a relationship between stone sulphur content and desulphurising
efficiency, there being a fall in desulphurising efficiency of approximately
10 percentage points when the sulphur content of the stone rose from
4% to 6% by weight.

Increasing the bed depth appears to be advantageous but there is
evidence which suggests that the effect is indirect and results from a
higher concentration of fines in the freeboard above the bed. Similar
effects can be obtained by increasing the superficial gas velocity in some
circumstances. Bed depths greater than 1 metre are unlikely to show an
appreciable advantage and economic penalties are incurred if the total
pressure drop exceeds the capacity of a single fan. Desulphurisation

efficiencies exceeding 80% can be obtained with beds 50 cms deep.

The particle size of the bed material has a relatively minor effect on performance, no difference was observed for example when a stone with a size range of 300-2000 microns was substituted for one having a size range of 600 - 3200 microns. The type of stone employed does however have an effect on the operability of the gasifier since calcites and cement rocks for example decrepitate badly, producing large quantities of fines, whereas fine grained pure sedimentary limestones do not. It was soon established that the rate at which stone was replaced did not have an appreciable effect on desulphurising efficiency at values greater than the Ca/S stoichiometric ratio. More recent work has shown that stone replacement rates as low as half the Ca/S molar rate can be used without ill effects.

The regenerator air rate is related to the sulphur throughput of the gasifier. Too small an air rate will result in a build-up of sulphur in the bed material and too high an air rate will lower the sulphur content of the stone until oxygen appears in the regenerator tail gas and the selectivity for the regenerating reaction falls to a level which restores the balance, the excess oxygen forming calcium sulphate. It should be possible in practice to control the regenerator air rate to give trace level O_2. Provided that the regenerator temperature exceeds 1050°C and the air rate is not excessive then the concentration of SO_2 in the tail gas is found to be in the range 6-8% by volume. Increasing the regenerator temperature does not have much effect up to 1150°C ,though since the amount of stone passing through the regenerator diminishes as the temperature rises, the difference between the sulphur content of the stone entering and leaving the regenerator is increased in proportion. It follows that increasing the temperature of the regenerator will tend to increase the sulphur content of the gasifier bed material.

Since it is difficult to bring about major variations in the
desulphurising performance of the gasifier it follows that the device is
inherently stable and is easily controlled.

Desulphurisation efficiencies exceeding 90% are possible provided that
attention is paid to the environment within the freeboard volume above
the gasifier fluidised bed.

MATERIALS OF CONSTRUCTION

The gasifier monolith was cast in G.R. Stein Durax C1600 refractory
concrete. This contains approximately 50% Al_2O_3, 42% SiO_2, 5% CaO and
traces of Fe_2O_3 and MgO. The maximum operating temperature for this
material is 1600°C and it melts at 1700°C. This refractory proved to be
reasonably durable, the walls of the gasifier and of the regenerator
remained in good condition throughout the test programme, and though a few
cracks did develop these did not have any effect on the operability on
effectiveness of the unit. The internal surfaces of the gasifier cyclones
however suffered considerable erosion damage and a better quality refractory
is required for this duty. Durax C 1850 appears to give satisfactory
service in this location. This is a high purity high alumina concrete
containing 96.8% Al_2O_3, 2.7% CaO and traces of Fe_2O_3, S, O_2 and alkalies.
Its maximum continuous service temperature is 1800°C.

The most difficult constructional problem was posed by the outlet
tubes of the gasifier cyclones. The wall thickness that could be accommodated
was insufficient to allow normal refractories to be used, but tubes of self-
bonded silicon carbide, supplied by the Springfield laboratories of the UK
Atomic Energy Authority, were highly satisfactory.. Although this type of

material could also have been used for thermocouple sheaths it was found convenient to use instead an iron/chromium alloy supplied by the Land Pyrometer Co. under the trade name Firebird Blue. Silicon nitride was tested but became very brittle after exposure to gasifying conditions.

It would appear that the selection of constructional materials does not present a serious problem and it is likely that a large scale unit may be constructed using normal commercial refractories.

REACTOR CONFIGURATION FOR LARGE SCALE PLANTS

In addition to the information which has already been given concerning the operability of the process, operating conditions and materials of construction, the designer of a large scale plant will also want to know what shape the reactor should be, how many fuel injectors should be used, the size and number of air distributor nozzles and so on. The answers to some of these questions may be obtained from previous publications (1) but experience gained during the test programme has enabled answers to be formulated for most of the rest. For example, although good results were obtainable with the pilot scale gasifier when one fuel injector was used, this had to be situated centrally, therefore a large scale unit will probably require one injector for every five square feet of bed. Horizontal fuel injectors are preferred since it is possible to either withdraw them or to rod them out should they become blocked. In the case of a large reactor some of these nozzles would need to penetrate several feet into the bed and this raises questions concerning heat transfer and support. Suitable arrangements have been devised, but an alternative possibility is to use a relatively narrow annular reactor as shown in figure 4 and to insert fuel injectors through both containing walls. This type of construction

G. MOSS

allows horizontal staging to be practised on the flowing bed. What this means is that the carbon content of the bulk of the gasifier bed can be made higher than in the section adjacent to the transfer duct leading to the regenerator, simply by adjusting the fuel distribution. In a similar fashion the temperature and calcium sulphate content of the stone in the regenerator can be reduced prior to its transfer back to the gasifier by injecting some fuel at an appropriate point along the boundary of the regenerator bed.

The possibility of using burn-back gas burner nozzles has already been mentioned whilst considerable experience has been gained of the techniques of cyclone drainage and fines injection. Sufficient information is available to enable a large scale demonstration unit to be built with a high degree of confidence in a successful outcome.

THE REDUCTION OF SO_2

The reduction of SO_2 to elemental sulphur did not form part of the Abingdon programme. Fortunately, however a suitable process has been developed by the Foster Wheeler Corporation (6). This process which has been given the trade name "Resox" reacts SO_2 with coal in order to produce mostly elemental sulphur and CO_2. Sulphur conversions of 90% have been claimed and a 20 MWe demonstration plant is installed at Chattahoocha in Florida.

REFERENCES

1.   Moss, G. & Tisdall, D.E. "The Design, Construction and Operation of the Abingdon Fluidised Bed Gasifier". Proceedings of the Third International Conference on Fluidised Bed Combustion. EPA-650/2-73-053. Dec. 1973.

2.   E.P.A. Study of Chemically Active Fluid Bed Gasifier for Reduction of Sulphur Oxide Emissions. Final Report O.A.P. Contract CPA 70-46,Esso Research Centre, Abingdon, Oxon June 1972.

3.   E.P.A. Chemically Active Fluid Bed Process for Sulphur Removal During
     Gasification of Heavy Fuel Oil-Second Phase. E.P.A. 650/2-73-039, Nov.1973.

4.   E.P.A. Chemically Active Fluid Bed Process for Sulphur Removal During
     Gasification of Heavy Fuel Oil - Second Phase EPA 650/2-74-109 Nov. 1974.

5.   Curran, G.P. et al United States Department of the Interior, O.C.R. Research
     and Development Report No.16, Interim Report No.3 on CO_2 Acceptor Process.

6.   Bischoff, W.F. and Steiner, P. "Coal Converts SO_2 to S"
     Chemical Engineering Jan. 6th 1975.

DEFLUIDIZATION CHARACTERISTICS
OF STICKY OR AGGLOMERATING BEDS

M. J. GLUCKMAN, J. YERUSHALMI, and A. M. SQUIRES

INTRODUCTION

Industry has compiled a catalogue of cases where fluidized beds were operated under conditions in which the bed material tended to agglomerate. The tendency to agglomerate is a direct result of stickiness of the bed material. This stickiness may be an inherent property of the bed particles themselves coming into play at some temperature level, or the stickiness may be due to a liquid deposited upon the bed material.

Agglomeration Due to Sintering

The advantages of fluid-bed processing have prompted, over the past two decades, a number of studies on direct reduction of iron-ore fines in a bed fluidized by hydrogen or by other reducing gases (1-6). Kinetic studies have shown that the rate of reduction is inversely proportional to the size of the particles, so that use of fines would afford high rates of reduction to metallic iron. Reduction rates are enhanced by higher temperatures as well. But small particles and higher temperatures often proved incompatible with stable fluidization of the bed. Stickiness of the fine ore or the reduced iron has been reported at temperatures as low as 700 — 900°F.

Development of the H-Iron process (fluid-bed reduction of ore fines with hydrogen) at Hydrocarbon Research, Inc., revealed that larger particle size, lower temperature, higher velocity, and higher gange content tended to favor stable fluidization (7). The impression was gained that a sharp boundary existed in the spece defined by these variables between a region in which stable fluidization can be reliably achieved and a region in which stable fluidization cannot be maintained.

For example, when a given iron powder was brought to active fluidization at a certain temperature, and the velocity was then reduced, a point was reached at which the bed "dropped dead", although the velocity at that point was above the theoretical minimum for fluidization of the powder. When the velocity was raised again, the bed recovered provided it had not been allowed to remain in the slumped condition a long time. When the experiment was repeated at a higher temperature, the bed slumped at a correspondingly higher velocity. The temperature-velocity pairs marking the boundary between stable fluidization and the slumped condition appeared precise for a given powder. When the bed was in the slumped condition, the gas tended to pass upward through the bed in well-defined chimneys or tracks, termed "rat-holes" at Hydrocarbon Research.

A more novel approach to the fluid-bed reduction of iron ores was studied on the bench scale at Battelle Memorial Institute (8). A bed of coarse, spherical pellets of reduced iron, typically -6 +8 mesh, is fluidized by hydrogen at velocities approaching 35 ft/sec and at room temperatures in the range of 1400-1600F. The coarse size of the bed material and the high fluidizing velocities ensure stable and smooth fluidization in a temperature range in which stickiness would have plunged a bed of fine iron into catastrophic agglomeration. The feed, in the form of pulverulent iron ore, -325 mesh, is injected into the fluid bed. Some of the ore particles are reduced having been captured upon the surface of the bed particles; others are reduced in the gas phase and the resulting tiny

particles of iron deposit on the bed material. Capture rates
of 75% have been achieved in beds only 6 inches deep operating
at 1550 to 1600F, and capture rates over 95% were recorded in
beds 3 ft. deep at 1500F.

Thus the agglomerating tendency of powdered iron, which
plagued the H-Iron development, was exploited by Battelle.
The stickiness of iron in the range of 1400-1600F permits
efficient capture of the fines, while the bed material is
kept in stable fluidization through the use of coarse particles
and associated high fluidizing velocities.

Fluidized beds gasifying coal have generally operated
below a temperature of about 1900F in order to avoid formation
of ash clinkers. It appears that the ash matter of most coals
begins to exhibit stickiness at this temperature and occasionally at lower temperatures.

In first attempts at Hydrocarbon Research to gasify
anthracite "silt", the fine wastes from anthracite mining and
cleaning, clinkers formed under circumstances which strongly
implied defluidization behavior very similar to that described
earlier for iron (9). Clinkers began to form at temperatures
as low as 1700F. Defluidization studies were subsequently
undertaken on the ash, and as in the case of iron, a sharp
boundary between the region of stable fluidization and the
slumped condition was observed. It should be noted that
boundary temperatures were below 2000F although the initial
deformation temperature of the ash was 2650F.

Raising the velocity was recognized as a cure to the
clinkering problem. This was done and successful gasification
trials were conducted in a large pilot unit (10).

Fluid bed gasification of coal at temperatures below
1900F has generally been attended by several disadvantages
(14). For example, gasification by air is limited to coals
of high reactivity, such as lignites. Also, the carbon level

of the ash purge is appreciable owing to the complete mixing
of solids and the need to maintain a sufficient carbon inventory
in the gaisification bed. In the absence of subsequent utilization, this purge represents serious carbon loss.

The late A. Godel obviated these and other difficulties
in his Ignifluid boiler, in which the agglomerating tendency
of the ash is exploited. The Ignifluid gasifies coal with
air and burns the resulting fuel gas to raise steam (11-15).

The Ignifluid operates at temperatures around 2200 to
2550F. At these temperatures the ash matter of all coals is
sticky, and one might expect that a catastrophically massive
clinker would form, but this does not happen. The Ignifluid
feeds upon coarse coal and operates at high fluidizing velocities, 30-50 ft/sec, typically. Godel discovered that under
these conditions, small ash agglomerates form throughout the
bed and remain fluidized, interspersed in particles of coke.
These ash agglomerates grow and finally reach the grate which
carries them to the ash pit.

The ash agglomerating gasification process of Union Carbide/
Battelle (16,17), the Pope, Evans and Robbins carbon burn-up
cell (18), and the Battelle fluidized bed combustor (19) all
exploit the stickiness of coal ash.

Agglomeration Due to the Presence of a Sticky Liquid

Attempts to carbonize a caking coal in a fluid bed have
mostly been beset by difficulties caused by uncontrolled formation of large agglomerates. Much of the U.S. Eastern coal
is caking, but under pressure and in an environment rich in
hydrogen, even Western coals may display caking tendencies.
To circumvent coal's agglomerating tendency, resort has often
been made to awkward dodges such as subjecting the coal to
pretreatment to reduce caking properties, carbonizing the coal

in stages operating at progressively higher temperatures, carbonizing the coal in dilute free fall region, or in the presence of large excess of recycled char. In general these dodges are expensive both in respect to recovery of valuable gaseous and liquid products of carbonization, as well as in respect to the additional hardware that is needed.

To exploit coal's caking propensity, a team at our institute has proposed a coke accreting fluid bed for flash hydrogenation of coal. Finely divided coal would be supplied to a bed of coke beads, about 1/4 inches in size, that is fluidized with hydrogen at around 1450F and at elevated pressure. The coal would be heated almost instantaneously, it would melt, and would react promptly to evolve vapor products and leave a sticky, semi-fluid residue. The latter would adhere to a coke pellet, forming a smear. Further polymerization and coking reactions would transform the sticky smear into dry coke. Studies at the Bureau of Mines (20) on carbonization of caking coals in a free fall dilute-phase reactor indicate that the coking reaction occurs in less than one second. Since heat transfer rates in the coke accreting beds are likely to be higher than those in a free fall dilute-phase reactor, faster coking reactions might be expected. Nonetheless, the coke pellets in the bed will at all times be coated with a sticky matter in an amount depending on the coal feed rate. Accordingly, apart from stoichiometric considerations, the fluidizing velocity of the hydrogen and the mean size of the coke pellets in the beds will have to be chosen to ensure operation in a region of stable fluidization.

The Fuller Company (21,22) carried the development of an accreting bed process for cement to the 500 bbl/day scale. A carefully sized bed of ordinary clinker is used in the startup of the Fuller process, but the bed quickly becomes altered to the bead form as the original startup material is gradually

discharged along with the product. Finely divided raw materials
for cement are introduced into the bottom of the bed. The raw
materials are heated almost instantaneously to the temperature
of the bed, and they quickly react to form cement. This adheres
to the beads of cement that make up the bed, which accordingly
grow in size. The basis of the process is the fact that intermediate products formed by the reacting solid materials are
molten. Blanks and Kennedy (23) state that a liquid phase
begins to form in a conventional cement-making rotary kiln as
the solids pass the point at which the temperature is about
2300F. In Fuller's process, which operates a little below
2400F, the raw materials apparently form a melt, which produces
smears upon beads that happen to be in the vicinity of the inlet
uniform, when the rate is viewed over several minutes of time.

Common to all accreting bed processes is the presence in
the bed of coarse beads bearing a burden of sticky matter.
The degree of stickiness in such beds obviously depends on
the amount and nature of the sticky matter. In the absence
of information on the defluidization limits of such systems,
it is hard to determine whether processes such as Fuller's
have approached their optimal operating capacities.

The presence of bed stickiness places operating limits
beyond which it is not safe to go. These limits may depend
in a complex way on many variables, i.e., particle size, size
distribution and composition, fluidizing velocity, temperature,
pressure, bed geometry, etc. But however complex the structure
of these limits is, it is the major thesis of this report that
these limits exist in sharp and reproducible form, and that
they are governed by precise rules. These limits and the
principles that underlie them have been virtually unexplored,
and the absence of knowledge in this area is reflected in the
measures adopted by industry to counter troubles attending

operation of fluidized beds under conditions where the
particles are sticky.

Under a second and a more novel approach, the agglomerating propensity is exploited rather than dodged. We strongly
advocate this approach. The very stickiness that many have
tried to avoid can be exploited to good advantage. This approach
has seen increasing application and would have undoubtedly become
more common a technique if not for the lack of knowledge on the
behavior and the operating limits of beds of sticky particles.
We have accordingly undertaken a study whose aims are

- To demonstrate that the defluidization of beds of sticky
  particles is a well-ordered phenomenon obeying precise
  rules, and to
- Elucidate underlying features of the defluidization
  phenomenon.

DEFLUIDIZATION EXPERIMENTS AT HIGH TEMPERATURE

Experimental Equipment

Figure 1 illustrates the directly heated equipment used
for the high temperature defluidization study. High temperature gas is generated in a furnace by the combustion of methane
in air (providing the freedom to control the nature of the
fluidizing gas, i.e., oxidizing or reducing). This gas is
passed through a conical distributor into a two inch diameter
quartz glass or stainless steel fluidized bed. Pressure and
temperature probes enter the bed from the top and are used to
record the condition of the bed via a two-pen recorder. This
equipment is capable of being heated to 2000F. For lower
temperature experiments (polymer particles) a two inch diameter
bed heated indirectly by electric globars was employed.

A defluidization experiment is conducted as follows: The

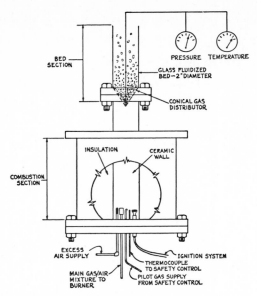

Fig. 1 Details of apparatus for studying defluidization limits of a directly heated fluidized bed.

particles to be studied are charged to the cold reactor and are
fluidized with air. Methane is added to the combustion chamber
and ignition takes place. Air and methane rates are set to
obtain the desired superficial velocity and gas atmosphere,
and the bed temperature is allowed to increase. As the temperature rises, the pressure drop through the bed remains substantially constant. As the defluidization temperature is approached,
the bed takes on a sluggish appearance and generally, within a
period of approximately five seconds the bed defluidizes.

Defluidization is a sharp phenomenon - the bed is well
fluidized one minute and dead the next. As the bed defluidizes
the particles become loosely stuck together, and the fluidizing
gas blows a hole through the mass of particles to escape. As
the hole appears the bed pressure drop decreases dramatically
as shown on the pressure drop plot in Figure 2. The exact
temperature of bed defluidization is determined by the simple
construction shown in Figure 2. The rapid rate at which the

bed pressure drop decreases is a good indication of the "sharpness" of the defluidization.

We have conducted defluidization trials on a number of different particle systems:

- Copper shot
- Polyethylene beads
- Polypropylene particles
- Polyethylene terephthalate (Dacron)
- Glass spheres

Experimental Results

Experiments were initiated with the copper particles to investigate agglomeration due to solid state sintering. A "pure" inorganic substance was chosen for these studies as it would exhibit a sharp and predictable melting point. As long as the melting point temperature was not reached in the bed we could feel confident that the melting process had not commenced, ensuring the absence of a liquid phase. Although

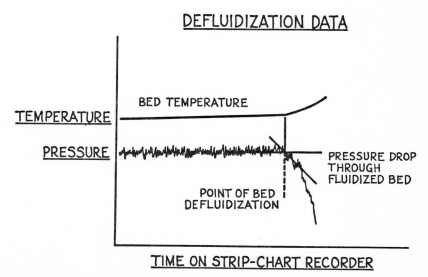

Fig. 2 Plot of bed pressure drop and temperature vs. time for a defluidization experiment.

pure copper particles were used, all of the defluidization
experiments were conducted in an oxidizing atmosphere leading
us to believe that the particles would be coated with a layer
of cupric oxide. The melting point of cupric oxide is 1879F
and our defluidization experiments were conducted over the
temperature range 1435F to 1880F, leading us to believe that
the copper particles had not melted. This was confirmed by
visual observation and photomicrographs to be shown later.

As the main thrust of our defluidization studies is
centered around coal ash, and as historical evidence points
to the fact that ash agglomeration occurs in all likelihood
by a mechanism of sintering in the presence of a liquid phase,
we decided to choose particles for our future studies that
exhibited high temperature behavior similar to coal ash. This
requirement indicated that materials with glass-like behavior
should be examined. We chose polymeric materials for our low
temperature defluidization studies and silicate glasses for
higher temperature investigations.

The defluidization results for two sizes of copper shot
are shown in Figures 3 and 4. The data for the larger copper
particles demonstrate a linear relationship between temperature and velocity as anticipated. A similar relationship for
the smaller particles indicates that smaller particles will
defluidize at lower velocities than larger particles. For
these smaller particles, defluidization was observed at a
temperature as low as 1435F.

Defluidization results for the polymer particles investigated are shown in Figures 5, 6, and 7. These results for
polyethylene beads, polypropylene beads, and dacron particles
all exhibit the same type of relationship between temperature
and defluidization velocity as the copper shot.

Finally, defluidization data for two sizes of low tem-

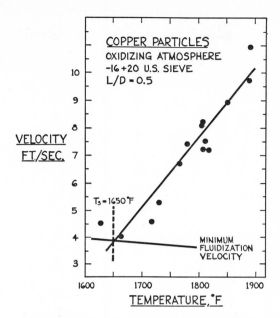

Fig. 3   Defluidization limits for − 16 + 20 sieve copper shot.

perature silicate glass spheres are presented in Figures 8 and 9. It is evident once again that the defluidization of glass particles exhibits the same velocity-temperature relationship as all of the other particles tested. Results of bed depth studied using glass spheres are shown in Figure 10.

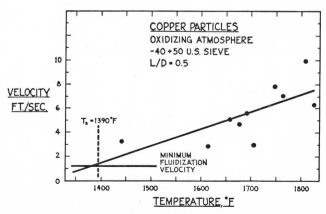

Fig. 4   Defluidization limits for − 40 + 50 sieve copper shot.

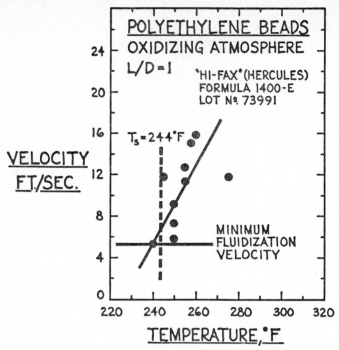

Fig. 5 Defluidization limits for polyethylene beads.

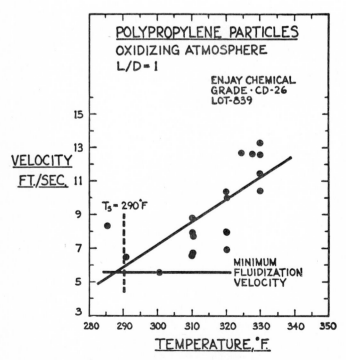

Fig. 6 Defluidization limits for polypropylene beads.

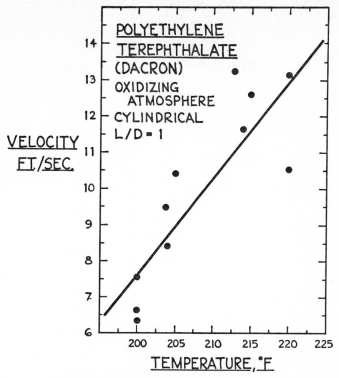

Fig. 7 Defluidization limits for Dacron particles.

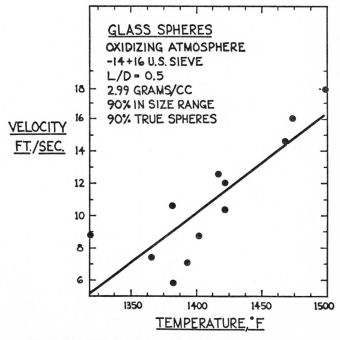

Fig. 8 Defluidization limits for − 14 + 16 sieve glass spheres.

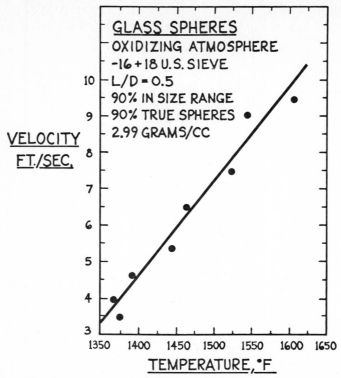

Fig. 9   Defluidization limits for − 16 + 18 sieve glass spheres.

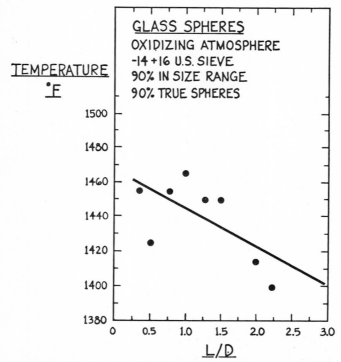

Fig. 10   Effect of bed depth on defluidization of − 14 + 16 sieve glass spheres.

All of the results plotted in Figures 3 through 10 were
correlated by linear regression analysis. The velocity-
temperature data were fitted by the following linear relationship:

$$\text{Velocity} = A\left(\frac{\text{Temperature}\,^{\circ}F}{1000}\right) + B$$

The bed depth studies were fitted by the following relationship:

$$\frac{\text{Temperature}\,^{\circ}F}{1000} = A\left(\frac{L}{D}\right) + B$$

$$\text{where,}\quad \frac{L}{D} = \frac{\text{Bed Height}}{\text{Bed Diameter}}$$

Table 1 lists the coefficients A and B as well as regression
coefficients for all of the data presented.

Discussion

Photographs of particle agglomerates taken from the fluidized beds after defluidization are shown in Figure 11. The
scale mark in all of these figures is one inch long with a
midpoint mark at one half inch. Careful investigation of
these multiparticle agglomerates indicate certain interesting
characteristics that they have in common, providing at least
a first clue to the mechanism of agglomeration. All of the
pictures in Figure 11 indicate very strongly that on a group
basis the particles - although stuck together - have maintained
their original shapes. This indicates that neither melting nor
substantial softening has taken place. In general, the forces
holding the particles together are very weak. Rolling an
agglomerate of copper particles between two fingers results in
large numbers of particles breaking loose and separating from
the parent agglomerate. This accounts for the ease with which
a bed that has been defluidized will be refluidized if the
velocity is increased.

Table 1 Correlation Coefficients and Regression Coefficients for Defluidization Experiments

Figure No.	Particles Used	Particle Size	L/D	Velocity ft/sec	Number of Data Points	A	B	Regression Coefficient
3	Copper shot	-16+20 sieve	1/2	---	14	25.275	-37.772	0.95
4	Copper shot	-40+50 sieve	1/2	---	10	13.407	-17.037	0.66
5	Polyethylene beads	0.3cm x 0.36cm	1	---	10	214.690	-44.033	0.57
6	Polypropylene beads	0.22cm x 0.34cm	1	---	18	127.701	-31.040	0.72
7	Dacron particles	0.14cm x 0.33cm	1	---	11	270.152	-46.355	0.86
8	Glass spheres	-14+16 sieve	1/2	---	12	60.504	-74.620	0.84
9	Glass spheres	-16+18 sieve	1/2	---	8	25.275	-31.007	0.98
10	Glass spheres	-14+16 sieve	--	12	8	-0.022	1.466	0.65

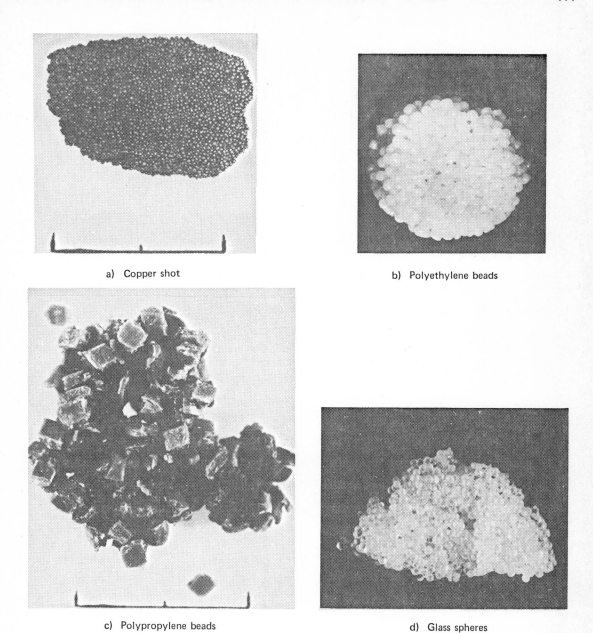

a) Copper shot

b) Polyethylene beads

c) Polypropylene beads

d) Glass spheres

Fig. 11 Photographs of particle agglomerates removed from defluidization equipment.

    The particles in the polymer and glass agglomerates were bonded together more securely than the copper but could still be readily separated by hand.

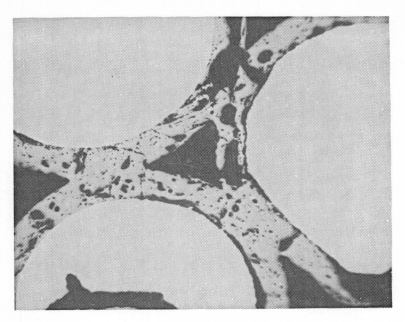

a) Copper particles 100X

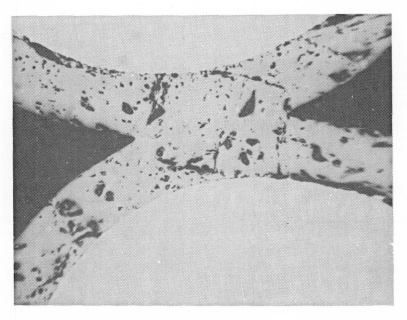

b) Copper particles 200X

Fig. 12 Photomicrographs of a section taken through the center of a three particle cluster of copper shot. (Courtesy of E. I. duPont de Nemours.)

Figure 12 shows two views of a section taken through the center of a three particle cluster of copper. The central white areas represent the pure copper and the shaded rings surrounding the copper represent cupric oxide. The black spaces between the cupric oxide rings are voids or air spaces. The contact between two cupric oxide layers is minimal in terms of prior sintering history. As the sintering process progresses, necking or densification becomes evident. In Figure 12 necking or densification is not apparent. Therefore, the sintering process holding the particles together must be in its infancy.

The micrographs of Figure 13 for polyethylene beads tell a somewhat different story. The fracture surface shown in Figure 13b indicates the existence of relatively strong forces at the points of contact of the particles. Such forces are indicative of a melting or viscous flow phenomenon.

One of the first indications that we had relating sintering to particle agglomeration in fluidized beds was obtained during a visit to the Babcock and Wilcox Research Laboratories in Alliance, Ohio. Dr. S. Vecci (24) explained how the sticking properties of coal ash materials to boiler steam tubes could be correlated with the sintering temperature of the ash. "Initial sintering temperatures" are defined by and measured on a dilatometer. The particles are placed in a cylindrical sample holder and a shaft with a piston at the end is pushed against the sample in the cylinder. A constant force is applied to the rod and the sample in the cylinder is heated at a programmed rate of temperature rise. The change in length of the sample in the cylinder is plotted as a function of temperature. Figures 14a and b show dilatometry studies conducted on our copper and polymer particles. Initially, the particles increase their length linearly with temperature.

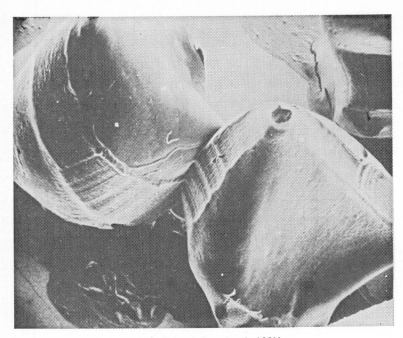

a) Polyethylene beads 100X

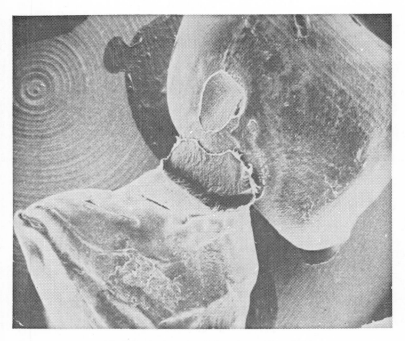

b) Polyethylene beads fracture surface 10X

Fig. 13 Micrographs of a two particle cluster of polyethylene beads. (Courtesy E. I. dupont de Nemours.)

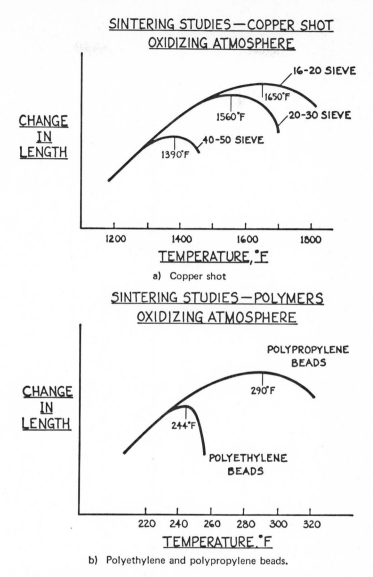

Fig. 14 Dilatometry studies on copper shot and polymer beads.

This is due to thermal expansion, the coefficient of which can be calculated from the slope of the linear portion of each curve. At some temperature which is different for each particle type and size, the slope of the length change curve begins to decrease. This indicates the onset of a phenomenon that is competing with thermal expansion, i.e., sintering or

densification. From this point onward the slope of the length
change curve decreases rapidly until it becomes zero and then
continues to decrease and becomes negative. When the slope of
the length change curve is zero, the expansion due to thermal
effects is just balanced by the contraction due to sintering.
This is the point for which we have defined the "initial
sintering temperature" or T_s.

Just as Babcock and Wilcox found sintering temperatures
for different ash materials to be a significant measure of ash
stickiness, we have found T_s to be a temperature of great
significance in defluidization. Referring to Figures 3 and 4
for copper shot and Figures 5 and 6 for polyethylene and polypropylene beads, we have plotted T_s for each of these particles
on the defluidization diagram. The results of this are dramatic
and significant. It appears that T_s is a measure of the temperature at which the high temperature defluidization curve
intersects the minimum fluidization velocity line.

The significance of this result lies in the fact that T_s
appears to define the maximum temperature for which the
traditionally accepted minimum fluidization velocity separates
non-fluidized from fluidized beds. For any temperature above
T_s, the minimum fluidization velocity for any type, shape and
size of particle is now defined by the curves presented in
this report. Once this observation has been made the logic
seems irrefutable. The traditional minimum fluidization
velocity represents a balance of drag, buoyancy, and gravity
forces acting on a particle in a large system of particles.
However, above T_s at least two other major forces must be
included in the balance, i.e., kinetic energy forces due to
the motion of the particles and adhesive forces due to the
sintering phenomenon. Therefore, our new high temperature
minimum fluidization velocity curves represent the balance

between drag, buoyancy, gravity, kinetic energy, and sintering forces.

A definite statement of what appears to be a fundamental result cannot be made on the basis of a limited number of experiments. However, the indications are strong that we have defined a quantity of fundamental significance to the understanding of an important aspect of fluidized bed behavior.

DEFLUIDIZATION EXPERIMENTS AT ROOM TEMPERATURE

Experimental Equipment

Figure 15 shows our arrangement for study of the defluid-ization of coarse solids upon addition of sticky matter. For bed material, we have used four size cuts of gravitating-bed thermal cracking catalyst (TCC beads) supplied to us by Mobil Oil Company, and carbon tablets of a single uniform size.

As the sticky matter, we have used Ace Plastic Coating, #3100, of Ace Glass, Incorporated. The fluid is highly viscous (41 poise) and is generally used in glass coating of laboratory equipment. It is white, odorless, and has a density of 1.198 gm/cc.

The solids are fluidized by air in a 3-inch glass column and the sticky matter is then added slowly by driving it from a reservoir with pressurized nitrogen. Our procedure is to set the fluidization velocity at a given level and slowly add the sticky matter until defluidization occurs. The contents of the bed are then dumped and weighed. The relative content of sticky matter is calculated by comparison with the initial weight of the bed material. This ignores sticky matter that adheres to the glass column. In general, this is small.

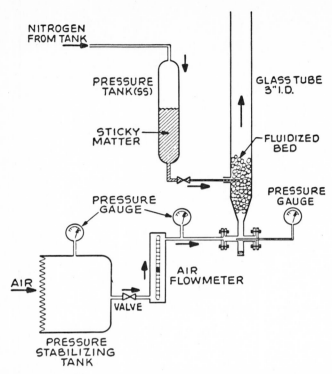

Fig. 15 Apparatus for adding sticky matter to a bed of coarse solids fluidized by air.

Experimental Results

Figure 16 depicts the defluidization limits for the four sized cuts of TCC beads. It should be noted that the defluidization lines extrapolate, at zero weight of sticky matter, to a velocity approximately equal to the minimum fluidization velocity of the particles in question.

In all the experiments with the TCC beads, the initial bed weight, W_B, was kept constant at 400 grams. That meant, of course, that the settled bed height was about the same in these experiments. To get some feeling for the effect of different bed heights on the defluidization limits, we ran two series of defluidization experiments with beds of carbon tablets of different initial inventory. The data are presented in Figure 17.

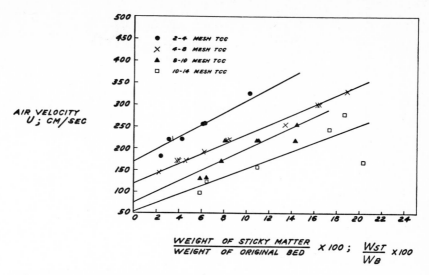

Fig. 16 Defluidization limits for four sized cuts of TCC beads. W_B = 400 grams.

DISCUSSION

As the sticky matter is added, the bed continues to fluidize nicely until the defluidization is reached, whereupon the bed "dies" all of a sudden. It should be noted that

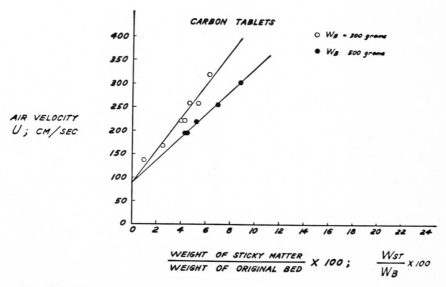

Fig. 17 Defluidization limits for beds of carbon tablets of different initial inventory.

s the defluidization limit is approached, the character of
the fluidization changes. The bed becomes somewhat more
sluggish, and clusters of particles can be seen where the
bed is thin, or in splashed material above the bed. Notwithstanding, there is no question but that the bed remains
fluidized up to the defluidization limit; one moment it is
alive, and an instant later, it is dead.

CONCLUSIONS

The experimental results outlined above have demonstrated
that the defluidization of beds of sticky particles is a well-
ordered phenomenon obeying precise rules. For the high temperature defluidization studies, the initial sintering temperature of the particles emerges as a quantity of fundamental
importance for predicting the temperature at which particle
sintering can affect the character of fluidization.

ACKNOWLEDGEMENT

This study is part of a program supported by Grant
GI-34286 from the RANN Program (Research Applied to National
Needs) of the National Science Foundation to The City College
of New York.

REFERENCES

)   A.M. Squires and C.A. Johnson, J. of Metals, 1, (April 1957).

)   A.M. Squires, Atomic Energy Commission Symposium Series,
    14, 181 (1969).

    R.J. MacMullan and C.A. Johnson, Mining Congress J., 47
    (June 1962).

    R.J. Priestley, Ind. Eng. Chem., 49, 62 (1957).

(5)   G. Tomasicchio, <u>Proceedings of Int'l Symp. on Fluidization</u>,
        Netherlands U. Press, Amsterdam, p. 725 (1967).

(6)   C.S. Cronan, <u>Chem. Eng</u>. 64 (April 1960).

(7)   Account of one of the authors (AMS) who participated in
        the development of the H-Iron process.

(8)   B.G. Langston and F.M. Stephens, Jr., <u>J. of Metals</u>, 312
        (April 1960).

(9)   Account of one of the authors (AMS).

(10)   A.M. Squires, <u>Trans. Inst. of Chem. Engrs. (London)</u>, <u>39</u>,
        3 (1961).

(11)   A. Godel and P. Cosar, <u>AIChE Symp. Series</u>, <u>67</u>, 210 (1967).

(12)   A.M. Squires, <u>Science</u>, <u>169</u>, 821 (1970).

(13)   A.M. Squires, <u>Sci. Am.</u>. <u>227</u>, 26 (October 1972).

(14)   J. Yerushalmi and A.M. Squires, <u>Conference Proceedings</u>,
        <u>Power Generation - Clean Fuels Today</u>, Electric Power
        Research Institute (EPRI-SR-1), p. 45 (April 1974).

(15)   J. Yerushalmi, et al., <u>Science</u>, <u>187</u>, 646 (1975).

(16)   W.M. Goldberger, paper presented at the <u>Fourth Synthetic</u>
        <u>Pipeline Gas Symposium</u>, (Chicago) (October 1972).

(17)   Corder, Batchelder, and Goldberger, paper presented at
        the <u>Fifth Synthetic Pipeline Gas Symposium</u>, (Chicago)
        (October 1973).

(18)   S. Ehrlich, J.W. Bishop, J.S. Gordon, E.B. Robison,
        A. Hoert, Pope, Evans and Robbins, paper presented
        at <u>AIChE</u>, (New York) (November 1972).

(19)   W.M. Goldberger, paper presented at <u>ASME</u>, (Pitts-
        burgh) (November 1967).

(20)   R.W. Hiteshue, S. Friedman, and R. Madden, <u>Bureau of</u>
        <u>Mines RI6376</u>, (1964).

(21)   A.M. Sadler, <u>AIChE</u> Meeting, (New York) (November 1967).

(22)   R. Pyzel, <u>U.S. Patent</u>, 2,776,132 (Jan. 1, 1957).

(23)   R.F. Blanks, H.L. Kennedy, <u>The Technology of Cement and</u>

<u>Concrete</u>: Volume I, <u>Concrete Materials</u>, John Wiley and
Sons, (New York) (1955).

(24) S. Vecci, Babcock and Wilcox, Alliance Research Center,
Alliance, Ohio, discussions with M.J. Gluckman and
J.H. Siegell (August 1973).

DESIGN CONSIDERATIONS FOR DEVELOPMENT OF A COMMERCIAL FLUIDIZED BED AGGLOMERATING COMBUSTOR/GASIFIER

J. M. D. MERRY, J. L.-P. CHEN, and D. L. KEAIRNS

ABSTRACT

A multi-stage fluidized bed coal gasification process is being developed to produce low Btu gas for combined cycle electric power generation. The first stage is a recirculating bed coal devolatilizer and the second stage is an agglomerating combustor/gasifier. Design considerations for the development of a commercial agglomerating combustor/gasifier unit are presented which integrate requirements imposed by the process and the problems of scaling from small scale unit data. Design criteria and laboratory support data are reviewed and scaling criteria identified for relating the design and operation of a commercial vessel to that of simulating models.

INTRODUCTION

Westinghouse, in conjunction with ERDA and a joint industry team, is developing a fluidized bed coal gasification process for producing low Btu fuel gas for electric power generation.[1,2,3] A Process Development Unit (PDU) with a design capacity of 544 kg/hr of coal has been built,[4,5] and preliminary designs are being prepared for a Generating Pilot Plant with a coal feed rate of 45.4 Mg/hr. The reactor vessels in the Generating Pilot Plant will represent full size modules in a larger generating facility.

This work is being performed as part of the Westinghouse Coal Gasification Program. The project is being carried out by a six-member industry/government partnership comprising ERDA, Public Service of Indiana, Bechtel, AMAX Coal Co., Peabody Coal Co. and Westinghouse.

Paper prepared for the International Fluidization Conference, Asilomar Conf. Grounds, Calif., June 15-20, 1975.

This paper describes the development of the agglomerating combustor/gasifier vessel and considers the problem of scale-up to the size required in the Generating Pilot Plant.

FUNCTIONS OF AGGLOMERATING COMBUSTOR/GASIFIER

The agglomerating combustor/gasifier, Fig. 1, carries out the following functions in the process:

(i) Gasification: char particles produced in a recirculating bed coal devolatilizer vessel[1] and residual ash particles are fed to the gasifier section where the char reacts with steam in a fluidized bed to form CO and H_2 which, with some CH_4 evolved in the devolatilizer, are the fuel components of the product gas.

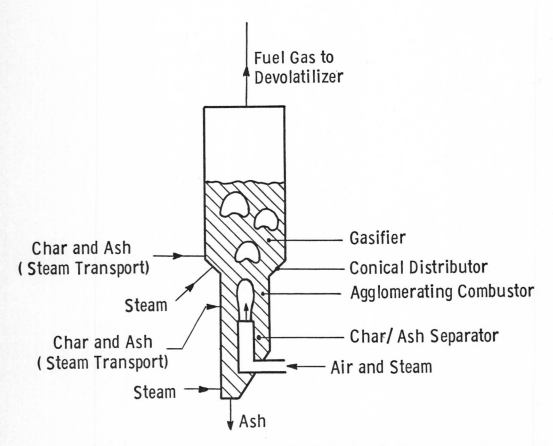

Fig. 1 Schematic layout of agglomerating combustor/gasifier.

(ii) Combustion for process heat generation: fine char and ash particles elutriated from the devolatilizer are collected and fed to the agglomerating combustor. The fine char is burned in vertical air jets in the combustor section and the heat generated is transferred to the gasifier section to support the endothermic gasification reaction.

(iii) Ash agglomeration: fine ash particles fed to the combustion flame are heated to above their softening temperature and stick to larger ash particles present in the bed to form ash agglomerates. These ash agglomerates are circulated around the air jets, being entrained into a jet near its base and carried to its top in the jet stream; they grow in size on each pass as fine ash sticks to them, until they reach a critical size and settle into the char/ash separator section adjacent to the air nozzles.

(iv) Char/ash separation and ash removal: steam is supplied to the char/ash separator section, the region of bed adjacent to the air nozzles, to fluidize the ash agglomerates so as to separate out any char particles carried down to this level and to return them to the combustor section. This stream also cools the ash agglomerates to facilitate their removal from the bottom of the vessel in a stream of low carbon content.

DESIGN AND OPERATION CONSIDERATIONS

The functions described above will be carried out in one composite vessel. The gasifier section is larger in cross-section than the combustor section and the two are connected by a conical section which also acts as the gas distributor for process steam supplied directly to the gasifier section. Reactant feed streams enter the vessel at the following locations:

Steam: (i) Steam is supplied directly to the gasifier through the conical distributor as above.

(ii) Steam transports coarse char and ash from the devolatilizer to the gasifier at points directly above the conical distributor.

(iii) Steam is supplied to the char/ash separator in the agglomerating combustor.

(iv)   Steam transports fine char and ash from the
devolatilizer cyclone to the agglomerating
combustor. With a single combustion jet, the
fines may be fed at the vessel wall, but with
multiple jets it is more convenient to feed the
fines through tubes internal to and co-axial with
the air feed nozzles.

Air: (i)   Air is fed to the agglomerating combustor through
vertical nozzles. Vertical jets develop in the
fluidized combustor section, in which char fines
are burned and ash particles agglomerate. The
products of combustion leave the jets in the form
of bubbles which rise through the bed of the
gasifier section.

(ii)   If the rate of heat transfer from the combustor
section to the gasifier section is not sufficient
to sustain the endothermic gasification reaction,
it may be necessary to supply air directly to the
gasifier section to burn some char there.

Heat transfer is achieved through the movement of solids
between the combustor and gasifier sections. Solids are carried upwards
from combustor to gasifier by the jets and by the bubbles which form
at the end of the jets. There is a consequent downflow of solids along
the vessel walls so that continuous solids circulation between the two
sections is induced.

The vessel will operate at 1720 kPa (250 psia) and will be
refractory lined. The agglomerating combustor will operate at 1366–1477°K
(2000–2200°F) and the gasifier will operate at 1255–1366°K (1800–2000°F).
Process air will be supplied to the vessel at 450°K (350°F) and steam
will be dry saturated at 477°K (401°F). The ash agglomerates will be
drained from the vessel at 533°K (500°F). The hot fuel gas from the
gasifier is fed to the devolatilizer vessel to drive the recirculating bed.

SUPPORT STUDIES

A program of theoretical and experimental work has been undertaken[1,3] to simulate the hydrodynamics of the combustor/gasifier and to study the fluidization characteristics.

A 100 mm (4 inch) diameter cold pressurized model, Fig. 2, has been used to study the mechanics of char/ash separation and of solids circulation. The results may be summarized as follows:

(i) Greater than 95% separation of ash agglomerates from char can be achieved by supplying steam to the char/ash separator at the minimum fluidizing velocity of the ash particles to be withdrawn.[6]

(ii) For optimum solids circulation, the air jet should penetrate to about the level of the conical section joining combustor and gasifier, and steam should be supplied to the conical distributor at the minimum fluidizing velocity of the char particles in the gasifier.

A two-dimensional fluidized bed, Fig. 3, in which lead shot is fluidized by water was used to study the flow of solid particles and fluid around a vertical jet in the bed. The conclusions of this work were as follows:

(i) Interstitial fluid as well as solid particles are entrained into a vertical jet in a fluidized bed.

(ii) The steam supplied to the char/ash separator will be swept into the air jet and will enter the gasifier in the bubble phase.

(iii) Char fines fed to the wall of the vessel around the level of the air nozzle will be swept into the jet along with particles and fluid in that region of the bed.

A correlation has been developed[7] for predicting the penetration depth, L, of a vertical jet into a fluidized bed, namely

$$L = 5.2 \ (\rho_f d_o / \rho_p d_p)^{0.3} \ (D_b - d_o) \tag{1}$$

where D_b is the initial diameter of the bubble leaving the jet, d_o is the nozzle diameter, d_p is the mean particle diameter, and ρ_f and ρ_p are fluid and particle densities, respectively. This expression was derived by relating the jet penetration depth to the initial bubble

Fig. 2 Cold pressurized model of agglomerating combustor/gasifier.

diameter and the jet half-angle. Harrison and Leung[8] give the initial
bubble volume, V_o, as

$$V_o = 1.138 \ (G^2/g)^{0.6} \qquad\qquad (2)$$

and by making the approximate assumption that the volume of the bubble wake is 1/3 of the bubble volume, we deduce

$$D_b = 1.422 \ (G^2/g)^{0.2} \qquad (3)$$

Fig. 3 Front view of liquid fluidized bed simulating flow patterns around central jet in combustor.

where G is the volumetric gas flow rate from the nozzle.

Theoretical studies of the mechanism of ash agglomeration, based on the findings of Jequier[9] and Goldberger[10] will be tested against data obtained from the Process Development Unit when that is in operation, as will a model of the total process used for material/energy balance calculations which predicts material flow rates and product gas compositions. Data from other experimental studies will also be assimilated into the analyses as the data becomes available.

AGGLOMERATING COMBUSTOR/GASIFIER DESIGN

Design criteria for critical process parameters have been developed based on the functional requirements for the combustor/gasifier, the design and operating considerations, and the available data from the support studies. These criteria are summarized in Table 1.

Table 1 Design Criteria

Agglomerating Combustor Nozzle Diameter	• nozzle velocity in the range 5-20 m/s (16-65 ft/sec) to give conditions favorable for ash agglomeration (velocity sufficient to establish jet and avoid bubbling), fine char combustion (velocity to permit sufficient residence time for complete combustion of particles < 0.5 mm in a single pass) and solids circulation between the combustor and gasifier (jet penetration to base of gasifier)
	• pressure drop sufficient to uniformly distribute gas (multiple nozzle case)
Agglomerating Combustor Diameter	• internal diameter sufficiently large to allow nozzle spacing such that the bubbles formed from each jet do not overlap
Agglomerating Combustor Height	• height should be the same order as the length of the jet(s) to avoid bubble coalescence in the combustor, encourage coalescence in the gasifier, and maximize the solid circulation rate and hence the heat transfer

Table 1 (cont.)

Char-Ash Separator Cross-sectional Area	• steam flow sufficient to cool agglomerated ash • incipient fluidization of agglomerated ash particles at the level of the end of the air nozzles
Conical Steam Distribution Area	• designed to give incipient fluidization of the bed char particles for steam requirement (permits maximum solid circulation rate between combustor and gasifier and increases steam-char contact in gasifier to enhance gasification
Gasifier Diameter	• obtained from combustor diameter and conical steam distribution area
Gasifier Bed Depth	• specified to give sufficient char residence time to permit complete gasification

A preliminary design of the agglomerating combustor/gasifier vessel for the Generating Pilot Plant is being drawn up. This indicates that the vessel will be comprised of a gasifier section of approximately 4 m dia. and a combustor section of approximately 3 m dia. The scale of this vessel is larger than any of the presently existing equipment, as is evident from Table 2. Thus, careful consideration must be given to the following design features.

(i) Gas bypassing: as much steam as possible should be supplied directly to the gasifier for reaction with bed char, but a substantial quantity of steam ($\sim$ 33%) must also be supplied to the agglomerating combustor for fines feeding and particle separation. This latter steam will be swept into the combustion jets and will enter the gasifier in the form of gas bubbles along with the products of combustion. With large discrete bubbles, contact between bubble steam and bed char will be severely restricted.[11] However, if an array of smaller bubbles is produced at the base of the gasifier (e.g., by employing a multiple nozzle design in the combustor), contacting between bubble gas and bed char could be improved due to the coalescence of these bubbles as they rise through the gasifier bed.[12]

Table 2 Dimensions of Existing Equipment and Projected Designs, for Simulated and Actual Operation of Agglomerating Combustor/Gasifier

Equipment	Cold Pressurized Lab Model	PDU	Projected for Demonstration Plant	Suggested Intermediate-Scale Cold Model
Combustor i.d. (D_c) m	0.05	0.3	3.0	1.28
Height of combustor section m	0.11–0.3	0.91–1.36	3.0	1.2
Gasifier i.d. (D_g) m	0.1	0.51	4.0	1.8
Height of Gasifier Bed m	0.15–0.6	1.8–3.6	8.0	3.6
Pressure kPa	100–1000	1500	1500	Ambient–1500
Temperature °K	Ambient	1350	1350	Ambient
Design Comments	Single Nozzle	Single Nozzle	Multiple Nozzle	Semi-circular

(ii) Jet stability: the PDU and the laboratory equipment described earlier are all designed with single nozzles in the combustor section whereas the concept above calls for a multiple nozzle design. The effect of the vessel wall on a single axial fluid jet is likely to be greater than on multiple jets since the latter will tend to veer away from the wall because of the tendency of the subsequently formed bubbles to coalesce. Before designing in detail the multiple nozzle combustor for the Generating Pilot Plant, one should study the characteristics of large closely spaced multiple jets in a fluidized bed.

Tests are being recommended to improve our understanding of these critical design features. The basis for developing further tests is an understanding of the scaling factors.

SCALING FACTORS

In order to project performance for a commercial unit from small scale test facilities and to carry out critical experiments in smaller scale test facilities, it is important to identify the correct scaling factors. Five basic criteria are suggested for relating the hydrodynamics of a commercial scale combustor/gasifier to that of simulating models and development units. Identification of these factors would permit scale-up to the commercial size unit, and scale down from a given design to give basic dimensions and operating conditions for intermediate scale simulation.

(i) Scale vessel diameters by factor β, so that

$$(D_{g_2}/D_{g_1}) = (D_{c_2}/D_{c_1}) = \beta \tag{4}$$

where D_g and D_c are the internal diameters of gasifier and combustor sections, respectively, and suffices 1 and 2 refer to the Generating Pilot Plant vessel and the simulating model, respectively.

(ii) Scale initial bubble diameters by the same factor, i.e., $(D_{b_2}/D_{b_1}) = \beta$, to give geometric similarity between bubble arrangements in the two beds. D_b is determined by the nozzle gas flow rate, G, through Eq. (3), therefore, we require

$$(G_2/G_1) = (D_{b_2}/D_{b_1})^{2.5} = \beta^{2.5} \tag{5}$$

(iii) Choose jet nozzle diameter, d_{o_2}, in model to give scaled penetration depth, i.e., $(L_2/L_1) = \beta$. In a cold model operating at atmospheric pressure, this may not be possible since fluid density (ρ_f) will be considerably less than in hot, pressurized plant. In this case, choose d_{o_2} to give maximum jet penetration depth, L_2, for given gas flow rate, G_2 (and so for given critical bubble diameter, D_{b_2}). From Eq. (1), we find that the required nozzle diameter in the model is given as

$$d_{o_2} = 0.23 \, D_{b_2} \tag{6}$$

To maintain the same mechanism of solids circulation, the height of the combustor section should be approximately equal to the jet penetration depth in each vessel, i.e., $H_2/L_2 = H_1/L_1 \sim 1$.

(iv) To simulate char/ash separation, choose the particle size for ash agglomerates to have the same minimum fluidizing velocity at the operating conditions of each vessel.

(v) Choose particle size in the gasifier section to give $(U_b/U_{mf})_2 = (U_b/U_{mf})_1$, where U_b is the bubble rise velocity and U_{mf} is the superficial minimum fluidizing velocity. This maintains the same pattern of gas circulation around the bubbles in each case.[11] For $U_b \gg U-U_{mf}$, we can take $U_b \propto D_b^{0.5}$ so that we require

$$(U_{mf})_2/(U_{mf})_1 = (D_{b_2}/D_{b_1})^{0.5} = \beta^{0.5} \qquad (7)$$

With most of the gas traveling in the bubble phase, this criterion is approximately the same as setting (U/U_{mf}) constant, where U is the superficial gas flow rate through the bed.

These criteria and scaling factors could be used for scaling-up to the Generating Pilot Plant design, or alternatively, for scaling-down from a given design to give the basic dimensions and operating conditions for simulation of bed hydrodynamics in a large-scale cold model.

CONCLUSIONS

Functional requirements for an agglomerating combustor-gasifier to produce low Btu gas for combined cycle power generation are identified and design criteria projected based on operating requirements and laboratory support data. Problems associated with scale-up to a commercial scale combustor/gasifier have been identified. Hydrodynamic scaling factors are presented to permit scale-up of data from test units to give basic dimensions for intermediate scale simulation and to identify critical tests on small and intermediate scale units.

SYMBOLS USED

d_o nozzle i.d.

d_p mean particle diameter

D_b initial bubble diameter

D_c combustor i.d.

D_g gasifier i.d.

g acceleration due to gravity

G   volumetric gas flow rate per nozzle

L   jet penetration depth

U_b bubble rise velocity

U   superficial gas velocity

U_{mf} U at min. fluidization

V_o initial bubble volume

β   scaling factor

ρ_f fluid density

ρ_p particle density

REFERENCES

1. Archer, D. H., E. J. Vidt, D. L. Keairns, J. P. Morris, J.L.-P. Chen, "Coal Gasification for Clean Power Production," Proceedings of the Third International Conference on Fluidized Bed Combustion, Hueston Woods, Ohio, 1972, issued as EPA 650/2-73-053, Dec. 1973.

2. Lemezis, S. and D. H. Archer, "Coal Gasification for Electric Power Generation," Combustion 45, 6 (1973).

3. "Advanced Coal Gasification System for Electric Power Generation," Annual Technical Reports 1973-1974 to Office of Coal Research, U.S. Department of the Interior by Westinghouse Electric Corporation, Contract No. 14-32-0001-1514.

4. Archer, D. H., D. L. Keairns, E. J. Vidt, "Development of a Fluidized Bed Coal Gasification Process for Electric Power Generation," Energy Communications, 1(2), 115-134 (1975).

5. Holmgren, J. D. and L. A. Salvador, "Low Btu Gas from the Westinghouse System," CEP 71 (4), 87 (1975).

6. Chen, J. L. P. and D. L. Keairns, "Particle Segregation in a Fluidized Bed", Can. J. Chem. Engr., 53, 395-402, August 1975.

7. Merry, J. M. D., "Penetration of Vertical Jets Into Fluidized Beds", A.I.Ch.E.J. 21(3), May 1975.

8. Harrison, D. and L. S. Leung, "Bubble Formation at an Orifice in a Fluidized Bed," Trans. Inst. Chem. Engrs., 39 409 (1961).

9. Jequier, L., L. Longchambon and G. Van de Putte, "The Gasification of Coal Fines," J. Inst. Fuel, 33 584 (1960).

10. Goldberger, W. M., "Collection of Fly Ash in a Self-Agglomerating Fluidized Bed Coal Burner," ASME Paper 67-WA/FU-3, Winter Annual ASME Meeting, Nov. 1967, Pittsburgh, Pa.

11. Pyle, D. L. and P. L. Rose, "Chemical Reaction in Bubbling Fluidized Beds," Chem. Eng. Sci. $\underline{20}$ 25 (1965).

12. Whitehead, A. B. and A. D. Young, in A.A.H. Drinkenberg (Ed.) "Proceedings of the International Symposium on Fluidization," Eindhoven, 1967, p. 294 (Amsterdam: Netherlands University Press).

PRODUCTION OF GASEOUS FUELS FROM COAL IN THE FAST FLUIDIZED BED

JOSEPH YERUSHALMI, MICHAEL J. GLUCKMAN, ROBERT A. GRAFF,
SAMUEL DOBNER and ARTHUR M. SQUIRES

INTRODUCTION

In many sectors of the U.S. economy, the sober realities that Figure 1 conveys are already in evidence. Figure 1 traces the U.S. production and consumption of natural gas (1). During 1968, gas production sailed past the 10-year running average of discoveries, and early in 1972 growing demand collided with declining production capability. An increase in gas well drilling notwithstanding, new discoveries in the past few years have been decreasing and domestic production could fall below 15 trillion cubic feet in 1980 (2).

The situation is compounded by the price control that the Federal Power Commission exercises over interstate gas. Not unexpectedly, committments of newly discovered gas to pipeline companies have been negligible since 1969. Interstate gas peaked at ca. 15 trillion cubic feet in 1972, and could fall below 8 trillions in 1980, 5 trillions in 1985 (2). It is no wonder that Texas and Louisiana are enticing industry from the Northeast with offerings of intrastate gas. At the same time, as their old contracts expire, power companies in the gas-producing states find renewal difficult in the face of increased competition from industry vying for the intrastate gas. It should be appreciated that over half of the utility boilers now firing intrastate gas have been designed to handle natural gas only (3).

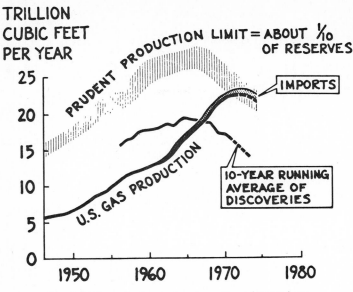

Fig. 1 U.S. production and consumption of natural gas.

Trends in U.S. and World production and consumption of oil portend an approaching crisis of even larger proportions. The chokes are already off the U.S. wells, and imports are rising. At the same time, as Figure 2 illustrates, the free world consumption of oil is just now sailing past the 15-year average of discoveries. Demand continues to grow both in Europe and

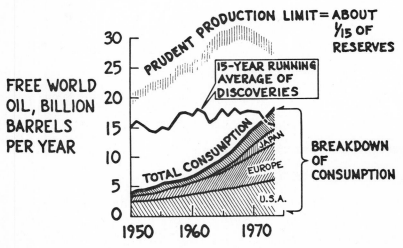

Fig. 2 Free world production and consumption of oil.

Japan and particularly in the developing countries. A scenario paralleling
that for U.S. oil appears to be unfolding for world oil.

When will the world's marginal production capability for oil vanish?
Figure 3 depicts a projection made by Warman (4) in September 1973. It is
a composite of data of widely varying degrees of reliability, yet it must be
essentially true in outline.

Figures 1 to 3 convey an urgent message: Industry must soon plan to
rely upon coal for its basic energy needs.

Gaseous fuels from coal can capture many of the roles of oil and gas in
industry. It is unfortunate that as the fourth quarter of the century begins,
the options available commercially to a potential client in industry boil down
to essentially three old gasification processes, all poorly suited to the
tasks ahead. On the whole, these processes are geared to gasifiers of low
capacity to process coal (typically 500 ton/day for the largest units) and
they produce a gas that requires extensive cleaning, on account of either dust

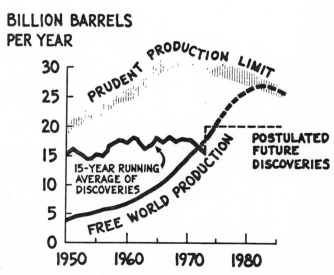

Fig. 3 Warman's projection of free world oil production, with future discoveries postulated at 20 billion barrels per
year.

or tars. These processes are expensive. It is accordingly not surprising that, in the face of diminishing supplies of natural gas, industry continues to replace gas with oil, or is opting for intrastate gas at prices that have hitherto been considered exorbitant.

In a large measure, industry makes its judgments upon economics. But to the engineer, otherwise novice in the art of costing, a process can be perceived attractive upon examination of its design and operating characteristics. What ingredients might combine to make a gasification process attractive?

Above all the process must be reliable. Simplicity marks its design, and easy maintenance attend its operation. In addition we may note the following desired characteristics:

- Large coal-treating throughputs per unit gasifier.
- Minimum coal preparation steps upstream, and minimum gas cleaning downstream of the gasifier. This entails,
    == Absence of tars in product gas.
    == Low dust loadings in gasifier effluent.
    == Capability for feeding a caking coal directly into the gasifier.
- Capability for processing a wide variety of coals (and cokes).
- Capability for operation at elevated pressure.

Production of gaseous fuels from coal in the fast fluidized bed has the potential to fill this bill. In this report we focus upon three schemes for gasifying coal at elevated pressure, each aiming at a different objective:

(1) Complete gasification in an air-blown fluidized bed operating at 1200° to 1400°C, with discharge of ash in the form of ash agglomerates.

The product is a low-Btu gas at about 125 Btu/SCF for use in large
power systems or as a heating gas in industry.

(2) Partial gasification by steam at temperatures around $800°$ to $1050°C$.
The product is an industrial gas at about 250 Btu/SCF which may serve
as fuel for industrial boilers (including retrofit of existing gas-
and oil-fired units), or for direct-heat applications. A coke

byproduct can be burned for process steam or power, or it may
be gasified by (1) above.

(3) Partial gasification by hydrogen in a flash hydrogenation at $650°$ to
$800°C$. The products are methane, ethane and benzene. A coke byproduct
might contain little sulfur if the coal has only a small amount of
pyritic sulfur.

We begin, however, with a discussion of the phenomenon of fast fluidization
and a presentation of fluidization data obtained in a fast fluidized cold model
at our laboratory.

THE FAST FLUIDIZING BED

The Bubbling and Turbulent Regimes

Figure 4 depicts a phase diagram typifying the pressure gradient across
a bed of fine powder* as a function of upward velocity of the gas. The gradient
first increases sharply due to the rise in pressure across the fixed bed of
powder. At minimum fluidization velocity, the powder begins to expand in a
particulate-like fashion and the pressure gradient becomes essentially equal
to the fluidized density of the bed. Beyond the minimum bubbling velocity,
bubbles appear, and grow as the velocity is raised further.

---

* Geldart's Group A solids (5).

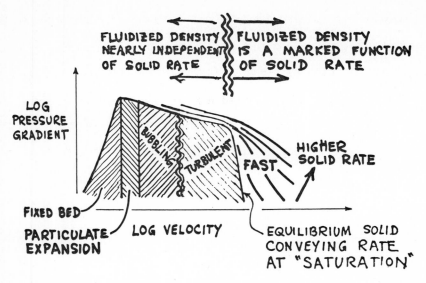

Fig. 4 Fluidization phase diagram for a fine powder, showing different regimes of fluidization.

Over the years, "fluidization" has become virtually synonimous with operation in the bubbling regime - corresponding typically to gas velocities around 1 ft/sec. The underlying function of fluidization has been to afford contact between a gas and a large inventory of solid surface per unit bed volume. Little attention was paid to opportunities of achieving the same end at gas velocities beyond the bubbling regime.

An important paper by Lanneau (6), appearing in 1960, provided strong evidence for such opportunities. Lanneau studied the fluidization characteristics of a fine powder in a bed 15 feet deep and 3 inches in inside diameter at gas velocities that ranged to about 5 ft/sec, and at two pressure levels, 10 and 60 psig. The character of the bed at a given gas velocity was manifest in recordings obtained from small-point capacitance probes inserted inside the column. Lanneau noted that as the velocity was raised beyond 1 ft/sec, the heterogeneous, two-phase character of the bubbling regime gave way to a condition of increasing homogeneity. At velocities in the range of 3 to 5 ft/sec, the tracings from the probes indicated that a condition of "almost a uniform or 'particulate' fluidization was approached". Lanneau provided an argument that a velocity around 1 ft/sec, where bubbles can clearly be seen, is the worst velocity to use from the standpoint of efficiency of contact between gas and solid. Lanneau provided a rationale for what the practical man already knew (7), that contacting is better at higher velocity.

It is now of course appreciated that in a bed operating in the bubbling regime:

(1) Considerable portion of the gas residing in the rising bubbles bypasses the solid in the bed.

(2) The rapid mixing of the solid produces appreciable backmixing of gas. This cuts conversions, and might promote undesirable secondary reactions.

In addition:

(3) Experience has taught that bubbling fluidized beds are not easy to scale up.

(4) The low-velocity characteristic of the bubbling regime translate to low processing capacities per unit cross sectional area of the bed.

(5) Solid mixing at these velocities is not sufficiently vigorous to allow processing of solid that do not flow freely or that tend to agglomerate.

Because Lanneau worked in steel equipment, it remained for Kehoe and Davidson (8) to describe the transition from bubbling to what they called the "turbulent regime" of fluidization, and it remained for Massimilla (9) to provide a laboratory research demonstrating the higher contacting efficiency in a turbulent fluidized bed.

What Kehoe and Davidson witnessed, as the velocity was increased through a narrow bed of fine powder, was a breakdown of the slugging

regime "into a state of continuous coalescence -- virtually a channelling
state with tongues of fluid darting in zig-zag fashion through the bed".
The velocity of the breakdown of the slugging regime was not sharp, but
usually occurred between 1 to 2 ft/sec. In Lanneau's experiments, the
transitions to the turbulent regime occurred roughly around 3 ft/sec. At
a pressure of 60 psig, the transition occurred at velocities somewhat lower
than those recorded at 10 psig.

A bubbling fluidized bed displays a distinct upper surface level; in
a turbulent fluidized bed, the upper surface is present but is considerably
more fuzzy. Both beds are characterized by a definite carryover of solid
that depends upon the fluidizing gas velocity and the distance above the
bed at which carryover is measured. If this distance is beyond the Transport
Disengaging Height, the carryover is a constant as if the gas were "saturated"
with solid. The solid inventory in the bed would of course remain constant
only if the solid carried over is returned to the bed.

Carryover increases of course with gas velocity. If, near the end of
the turbulent regime, the experimenter maintains a given solid inventory
in the bed, then, as the velocity is slowly raised further, , there would be
a sharp drop in bed density over a narrow velocity range as seen at the right
for the lower curve in Figure 4. It will be appreciated that to preserve a
constant solid inventory, the experimenter must return solid at precisely the
same rate that material is carried over from the apparatus. That is to say,.
the solid returns must equal the carryover at saturation, as indicated in the
figure.

Fast Fluidization

If, on the other hand, the experimenter causes solid to flow into the
bottom of the fluidized bed at a rate well beyond the saturation carryover,

he will increase the inventory of solid in the bed. The effect upon bed densities is not marked at velocities below those associated with the aforementioned sharp drop in density. Addition of solid to a bubbling or a turbulent bed at a rate above the saturation carryover will simply cause the vessel to fill up continually, but densities will remain substantially the same.

At higher velocities, however, the fluidized density becomes a strong function of solid rate into the bottom of the bed.

W.K. Lewis and E.R. Gilliland (10) clearly recognized in 1940 what we at the City College have dubbed the fast fluidized bed condition:

"If one will operate at a gas velocity sufficient to blow all or substabtially all of the solid material out of the reactor in a relatively short time, provided no fresh solid material be introduced during this time, but will feed into the reactor simultaneously solid material at a sufficiently high rate, one can maintain in the reactor a high concentration of solid granules approaching that of the 'liquid state' . . . [of the bubbling fluidized bed], and yet be blowing the solid particles out of the top of the reactor at a corresponding rate."

They reported a fast fluidized density as high as 8 lbs/ft^3 for a pulverized clay catalyst at a superficial fluidizing gas velocity of about 8 ft/sec.

Those who have practiced pneumatic transport of solid are likely to be familiar with instances where powders where conveyed under essentially fast fluidized conditions. But examples of use of fast fluidized bed reactors are relatively rare.

Although the Kellogg reactors at Sasolburg, South Africa, for Fischer-Tropsch synthesis may be fast beds, it remained for Lurgi Chemie und

Huttentechnik GmBh to appreciate the broad commercial potential of the fast fluidized bed and to realize this potential in two successful commercial processes (11). Lurgi's design for calcining aluminum hydroxide to provide cell-grade alumina is shown schematically at the upper right in Figure 5. In contrast, Figure 5 shows equipment for bubbling and turbulent beds having approximately the same gas-treating capacity. The fast bed installation is characterized by a large external cyclone and a standpipe of large diameter for circulation of solid at the rate required to maintain the fast fluidized condition.

The City College Studies of Fast Fluidization

Our stand for study of fast fluidization includes a two-dimensional fast bed built in plexiglas, 2 by 20 inches in cross section and 23 feet in height; a 3-inch and a 6-inch round fast beds, each 24 feet tall.

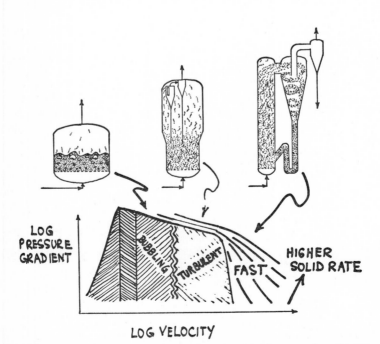

Fig. 5 Fluidization phase diagram for a fine powder, showing schematic diagrams of equipment suitable for use in the bubbling, turbulent, and fast fluidization regimes.

The beds are blown with air supplied from a compressor that delivers 1200 CFM at 10 psig. The present program explores the fluidization characteristics of several solids of different size, size distribution and density. Here we report observations and data on fluid cracking catalyst (FCC) with a size range of 20 to 130 microns (volume-surface average diameter is about 60 microns) particle density of 55 lbs/ft^3 and a settled density of 31.9 lbs/ft^3.

We have viewed the operation of the two-dimensional fast bed by means of high speed photography. The air velocity was held fixed at 12 ft/sec, and the solid throughput was varied. At low solid throughputs, the solid is conveyed upward in dilute-phase transport, but contrary to the impression created by many discussions of this subject, fine particles are not streaming upwards discretely. Even at very low solid loadings (0.1 lbs/ft^3), some segregation is apparent. Relatively denser clouds of particles go up surrounded by a more dilute environment. At higher solid loadings, though still within the dilute phase regime, solid segregation becomes more pronounced: Particles throng into vertical streamers which move upwards surrounded by a faster moving leaner phase. Some of the streamers or strands weave a bit or even halt momentarily. At a solid loadings around 2 lbs/ft^3, solid backmixing gradually comes into play, and the fast bed is established.

The fast bed can be regarded as essentially a dense suspension marked by vigorous and intensive backmixing of solid. At loadings between 3 to 5 lbs/ft^3, the solid at any moment appears distributed in two phases. Dense strands and ribbons rise and fall and drift from side to side at high speeds, while the bulk of the colume of the bed is taken by particles moving rapidly upwards in a more dilute environment. Solid interchange between the two phases appears rapid and intensive; dense strands of solid break apart, some gradually and some in an explosive fashion, as new strands form.

At loadings beyond 5 lbs/ft^3, observations of the details of the structure of the fast bed become difficult, but suggest that both the dense (ribbons and strands) and the lean phases become on the whole continuous. Strands and ribbons become linked in a system of rapidly circulating material that includes many vortices resembling tiny tornadoes. The impression is that gas-solid interaction in on a fine scale. The fast bed thus affords intimate contact between a high velocity gas and a large inventory of solid surface per unit bed volume.

In the remainder of this section we present some data obtained in our 3-inch fast fluidized bed. The set up is shown schematically in Figure 6. On the left is a 1-foot companion, low-velocity, bubbling bed that serves for storage of solid and control of solid circulation. Normally, solid flows by gravity from the bottom of the companion bed into a well-aerated 3-inch I.D. U-tube that extends without contraction or expansion to the bottom of the 3-inch fast bed seen on the right. Solid rate can be controlled by a buttefly valve installed in the 3-inch tube just below the bottom of the 1-foot companion bed. Having traversed the fast fluidized bed, the solid is returned via cyclones to the companion bed. In few experiments, solid was fed by a paddle-wheel through a line that extended diagonally from the bottom of the companion bed to the 3-inch column (see Figure 6).

Solid circulation rates were measured with the aid of a sintered-plate butterfly valve installed in the middle of the 1-foot companion bed as follows: At a given moment, the valve was closed, and the rate of descent of the upper level of the bed was timed. Solid returning to the companion bed through the cyclone diplegs formed a fluidized bed on top of the butterfly valve which accordingly acts as a distributor for this upper bed. It should be appreciated that, apart from the additional pressure introduced by the

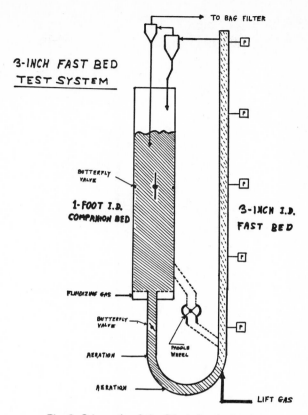

Fig. 6 Schematic of the 3-inch fast bed system.

sintered butterfly valve, the solid head in the companion bed remains constant throughout this procedure. Pressure readings were taken at many locations around the solid circulation loop, and particularly along the fast bed at elevations of 2, 7.5, 13, 18.5, and 24 feet from its bottom.

Data are shown in Figures 7 to 11. Figure 7 shows the pressure gradient across a section of the fast bed extending from an elevation of 7.5 to 18.5 feet from the bottom of the bed, as a function of the superficial fludizing gas velocity and the solid rate. Solid hold ups in the column were not measured, but at the range of conditions shown in Figure 7, the pressure drop is almost entirely due to the static head of the solid, and the pressure gradient accordingly can be regarded essentially as the fluidized density in

PRESSURE GRADIENT, 7.5 FT TO 18.5 ELEVATION

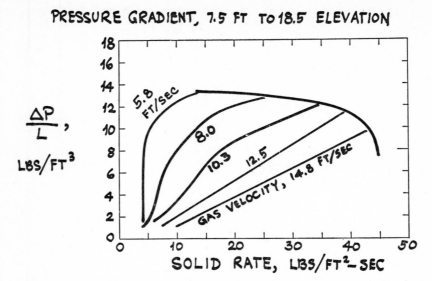

Fig. 7 Pressure gradient (across a section extending from 7.5 to 18.5 feet above the bottom of the 3-inch fast bed) vs. solid rate at different fluidizing gas velocities. The slumped bed height in the companion bed was 121 inches.

PRESSURE GRADIENT AT AIR = 8.0 FT/SEC

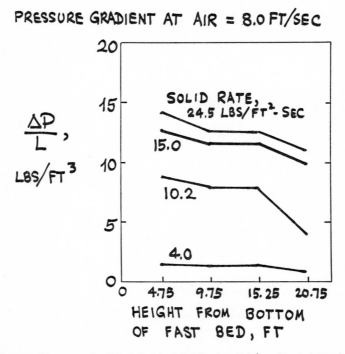

Fig. 8 Pressure gradient profile across the 3-inch fast bed fluidized at 8 ft/sec. The height of the slumped solid in the companion bed was 121 inches.

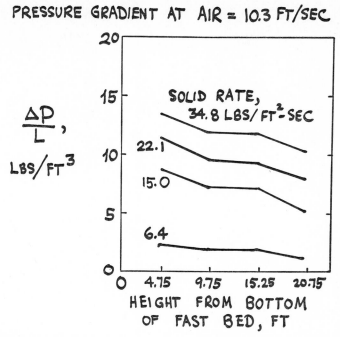

Fig. 9 Pressure gradient profile across the 3-inch fast bed fluidized at 10.3 ft/sec. The height of the slumped solid in the companion bed was 121 inches.

the section of the fast bed in question. Figure 7 thus attests to the high solid loadings that can be achieved in the fast fluidized bed. The upper curve in Figure 7 represents the maximum density that could be achieved in our apparatus under the given conditions and at the indicated section of the bed. Specifically, the upper curve shows data obtained with the solid control valve fully open. Solid densities as high as 13 lbs/ft^3 were recorded, corresponding to a gas voidage of about 76%.

Pressure gradients are normally higher at the bottom section of the bed, and, within the range of gas and solid rates used in our experiments, are lower toward the top of the bed. Figures 8 to 11 depict pressure gradients as a function of height from the bottom of the bed for four different fluidizing gas velocities. Figure 12 compares the pressure gradients at two different gas velocities.

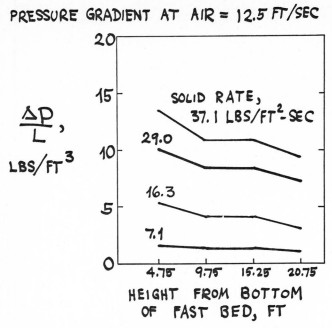

Fig. 10 Pressure gradient profile across the 3-inch fast bed fluidized at 12.5 ft/sec. The height of the slumped solid in the companion bed was 121 inches.

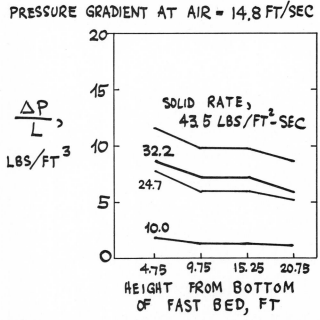

Fig. 11 Pressure gradient profile across the 3-inch fast bed fluidized at 14.8 ft/sec. The height of the slumped solid in the companion bed was 121 inches.

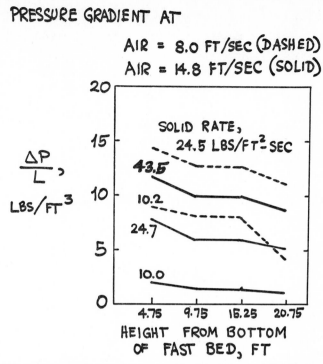

Fig. 12 A comparison of the pressure gradient profiles across the 3-inch fast bed fluidized at 8 and 14.8 ft/sec. The height of the slumped solid in the companion bed was 121 inches.

Slip velocities are high in the fast fluidized bed. Those corresponding to the data of Figure 7 are shown in Figure 13.

As our investigation proceeded, it soon became apparent that the fast fluidized bed possesses, as it were, its own personality or individuality. Changes in the design of the solid circulation loop, and, in particular, the manner in which the solid is conveyed and fed into the fast bed itself will affect, often appreciably, the fluidization characteristics in the fast bed. The height of the solid level in the 1-foot companion bed determines directly the maximum density gradient that may be achieved in the fast bed, and feeding solids with the paddle wheel produced results quite different from those obtained under the normal practice of conveying the solid via the U-tube. These differences are illustrated in Figure 14. The fluidization character in

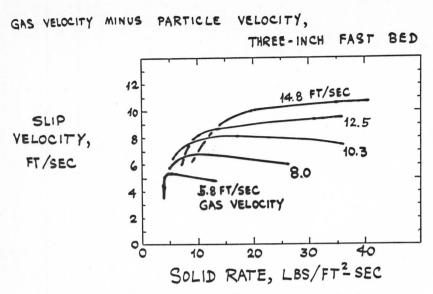

Fig. 13 Slip velocities corresponding to data shown in Fig. 7.

the companion bed, and the degree of aeration in the U-tube also caused
changes in the pressure gradients measured at a given elevation at the same
solid rate and fluidizing gas velocity.

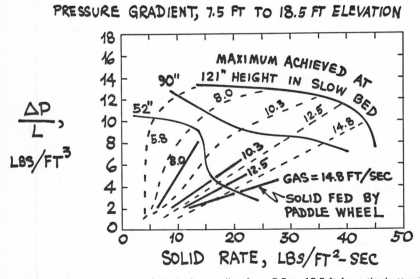

Fig. 14 Pressure gradient (across a section of the bed extending from 7.5 to 18.5 ft from the bottom) vs. solid rate. The three curves marked 121″, 90″ and 52″, represent the maximum gradient achieved with the indicated slumped solid height in the companion bed. The dashed curves are gradients reproduced from Fig. 7, and solid lines show data obtained when solid was fed by means of a paddle wheel.

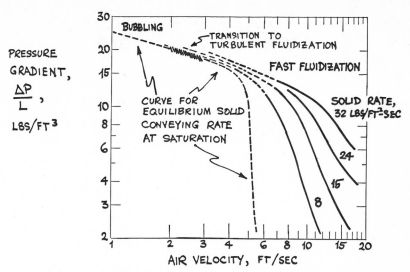

Fig. 15 Tentative phase diagram showing fluidization regimes for the fluid cracking catalyst. Gradients were measured across the bottom section of the bed, and the height of slumped solid in the companion bed was 121 inches.

In our 3-inch set-up we were able to generate data in the bubbling and the turbulent regimes. Combining these data with the results for the fast bed regime, we formed a portion of the phase diagram showing these regimes of fluidization (Figure 15). Our 6-inch fast bed is better equipped to measure gradients at low solid flow rates, and should in the near future provide an improved version of Figure 15.

The 3-inch fast bed is constructed in alternating glass and metal sections, and we could accordingly observe the transition from slugging (bubbling) to the turbulent regime. What we saw resembles very closely the description provided by Kehoe and Davidson.

Advantages of the Fast Fluidizing Bed

Vessel diameter is a far more important cost factor than vessel height, especially in an application at elevated pressure. In light of this fact, the capacity advantage of the fast bed is obvious from a glance at Figure 5.

The fast bed enjoys all of the bubbling bed's advantages of temperature uniformity, ease of introducing or withdrawing material or heat, capability of bringing a cold solid or gas feed almost instantaneously to temperature, etc'.

In addition, our strong impression is that the fast bed ought to afford excellent gas-solid contacting efficiency. Lanneau's argument (6) and Massimilla's data (9) support this impression, as do Wainwright and Hoffman's observation of excellent contacting for the oxidation of ortho-xylene in what was probably a fast fluidized bed (12). Lurgi also reports that absorption of HF issuing from Hall cells showed that the fast bed produced an effluent containing on the order of one-tenth of the HF found in effluents from a conventional bubbling fluidized bed.

Despite the backmixing of solid in the fast bed, we believe that the high gas velocities preclude any appreciable backmixing of gas, so that the operation approaches a plugflow condition.

Finally, the fast bed might prove easier to scale up than a bubbling fluid bed. At any rate, becuase of its high capacity, it would not be imperative to scale it up to uncomfortably large dimensions.

Advantages for the Gasification of Coal

The relatively high gas velocities characteristic of the fast bed would afford large gasification throughputs per unit cross sectional area of the bed. In addition, there are several factors that might contribute to high gasification rates per unit bed volume: First, the fast bed, as noted above, affords contact with a suprisingly large inventory of solid per unit bed volume; gas-solid contacting is expected to be excellent, and gas flow to approach a condition of plugflow.

The fast bed should be able to receive a feed of a highly caking

bituminous coal without pretreatment. This would be an important advantage.
Much of the Eastern coal is caking. Gasification in fixed or in a bubbling
fluidized bed at temperatures below 1000°C often resorts to awkward dodges to
circumvent the coal's caking propensity. Such schemes as preoxidation exact
serious penalties both with respect to the amount and quality of the product
gas as well as with respect to costs of additional hardware. The
fast bed closely resembles a draft tube shown by Consolidation Coal
Company (13) to be capable of receiving a caking coal. Consol's work also
suggests that feeding a caking coal directly into the fast bed could produce
a char of greater density and strength, a possible advantage not to be overlooked.

Review of the literature (see, for example, ref. 14) suggests that
a fast bed gasifier operating around or above 900°C could produce a gas free
of tars.

Other advantages will be discussed within the context of the
gasification schemes presented in subsequent sections.

COMPLETE GASIFICATION OF COAL BY AIR

The Ignifluid Process

The Ignifluid process is described in details elsewhere (15-17).
Briefly, the Ignifluid gasifies coal with air and burns the resulting fuel gas
to raise steam. Air is introduced at velocities ranging from 30 to 50 ft/sec
(at conditions) through a travelling grate into a bed of coarse particles
of coke arising from the devolatilization of the coal (Figure 16).
Temperatures in the Ignifluid gasification bed are around 1200° to 1400°C. At
these temperatures the ash matter of all coals is sticky, and one might expect
that a catastrophically massive clinker would form, but this does not happen.
Albert Godel, the inventor of the Ignifluid process, discoevered that small ash

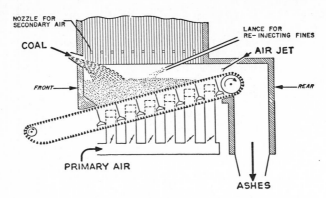

Fig. 16 Schematic of the Ignifluid gasification fluidized bed.

agglomerates form throughout the bed and remain fluidized, interspersed in particles of coke. The agglomerates grow in size at a controlled rate. The ash agglomerates typically constitute 10 to 20 per cent of the weight of the bed, and their carbon contents average about 5 per cent. Ultimately, the ash agglomerates reach the grate and are carted into the ash pit.

The mechanism by which the inorganic ash gathers into ash agglomerates in the Ignifluid bed - The Godel Phenomenon - has been described recently (18).

Fives-Cail Babcock of Paris, France, and Air Products & Chemicals,Inc. are working together to develop a revamp of the Ignifluid for "pure" gasification to supply a low-Btu gas for industrial use. Recent tests in a small unit at La Corneuve, France (Figure 17), lead us to believe that a gas of about 125 Btu/SCF could be produced in a commercial size unit. It should be remarked at the largest Ignifluid boiler in commercial operation can process about 400 tons of coal per day. Scale up to about 1000 tons/day appears feasible for the boiler unit as well as for a unit designed to operate as a gasifier per se.

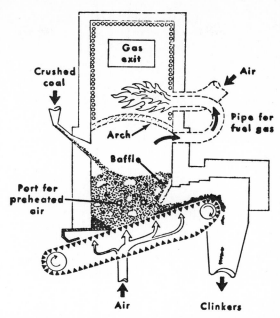

Fig. 17 Experiment on "pure" gasification in a revamp of a small test Ignifluid boiler at La Corneuve, France.

The City College Ash-Agglomerating Fast Fluidized Bed Gasifier

We have proposed a version of Godel's ash agglomerating gasifier to work at pressure (19). Our version would incorporates a fast fluidized bed to gasify coke fines, as shown in Figure 18. Development of this gasifier is

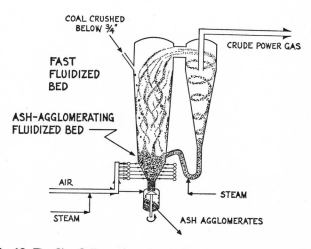

Fig. 18 The City College ash agglomerating, fast fluidized bed gasifier.

the subject of a proposal from Hydrocarbon Research, Inc. to ERDA.

An advantage of the design is that there are two controls on bed temperature: (1) a rapid response in temperature can be obtained by changing the flow of steam, normally fairly small relative to the flow of air; and (2) the inventory of carbon in the fast bed zone can be adjusted by regulating the rate of recirculation of fine carbon particles in the return leg from the cyclone. As we have pointed out earlier (19), the proposed gasifier should operate well at partial loads.

The proposed gasifier can afford a high capacity for processing coal, as the following example for New Mexico coal (7% ash) illustrates:

Inside diameter of gasification vessel = 10 ft

Fluidizing gas velocity = 12 ft/sec

Pressure = 20 atm;   temperature = $1200^{\circ}C$

Coal rate = 1,959 tons/day, moisture-and-ash-free (m.a.f.)

Gas make = 40,840 MMBtu/day

Higher heating value of the gas = 128.3 Btu/SCF (sulfur-free but not dried)

Electricity generating capacity that could be served at 40% efficiency = 198 megawatts (Mw).

PARTIAL GASIFICATION OF COAL BY STEAM

A surprise of the Synthane development of the U.S. Bureau of Mines was the production at $925^{\circ}C$ and 27 atm of a gas containing methane at an equilibrium ratio indicated by the black square near the top line of Figure 19, a figure published by one of the authors in 1961 (14). The top line reflected data for beds of coke fluidized by steam at atmospheric pressure and fed continuously with a caking bituminous coal (the star), a subbituminous coal (the open squares), and peat (the triangles). Looking at the line for batch gasification of cokes and anthracite coal at atmospheric pressure (the dot-dashed line in the figure) and data for gasification of bituminous char at 9.3 atm (the closed circles) and anthracite at 12 to 17 atm (the X's), a trend appeared toward lower methane equilibrium ratio with higher pressure for batchwise steam gasification of such materials in fluidized beds, without continuous feed of raw coal. There was accordingly the possibility that a similar trend would be seen for raw coal. The surprising data represented by the black square in Figure 19 was from a test at 27 atm conducted with continuous feed of lightly pretreated bituminous coal into a fluidized bed gasifier, about 18 inches in inside diameter, at Trenton, New Jersey, laboratory of Hydrocarbon Research, Inc. (20). The gasification medium was steam and oxygen.

The historic data at atmospheric pressure defining the top line in Figure 19, as well as the new data at 27 atm , were for a relatively small

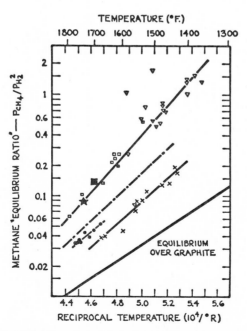

Fig. 19 Methane equilibrium ratios (14, 20) reckoned from composition of gases emerging from steam gasification of coals and chars in fluidized beds. See text for explanation of symbols.

ilization of fixed carbon in the coal. However, a commercially attractive operation can be visualized with essentially no extrapolation from the known experimental facts.

Figure 20 is a new concept for a fast fluidized bed gasifier with the advantage that the reaction with steam and raw coal is conducted in a fast fluidized bed zone that is free of products of combustion. The new concept takes advantage of the fast bed's relative freedom from massive backmixing of gas that characterizes the low-velocity, bubbling fluidized bed.

The gas emerging from the top of the steam-reaction zone in Figure 20 ought to have methane content according to the top line of Figure 19. For operation at an elevated pressure, this can amount to a substantial quantity of methane, and the gas at the top of the steam-reaction zone will be rich in fuel gas. In the upper part of the vessel, this rich gas is admixed with products of the combustion of coke with air that occurs in the annular space

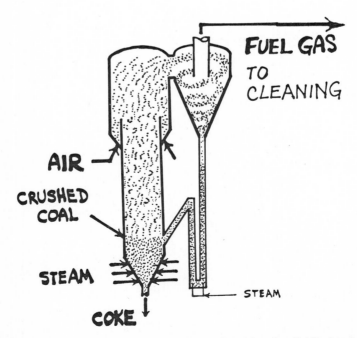

Fig. 20 New concept for partial gasification of coal in a fast fluidized bed.

provided around the upper part of the steam-reaction zone. The arrangement insures that no methane will react with oxygen in the air. The decomposition of methane is slow at temperatures of interest, and so methane will survive passage through the top part of the vessel in Figure 20.

An example for Illinois No. 6 coal (10% ash) illustrates the possibilities of Figure 20. The example is for a preheat temperature of 760°C, as might be achieved, for example, by fluidized bed combustion.

Inside diameter (top section) = 10 ft

Fluidizing gas velocity = 12 ft/sec

Pressure = 20 atm; temperature = 900°C

Coal feed, tons/day (m.a.f.)	3,897
Coke product, tons/day (m.a.f.)	2,003
Steam flow, lbs/hr	426,000
Gas make, MMBtu/day	45,026
Higher heating value of gas	
Dry, sulfur free (Btu/SCF)	226
Dry, sulfur- and CO_2-free	274

The effect of air preheat temperature is illustrated:

Air and Steam Temperature	Higher Heating Value of Gas Dry, sulfur-free	Also CO_2-free	Gas Make MMBtu/day
649°C	206	248	42,693
760°C	226	274	45,076
871°C	249	302	47,466

The fuel gas will be well suited for many industrial uses. For a distribution to a region, a version of Figure 20 blown with steam at the bottom and an equimolar mixture of steam and oxygen at the top might be

preferred , as in the example for Illinois No. 6 coal that follows. Oxygen
and steam are supplied at 450°C. In other respects, the arrangement and
gasification conditions are like those listed in the example above.

Coal feed, tons/day (m.a.f.)	8,509
Coke product, tons/day (m.a.f.)	4,332
Oxygen , tons/day	1,810
Steam flow, lbs/hr	333,000
Gas make , MMBtu/day	101,000
Higher heating value of gas	
Dry, sulfur-free (Btu/SCF)	409
Dry, sulfur- and CO_2-free (Btu/SCF)	536

It may be noted that the Koppers-Totzek requires, for production of
101,000 MMBtu/day of gas of far lower heating value, more than 4000 tons/day
of oxygen. Also, to process the amount of coal shown above several K-T units
will be needed in place of the single fast bed gasifier.

The proposed gasifier manufactures a coke byproduct, and we might
accordingly offer at this point a rationale for the partial gasification of coal:
As industry prepares to rely more upon coal for its future energy needs, it
will inevitably turn first to direct combustion of coal. Coals and chars
will be burned in increasing amounts in equipment of conventional or upgraded
design. As fluidized bed combustion matures to commercial status, it will
offer new opportunities.

With an opportunity to burn coal or char for process steam or power, it
will prove considerably cheaper to practice   partial instead of total
gasification of coal. Indeed, the preoccupation with complete gasification
of coal often lent unnecessary complexities to existing and proposed
gasification processes.

PARTIAL GASIFICATION BY FLASH HYDROGENATION

We have explored experimentally the flash hydrogenation of Illinois No. 6 coal at 100 atm and at temperatures ranging from 600° to 1000°C. The residence time of vapor products was controlled between 0.6 and 5 seconds. In a typical experiment, our coal sample is heated electrically to reaction temperature in about 1 second, and is maintained at temperature for about 10 seconds. We collect the vapor reaction product in a vessel and analyze the total mixture by gas chromatography. In a subsequent experiment, we burn the coke residue and analyze for carbon oxides.

Figures 21 and 22 give the yield structure based upon 29 runs (23 with carbon closure) made with a mine-mouth sample of coal. Over the range of 600° to 1000°C, the only light products observed in more than

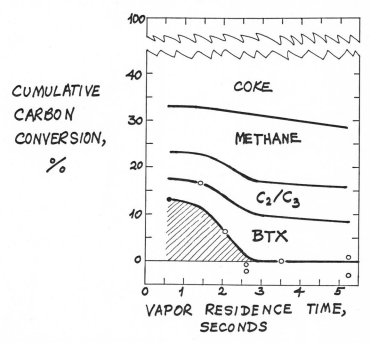

Fig. 21 Yield structure from flash hydrogenation of Illinois No. 6 coal at 100 atm and about 700°C vs. vapor product residence time. C_{2+} = yield of ethane and propane (with trace of butane), BTX = yield of benzene, toluene, and xylenes. The shaded area in the lower left is carbon unaccounted for by the carbon balance. A plotted point is the carbon not found in vapor products by hydrogeneation or carbon oxides from a subsequent combustion; a point falling on the bascissa represents a perfect carbon balance.

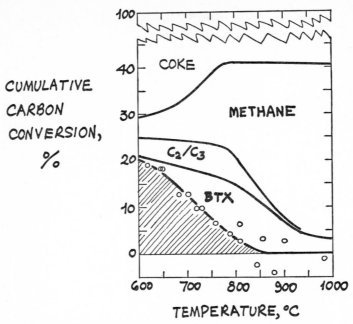

CUMULATIVE CARBON CONVERSION, %

TEMPERATURE, °C

Fig. 22 Yield structure at a product residence time nominally 0.6 seconds vs. temperature. (For significance of the symbols, shaded area, and plotted points, see legend for Fig. 21.) Propane yield falls linearly from 2% at 620°C to nearly zero at 780°C. The ratio of benzene to total BTX rises nearly linearly from 0.41 at 620°C to substantially unity beyond 790°C.

trace amounts (above 1%) are methane, ethane, propane and BTX (benzene toluene and xylenes). Direct tests show that xylene and lighter species introduced ahead of the reactor are transmitted quantitatively to our collection reservoir. Carbon balances suggest that little if any material heavier than xylene appears in products from runs beyond about 3 seconds in Figure 21 and beyond about 850°C in Figure 22. The precision of the carbon balance for these runs (95% confidence limit) is ± 3%, placing an upper limit to heavy tar, if any is present.

We believe that carbon unaccounted for in runs below about 3 seconds in Figure 21 and below about 850°C in Figure 22 reflects the presence of

species heavier than xylene.

Noncatalyzed flash hydrogenation of coal could supply needs of the pipeline gas, petrochemical, liquid fuel, and electric power industries. Our data are for a supply of hydrogen far too large relative to the coal sample to be economic (i.e., far too low hydrogen conversion). Data following the reaction path at a realistic supply of hydrogen are needed before flash hydrogenation can be evaluated economically. That such data are worth getting is illustrated by our yields from Figure 21 at 3 seconds: A plant consuming 27 million tons of coal could supply 0.22 trillion standard cubic feet of methane, about 1.4 million tons of ethylene (obtained from cracking ethane product), and 2.7 million tons of chemical benzene. These are respectively 1%, 13% and 50% of the 1973 U.S. consumption.

The fast fluidized bed is eminently suitable for the flash hydrogenation of coal. A raw coal introduced into the fast bed would be heated almost instantaneously to reaction temperature, and the high fluidizing velocities characteristic of the fast bed would permit short residence time for the vapor product. Proper choice of operating conditions in light of available data would ensure a product gas free of heavy tars.

We plan to use a mass spectrometer to analyze for sulfur species in combustion gases from the coke residue in our laboratory experiment. Such analysis can confirm our expectation that the coke residue will be low in sulfur content, provided the coal contains little pyritic sulfur. A coke byproduct low in sulfur would of course be a desired product.

ACKNOWLEDGEMENT

Work on fast fluidization and flash hydrogenation of coal is supported by Grant GI-34286 from the RANN Program ("Research Applied to National Needs") of the Unites States National Science Foundation. David Turner conducted nearly all the experiments on fast fluidization; Alan E. McIver assisted in gathering the fluidization data using the paddle wheel, and in the

installation of the test stand for study of fast fluidization. Eli Gilbert
assisted in developing the experiment on flash hydrogenation of coal ;
Anthony Chiaravalotti performed most of the experiments. We thank Richard D.
Harvey and his colleagues at the Illinois State Geological Survey for their
cooperation.

REFERENCES

(1) .H.E. Risser, Environmental Geology Notes, No. 64, Illinois Geological
     Survey, July 1973.

(2) Federal Power Commission, Bureau of Natural Gas, "A Realistic View of
     U.S. Natural Gas Supply", Washington, D.C., December 1974.

(3) William J. Amos, Air Products and Chemicals, Inc., private communication.

(4) H.R. Warman, Environment and Change, Vol. 2, No. 3, pp. 164-172, Nov. 1973.

(5) D. Geldart, Powder Technology, 7, 285 (1973.

(6) K.P. Lanneau, Trans. Inst. Chem. Engrs., 38, 125 (1960).

(7) R.M. Braca and A.A. Fried, "Operation of Fluidization Processes", in
     Fluidization, D.F. Othmer, Editor, Rheinhold, New York, 1956.

(8) P.W.K. Kehoe and J.F. Davidson, Inst. Chem. Engrs.(London) Symp. Series.,
     No. 33, 97 (1971).

(9) L. Massimilla, A.I.Ch.E. Symp. Series, 69, No. 128, 11 (1973).

(10) W.K. Lewis and E.R. Gilliland, U.S. Patent 2,498,088 (Feb. 21, 1950); a
     division of an application filed in 1940.

(11) L. Reh, Chem. Eng. Progr., Vol. 68 (February 1971), p. 58.

(12) M.S. Wainwright and T.W. Hoffman, Advances in Chemistry Series, 133,
     669 (1974).

(13) G.P. Curran and E. Gorin, paper presented at Third International Confe-
     rence on Fluidized Bed Combustion, EPA, Houston Woods, Ohio, 1972.

(14) A.M. Squires, Trans. Inst. Chem. Engrs., 39, 3 (1961).

(15) A.M. Squires, Science , 169, 821 (1970).

(16) A.A. Godel, Rev. Gen. Therm. 5, 349 (1966)

(17) J. Yerushalmi and A.M. Squires, In Conference Proceeding: Power Generation -
      Clean Fuels Today, EPRI ("Electric Power Research Institute") -SR-1,
      Monterey, California, 1974, p. 45.

(18) J. Yerushalmi, M. Kolodney, R. A. Graff, A.M. Squires and R.D. Harvey,
      Science, 187, 646 (1975).

(19) S. Dobner, M.J. Gluckman and A.M. Squires, A.I.Ch.E. Symp. Series,
      Vol. 70, No. 137, 223 (1974).

(20) S. Dobner, R.A. Graff and A.M. Squires, "Analysis of Trials of the
      Synthane Process", Report prepared for Hydrocarbon Research, Inc. under
      contract with U.S. Bureau of Mines, March 1974.

HEAT TRANSFER IN FAST FLUIDIZED BEDS

K. D. KIANG, K. T. LIU, H. NACK, and J. H. OXLEY

ABSTRACT

Preliminary heat transfer measurements in a fast fluidized bed are presented in this paper. Results indicate heat transfer coefficients as high as those of a classical dense fluidized bed can be obtained. The study also shows that (1) heat transfer coefficient in the fast bed increases with increasing solids recirculation rate of the bed material, and (2) heat transfer coefficient is very uniform in the fast bed.

INTRODUCTION

Ever since Lothar Reh of Lurgi Chemic und Huttentechnik GmBh first observed the fast fluidization phenomenon, this technique has been successfully applied to several commercial processes such as calcination of aluminum hydroxide[1,2]. The potential for widening the application of this technique to energy-related processes seems to be significant because of its capability for improving the throughput of several processes.

Little effort has been made to characterize the transport behavior of the system due to its extreme complexity. Presented in this paper are the results of our preliminary investigation on the heat transfer behavior of the fast fluidized bed.

EXPERIMENTAL RESULTS

Figure 1 shows a schematic diagram of a cold model fast fluidized bed system. The fluidized bed column is 4 inches in diameter and 12 feet in height. An 8-inch cyclone and a hopper were used for recirculation of bed

*The work reported in this paper was supported by the Battelle Energy Program.

material. For control of solid recirculation rate, a ball valve was installed between the cyclone dip leg and the bed. The entire setup was made of
transparent plexiglass for convenience of visual observation. Several minia-
ture heaters (3/4" OD x 2-1/4") were used for measurement of heat transfer
coefficients. These heaters were located 21" (Heater 1), 51" (Heater 2),
93" (Heater 3), and 123" (Heater 4) above the distributor plate corresponding
to L/D ratios of 5.25, 12.75, 23.75 and 30.75 respectively. The bed material
selected for this study was a cracking catalyst (Houdry Mineral HFZ-20) with
the following physical properties:

Bulk density (packed), g/cc   0.86-0.90
Particle size analysis, weight percent
    0-20 microns   2.0 max.
    0-40 ditto   10-15
    0-80   "   78-85
    0-105 "   93-97
    0-149 "   99 min.

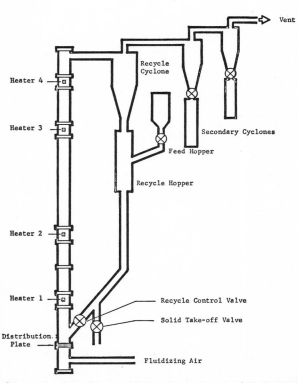

Fig. 1 Schematic of the 4-inch cold flow fast-bed model (4″ dia X 12′ long).

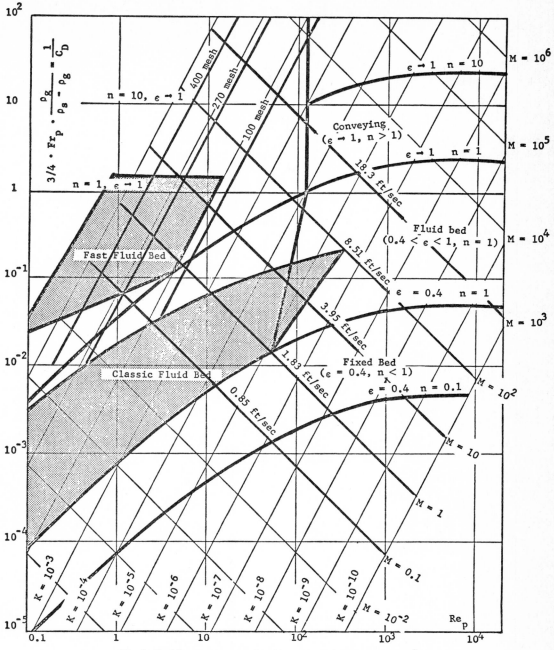

Fig. 2 Fluidized-bed status diagram for cracking catalyst at 70°F.

In typical operation, predetermined amounts of bed material (bed inventory) is charged to the bed. At a given gas velocity, the heat transfer coefficient is measured after the system reaches a steady state. The heat transfer coefficient, h, is then calculated by the following formula:

$$h = \frac{Q}{A(T_h - T_b)}$$

where Q is the electrical power input through the heater; A is the total surface area of the heater; T_h is the surface temperature of the heater; and T_b is the bed temperature.

Shown in Figure 2 is Reh's fluidization status diagram. The constant particle size lines and constant gas velocity lines are given for the case when the cracking catalyst is fluidized with air at room temperature. One hundred to 400 mesh lines correspond to the particle size range of the cracking catalyst used in this study. The diagram indicates that fast-bed mode fluidization should exist when superficial air velocity is greater than 2 ft/sec. It should be mentioned that the minimum fluidization velocity is less than 0.05 ft/sec and the terminal velocity is about 0.85 ft/sec for this bed material.

Comparision of Heat Transfer Behavior between Fast and Dense Fluidized Bed

Comparison of heat transfer coefficients of a classical dense fluidized bed and a fast fluidized bed is shown in Figure 3. It is clear that high heat transfer coefficients can be maintained at superficial gas velocity several times that of the terminal velocity of the particles, whereas for the case of dense fluidized bed, it is limited by the terminal velocity (about 0.85 ft/sec in this case) of the particle; i.e., the operability range is much narrower.

Effect of Solid Recirculation Rate

During this study, no attempt was made to determine the solid recirculation rate quantitatively. However, the solid recirculation rate was varied by adjusting the opening of the ball valve located between the recycle dip leg and the bed. Therefore, opening position of the valve should represent the magnitude of solid recirculation rate; larger valve opening corresponds to higher solid recirculation rate.

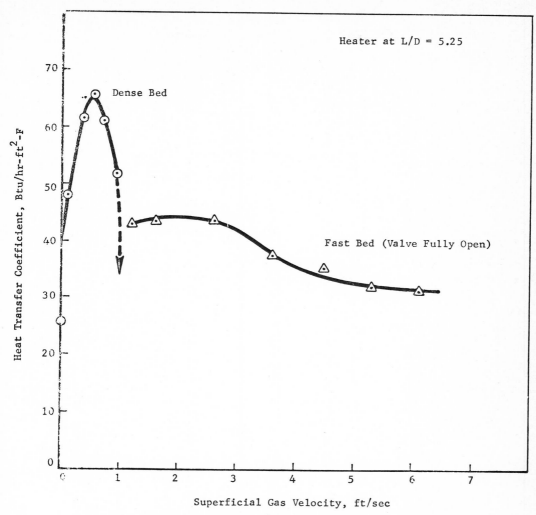

Fig. 3 Comparison of heat transfer behavior of dense and fast fluidized beds.

It is conceivable that at a given high gas velocity (e.g., 6 ft/ sec) the average bulk density of the bed increases with increasing solid recirculation rate. In other words, the bulk density of the bed should be higher at larger valve openings. This is demonstrated in Figure 4.

Heat transfer coefficients measured at position of Heater 1 (L/D = 5.25) for various valve opening positions and gas velocities are shown in Figure 5. With the valve fully closed, it is observed that the system became inoperable as gas velocity exceeded terminal velocity. This condition is in fact a dense fluidized bed operation. With the valve opening increased to 25

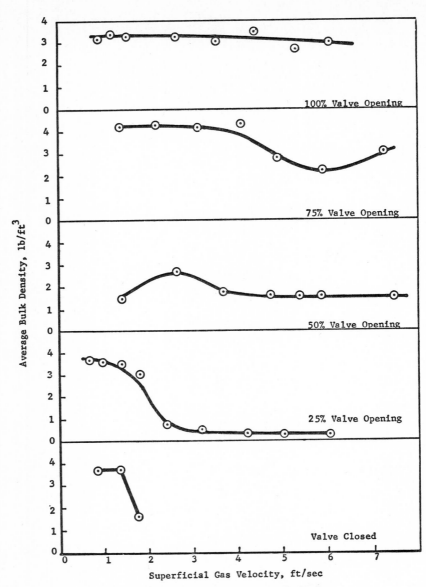

Fig. 4 Effect of recirculation rate on bulk density.

percent, heat transfer coefficient decreases sharply with increasing gas velocity. Although recirculation of solid can be maintained, it is believed that the
system is in the transport regime at this condition, due to insufficient
solid recirculation rate. The low heat transfer coefficient at higher
velocities can be explained by low bulk density of the bed (see also Figure
4).

For valve openings larger than 50 percent, the heat transfer coefficients decrease from about 45 Btu/hr-ft^2-F at low gas velocity to about 32 Btu/hr-ft^2-F at high gas velocity and level off. These observations suggest that high heat transfer rates can be obtained at high solid recirculation rate. It also seems that at high gas velocity (>5ft/sec) the heat transfer coefficient is independent of solid recirculation rate when the solid recirculation rate exceeds a certain value.

Similar results were observed at position of Heater 2 (L/D = 12.75) as shown in Figure 6.

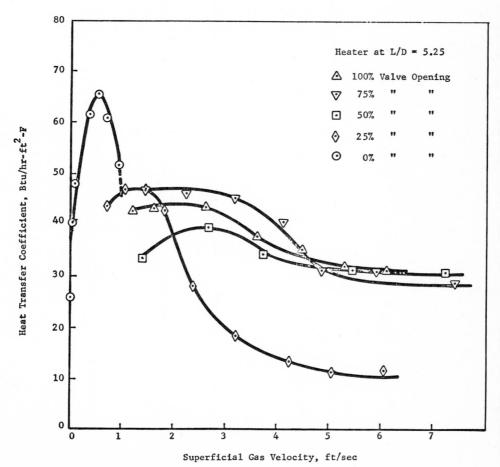

Fig. 5 Effect of recirculation rate on heat transfer.

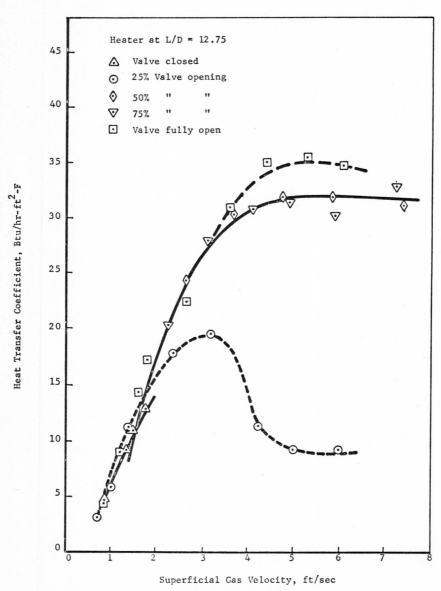

Fig. 6 Effect of recirculation rate and bed height on heat transfer.

Effect of Axial Position on Heat Transfer

Heat transfer coefficients at several axial positions in the bed are shown in Figure 7. The results seem to indicate that relatively uniform heat transfer coefficient can be obtained in the fast bed. The slight

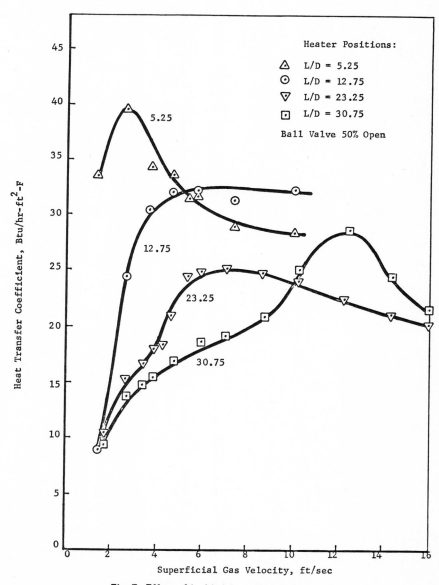

Fig. 7 Effect of bed height on heat transfer.

variation might be due to the large particle size distribution that results
in some inhomogeneity of bed density in the bed and end effects.

Effect of Solid Inventory on Heat Transfer

The effect of solid inventory on heat transfer coefficient is shown
in Figure 8. Heat transfer coefficients are generally higher for high solid
inventory (e.g., 18 lbs) than that of low solid inventory (e.g., 10 lbs).
However, higher initial bed inventory than necessary (e.g., 28 lbs) appears
to result in loss of bed material until a steady condition is reached.

Proposed Heat Transfer Correlation in Fast-Fluidized Bed

It is the authors' opinion that the correlation of heat transfer
coefficient in the fast-fluidized bed should be similar or identical to that
of a dense fluidized bed. For example, the heat transfer coefficient between
a dense bed and a vertically immersed tube has been correlated as[3]

$$\frac{h_w d_p}{k_g} = 0.01844 \, C_R \, (1 - \epsilon_f) \left(\frac{C_{pg} \, \rho_g}{k_g} \right)^{0.43} \left(\frac{d_p \, \rho_g \, U_o}{\mu} \right)^{0.23} \left(\frac{C_{ps}}{C_{pg}} \right)^{0.8} \left(\frac{\rho_s}{\rho_g} \right)^{0.66}$$

Similar correlation can be applied to a fast bed system, i.e.

$$\frac{h_w d_p}{k_g} = \text{const} \, (1 - \epsilon_f)^a \left(\frac{C_{pg} \, \rho_g}{k_g} \right)^b \left(\frac{d_p \, \rho_g \, U_o}{\mu} \right)^c \left(\frac{C_{ps}}{C_{pg}} \right)^d \left(\frac{\rho_s}{\rho_g} \right)^e$$

where a, b, c, d, e are parameters to be determined experimentally.

In the fast bed system, the void fraction term (ϵ_f) is not only
a function of the superficial gas velocity, but also strongly dependent on
recirculation rate of bed material ($\dot{W}_s$, in lbs/ft^2– sec), solid density and
other physical properties of gas and solid. It can be postulated that the
physical properties of gas and solid can be characterized by a single parameter;
minimum fluidization velocity (U_{mf}) is selected for convenience. Therefore

$$\epsilon_f = \epsilon_f \, (\dot{W}_s, \, \rho_s, \, U_o, \, U_{mf})$$

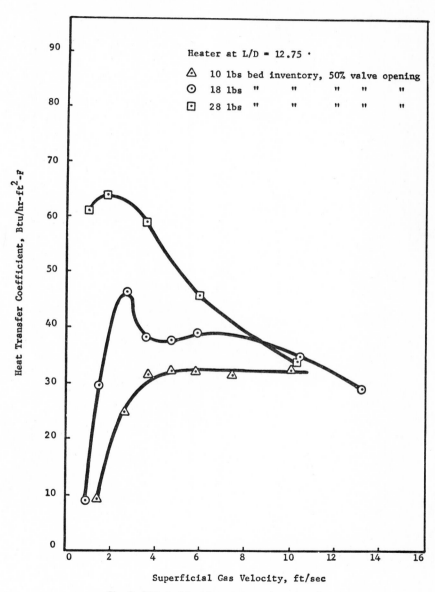

Fig. 8 Effect of bed inventory on heat transfer.

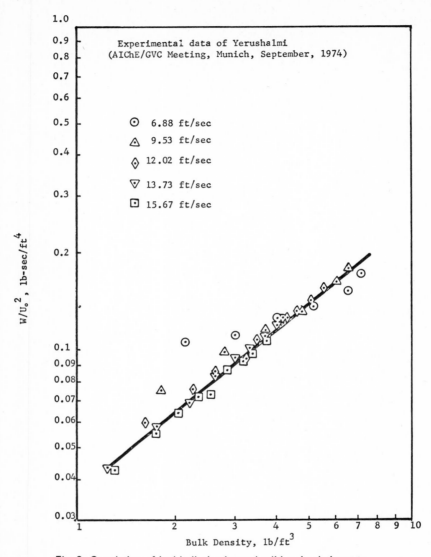

Fig. 9 Correlation of bed bulk density and solid recirculation rate.

Dimensional analysis shows that ϵ_f is a function of a dimensionless group

$$\epsilon_f = \epsilon_f \left(\frac{\dot{W}_s \, U_{mf}}{\rho_s \, U_o^2} \right)$$

In other words, the bulk density in fast bed is also dependent on the dimensionless group $\left(\frac{\dot{W}_s \, U_{mf}}{\rho_s \, U_o^2} \right)$. This relationship agrees with the experimental data reported by Yerushalmi et al[4] (shown in Figure 9) in that a straight line is obtained by plotting log $\left(\frac{\dot{W}_s}{U_o^2} \right)$ vs. log (ρ_b), where ρ_b is the bulk density of the bed.

CONCLUSIONS

Based on the above experimental observation, the following conclusions are drawn regarding heat transfer in a fast fluidized bed:

(1) High heat transfer coefficient can be maintained at gas velocity several times the terminal velocity of the particles

(2) High solid recirculation rate enhances heat transfer

(3) Heat transfer rate is very uniform throughout the bed.

SUGGESTED FURTHER WORK

The following studies are suggested for further development of this area of fluidization:

(1) Develop reliable methods of measuring solid circulation rate

(2) Conduct more extensive experiments to develop suitable heat transfer correlations

(3) Apply technology to practical systems such as fluidized bed combustion and gasification.

REFERENCES

(1) Reh, L., "Calcinating Aluminum Trihydrate in a Circulating Fluid Bed, A New Technique" Aust. Inst. Min. Met. No. 241, March 1972.

(2) Reh, L., "Fluidized Bed Processing", Chem. Eng. Prog., 67, No. 2, Feb. 1971.

(3) Kunii, D., and Levenspiel, O., "Fluidization Engineering", Wiley & Son, Inc., 1969.

(4) Yerushalmi, J., et.al., "The Fast Fluid Bed", AICHE/GVC Meeting, Munich, Germany, September 17-20, 1974.

FLUIDIZED-BED PERFORMANCE WITH INTERNALS HEAT EXCHANGER ABOVE THE SLUMPED BED

CALLIXTUS J. AULISIO, SHELTON EHRLICH, RICHARD W. BRYERS, and JOHN BAZAN

SCOPE OF THE WORK: A heat exchanger made up of horizontal tubes in a staggered array is installed within an 8 square foot fluidized-bed combustor and a 36 square foot cold model. The major fraction of the bed is below the bottom tube of the array. The paper describes the results of experiments.

OBJECTIVES: A relatively shallow fluidized bed, 2 feet slumped height is adequate for combustion of coarsely crushed coal, -¼". Control of bed temperature requires approximately 6.5 square feet of tube surface per square foot of bed cross section at peak firing rate, ~900,000 Btu/hr,ft^2 of grate. Mechanical considerations require that most of the heat exchanger be located above the level of the slumped bed. The objective of the cold experiments was to determine if the bed would expand to fill the heat exchanger and if the violent bubbling experienced with the open furnace would be moderated by tubes beginning 18" above the air distributor.

The objective of the hot experiments was to determine if the heat exchanger would remove heat generated in the bed despite the relatively low density of particles within the expanded bed.

RESULTS: Figure 1 shows the heat exchanger as installed
in the hot model. The cold model internals are identical in
all essential features except that 127 tubes are used.

Table 1 describes the particles used in the models.

The Ergun equation was used to estimate the minimum
fluidizing velocity. Cold tests in the hot model showed that
the Ergun equation accurately predicts the minimum fluidizing
velocity for these coarse solids. The cold model cannot now
be operated at close to minimum fluidization.

Tests without internals showed that bed expansion, at
minimum fluidization, was only 6%. This was determined by
visual observation. At high superficial velocities, to
12 ft/sec, the bed bubbled violently and a top surface could
not be defined. However, pressure drop values across 12"
of bed showed that the bed density did not decrease markedly
at the high velocities indicating that bed expansion was still
relatively low. With the tube bundle placed in the beds as
shown in Figure 1 bed expansion increased and pressure oscillations due to bubbling decreased. Figure 2 shows the pressure
transducer output for the 36 square foot cold model with and
without tubes at a superficial velocities of 5 to 12 ft/sec.

The tubes performed well as a heat exchanger. The single
experiment completed at the time of this writing showed a
bed-to-tube heat transfer coefficient of 40 Btu/hr,ft^2,°F.
This may be compared with 50 Btu/hr, ft^2, °F for a vertical
heat exchanger in the same furnace. The coolant was pressurized
water at an average temperature of 285°F. The integrated
heat flux to the entire bundle was found to be proportional

to the superficial velocity and bed mass over the 2:1 range in both parameters studied. That is, a constant bed temperature is maintained under varying load conditions by varying the bed mass. Local, tube-by-tube, heat transfer coeficients have been measured. Figure 3 shows that tubes in level 1 (at 18" above the air distributor) are not as effective as the tubes above this level in removing heat.

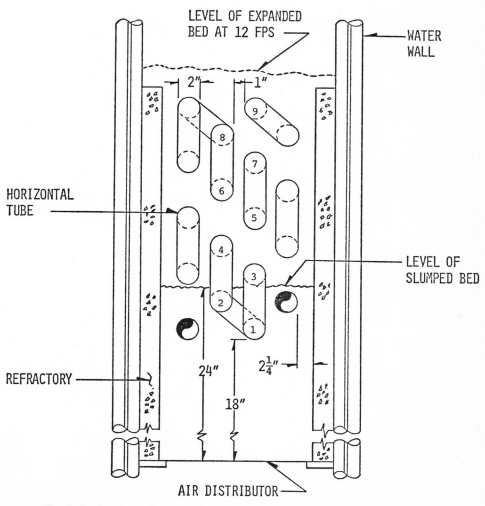

Fig. 1 Section thru tube bundle in hot model. No. indicates tube level (see Fig. 3).

Presently this is believed to be due to bubbles that grow
to relatively large size up to the bundle and blanket the
lowest row of tubes. The lower tubes reduce the bubble size
and tubes higher in the exchanger are thus more effective.
Bed temperatures are uniform throughout the heat exchanger
which indicates very substantial mixing.

SIGNIFICANCE AND COMMENTS: Internal surfaces that
are in the splash zone of a fluidized-bed of coarse particles will markedly reduce the pressure fluctuations caused
by bubbling.

The above-bed internals perform effectively as a heat
exchanger. This means that a heat exchanger may be designed

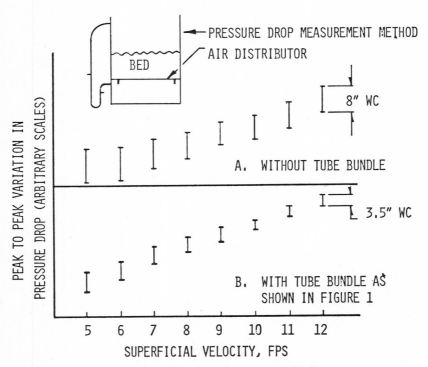

Fig. 2 Effect of internals on pressure oscillations.

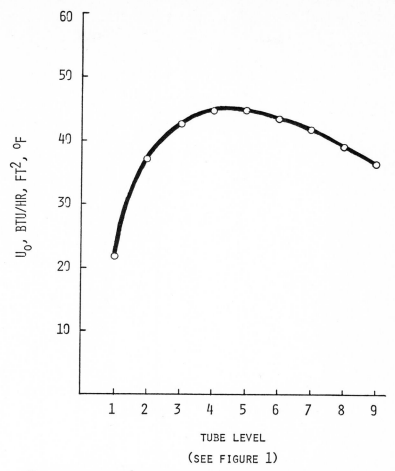

Fig. 3 Plot of local heat transfer coefficient vs. level of tube in bundle.

with relatively wide tube spacing. Because the heat flux to the bundle increases linearly with superficial velocity and bed mass the fluidized-bed boiler may be operated over a wide range of throughputs.

DESIGN AND OPERATING EXPERIENCE OF HIGH-PRESSURE AND HIGH-TEMPERATURE FLUIDIZED-BED REACTORS

BERNARD S. LEE

Two important design features must be kept in mind in the conversion of coal to high-Btu, pipeline-quality gas. First, because of the scale of operation involved in energy supply, the gasification reactors would be very large in size and require high capital cost. Second, the gasification reactions are either highly exothermic or highly endothermic; so the efficient transfer of heat becomes an important consideration. Fluidized-bed reactors, therefore, are commonly used in the coal gasification field because they meet these two criteria: ease of scale-up and outstanding heat transfer characteristics.

To achieve maximum conversion efficiency from coal to gas, high pressure is required to promote the formation of methane from carbon and hydrogen. In addition, high temperature is required to increase the reaction rate to an acceptable level. Thus, the fluidized-bed reactors have to operate at pressures up to 1500 psig and temperatures up to 2000°F. Fluidization data and design and operation guidelines are not available for these operating regions. This paper describes some design features, operating experience, and considerations for the scale-up of gasifier reactors to commercial size in the Institute of Gas Technology's (IGT) HYGAS program.

GAS DISTRIBUTOR

One of the key design areas in the fluidized-bed reactor is the gas distributor or grid. In the HYGAS program, three types of grid designs have been successfully tested. The first, a metal grid with simple perforations, has been used as a distributor for a coal slurry dryer. The temperature here, although high, permits the use of a metal grid. Hot gases up to 1500°F are distributed across the grid to fluidize a bed above into which is sprayed the feed slurry of coal and light oil.

The slurry dryer bed operates at 600°F or higher. The grid pressure drop is designed to be 0.8 psi, at a hole velocity of 70 ft/s. As a variation of the perforated grid design, special inserts with a "hockey puck" distributor that spreads the gas out horizontally and prevents solids weeping through the grid was tested. This design was found to be necessary at the time because we wanted to retain a starting bed in the slurry dryer. As we developed and improved the start-up procedure, it became possible to start the test with an empty reactor, thus the "hockey puck" design was replaced with a straight perforated grid that has been in operation for over 2 years. The typical bed height above the grid is 8 feet, representing a bed pressure drop of 1.5 psi.

The second type of grid is made up of cast refractory pieces built into a self-supporting arch. This grid is used to distribute the hot gases from the steam-oxygen gasifier into the high-temperature hydrogasification stage above. A 1900°F temperature is expected entering the grid, and 1600° to 1800°F above the grid. A number of designs were tested and are shown in Figures 1 to 3. A dense refractory material weighing 170 lb/cu ft was cast into brick shapes, which were fired in kilns and then fitted in circular courses inside the gasifier reactor. The distributor holes were cast into the refractory pieces, first on the side of the bricks, Figure 1, which made the casting job easier. However, in a small reactor with a few pieces of brick, any misalignment between two adjacent pieces would displace the two half holes. In later designs, the holes were cast in the center of each refractory piece, Figure 2. To reduce the pressure drop across this grid, we only kept a 1-inch-long orifice at the top of the grid, while a cone was cast into the lower half of the brick. Again, metallic inserts were tried when the starting bed was to be kept in the reactor during the heat-up period. The metallic inserts, however, did not survive the high temperature well: They warped and became greatly distorted. With the improvement on the start-up procedure, we can now operate with a refractory grid that has straight perforations.

The pressure drop across the refractory grid is 0.8 psi, with a hole velocity of 70 ft/s. The bed depth above the grid is 10 feet to give a bed pressure drop of 1.7 psi.

The difficulty with the original grid design was that the rise in the arch of 0.25 inch per foot of reactor diameter was too shallow. As a result, the pieces of refractory gradually dropped down during the heat-up and cooldown cycles of the pilot plant. With temperature cycling, the refractory pieces wedged in deeper and deeper in the center until the compressive forces on the bricks

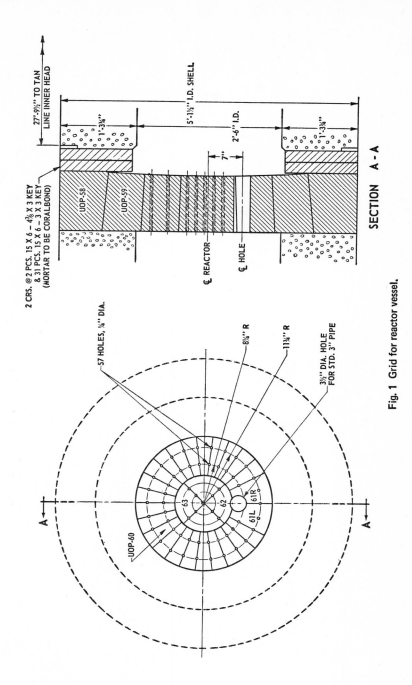

Fig. 1 Grid for reactor vessel.

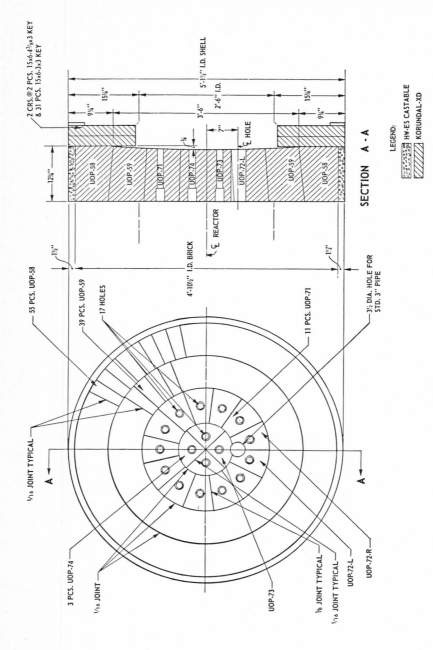

Fig. 2 Assembly for perforated dome.

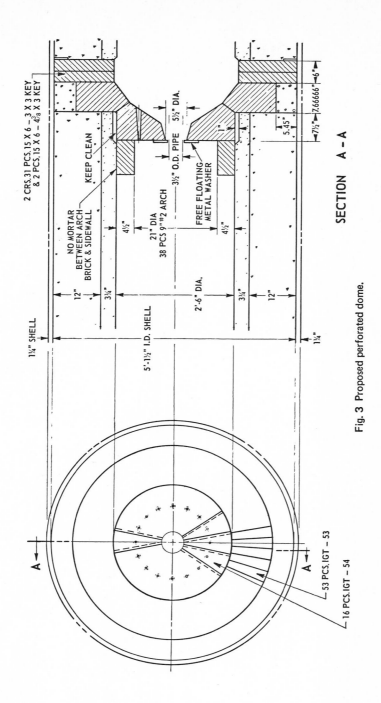

SECTION A – A

Fig. 3 Proposed perforated dome.

495

caused these to shear and spall. The solution to this problem was a redesigned
grid, supplied by Harbison-Walker Refractories Co., to give a much higher rise
in the arch, namely 2.5 inches per foot of reactor diameter. In addition, the solids
transfer pipe from the hydrogasification stage to the steam-oxygen stage below
was moved from the edge of the reactor to the center. This makes the grid design
completely symmetrical, as shown in Figure 3. Around the central transfer pipe
is a collar which rests, through high-temperature gaskets, on the refractory grid
itself. This arrangement has been in service since June 1974 and we have seen
no pinch spalling of the grid.

The third type of distributor is a ring with holes perforated around it and
with a conical nozzle attached to each hole to properly form the gas bubble as it
leaves the perforation. Such a grid was used on the steam-oxygen gasification
stage,where hot spots must be avoided if oxygen and coal are not to be randomly
contacted. The pressure drop across the ring is 17 psi with a hole velocity of
250 ft/s. The high pressure drop at full design flow rate was chosen to ensure
good gas distribution for operation at lower capacities. The bed height is 10 feet,
representing a bed pressure drop of 1.1 psi. Of particular interest is the selection
of the cone dimensions to help prevent the formation of clinkers and slagging, even
in the presence of direct oxygen contact with coal. The proper cone angle[3] prevents
solids weeping into the distributor hole and subsequent premature combustion in
the cone.

The diameter of the gasifier reactor in the pilot plant is 30 inches in all instances. In scaling up the grids to a commercial-size reactor of around 20 feet in
diameter, the standard perforated grid will be conventional in design, except for
a dished shape of the grid to withstand the grid pressure differential. The ring
distributor grid will become a pipe grid, with each hole having a cone nozzle. The
refractory grid can be built in large diameters; however, to keep the grid from be-
coming too thick and the weight of each individual piece from being too great, it
will be desirable if the top of the grid can be built in an arched shape, rather than
a flat surface. Depending upon the density of the grid material and the thickness
of the grid, the maximum design pressure drop across the grid should be no more
than 1.5 psi. Pressure surges must be avoided in the operation of a refractory grid,
otherwise the grid can be lifted, as we have experienced in the pilot plant.

SOLID TRANSFER BETWEEN STAGES

Another key design area is the transfer lines between fluidized beds and solids flow control valves. Many of the rules for the design of transfer lines or standpipes derived from experience with fine catalyst particles have to be modified when handling relatively coarse solids under high pressure. In a catalytic cracking unit, solids moved through the lines at many feet per second, which causes the solids to be self-fluidized.

The need to aerate a standpipe is minimal, because at high pressure the effect of the hydrostatic head compressing the interstitial gas is negligible. Under low-pressure conditions and with dense solids, aeration taps are normally located at a distance such that the hydrostatic head between the aeration taps is enough to compress the interstitial gas and reduce the relative velocity between the gas and solids to below the bubbling point. At high system pressures around 1000 psig, however, the small pressure change caused by the hydrostatic head of a leg of solids represents only a small fraction of the system pressure. This small fraction, therefore, has a small effect on the compression of the gas and hence on the reduction of the relative gas-solids velocity. Aeration taps are only installed in transfer lines to initiate solids flow in the pipe. Once the flow begins, even at a solids flow velocity of less than 1.5 ft/s, there is no need for aeration.

A number of solids flow control valves were developed and tested. To adapt conventional slide valves for high-pressure applications requires difficult design of the housing and purge connections to keep dust from entering the packing around the valve shaft. The packing design is made more difficult by the need to have in-and-out motion of the shaft through the packing as opposed to rotational motion. Therefore, in the HYGAS plant, we tested full-port ball valves, loose-fitting butterfly valves, and full-port rotating plug valves, all making use of rotational motion for flow control. The butterfly valve operated satisfactorily; however, the obstruction posed by this valve has to be eliminated if at all possible. On the other hand, the ball or plug valve presented a full opening with no obstruction; however, these valves' flow characteristics are far from linear. IGT has also developed its own high-temperature flapper valve, which can be positively opened or closed. The conventional counterweighted flapper cannot provide a positive

closure. Our own design, through the use of a minimum amount of linkage, allows
for the expansion of the transfer pipe caused by high temperature. All of the valves
tested can be scaled up to operate in a commercial reactor, where the transfer lines
might be 30 inches in diameter.

REACTOR PERFORMANCE

For a dense-phase fluidized bed, we found that at high pressure, the bed
length/diameter ratio is not a critical design parameter, as in low-pressure systems.
Fluidization is smooth at high pressure and can be fully developed at 3 times the
minimum fluidizing velocity, even for a 15-foot-deep bed. The formation of bubbles
is suppressed, resembling particulate fluidization represented by fluidization in a
liquid medium. For the reactor design the length/diameter ratio is not all-important,
but rather a minimum bed depth for ensuring good mixing. We continue to use the
Zenz criterion[4] of taking at least 30% of the pressure drop in the bed across the grid.
Thus, for adequate gas distribution in a deep bed a high pressure drop is taken
across the grid. Once this gas is distributed, bubbling and slugging is not noticed.

In the slurry dryer, a coal slurry is sprayed into the top of the fluidized bed,
operating at 0.5 ft/s superficial velocity. We have evaporated up to 50 gal of oil/
hr-cu ft of fluidized bed, a volumetric evaporation rate that is several times
higher than the highest known rate previously reported for fluidized-bed drying.

In the low-temperature hydrogasification stage, a lift-line reactor is employed.
A gas velocity of around 30 ft/s carries solids at a loading of about 1 lb of solids/CF
of gas. In this velocity range, the pressure drop in the lift line checks closely
with the Hinkle correlation,[1] provided an extra term is added to account for the
hydrostatic head of a column of dense gas.

In the high-temperature hydrogasification stage, the dense-phase fluid bed is
operated at a superficial velocity of about 0.5 ft/s. A superficial velocity of 1 ft/s
is maintained in the steam-oxygen gasifier stage to avoid slagging by ensuring
a high degree of mixing. We believe, however, that this requirement can be
reduced to about 0.8 or 0.7 ft/s.

In overall reactor performance, in any reactor system consisting of a number
of fluidized stages in series, the accumulation of fine particles, which are trapped
in the system, must be avoided through careful design. Such a situation was encountered in the HYGAS program in which the throughput of the reactor system
was limited. Only after a long period of diagnosis was the cause accurately

pinpointed[2]: By adjusting the velocity in the various stages, the dust was kept out of the reacting system.

REACTOR INSTRUMENTATION

Differential pressure is still the best means of obtaining data on the bed fluidization characteristics. The only additional correction that must be applied is the hydrostatic head of the column of dense gas between the two points of pressure measurement, a correction which is negligible at low pressure but significant at high pressure. Particularly for a relatively light solid such as coal char, which may have a fluidized density of 15 to 25 lb/cu ft, a column of cold nitrogen with a density greater than 5 lb/cu ft is a factor that must be accounted for in measurement.

For bed-level measurement, nuclear radiation gages, using gamma ray sources, were tested and found to be able to give only qualitative indications of bed level and bed density. It is difficult to sight through thick reactor walls and refractory linings that attenuate most of the radiation and to try to discern a small incremental change in radiation adsorption caused by changes in the level or density of the bed. The hostile environment in gasification reactors makes it impractical to place any sensing device inside the reactor. Probes that utilize principles of capacitance, conductivity, or ultrasonics for measurement cannot be checked because the sensors cannot withstand the high temperatures.

CONCLUSION

The coal gasification field represents a unique opportunity to apply fluidization techniques to a giant new industry. Because of the chemistry of the reactions involved, high temperatures and high pressures are required. Little is known about fluidization under these conditions, so that much design data have yet to be obtained and interpreted. Some actual operating experience and design criteria derived from the HYGAS program are presented here, including the design of gas distributors, solids transfer lines, solids flow control valves, the assessment of reactor performance, and reactor instrumentation.

ACKNOWLEDGEMENT

The work reported was sponsored by the Energy Research and Development Administration and the American Gas Association.

REFERENCES CITED

1. Hinkle, B. L., PhD Thesis, Georgia Institute of Technology, Atlanta, June 1953.

2. Lee, B. S. and Lau, F. S., "Results From HYGAS Development." Paper presented at 77th National Meeting of AIChE, Pittsburgh, June 2-5, 1974.

3. Punwani, D., Pyrcioch, E. J., Johnson, J. L. and Tarman, P. B., "Steam-Oxygen Char Gasification in a Nonslagging Fluidized Bed." Paper presented at the Joint American Institute of Chemical Engineers-Gessellschaft Verfahrenstechnik and Chemieingenieurwesen Meeting, Munich, September 17-20, 1974.

4. Zenz, F. A., "Bubble Formation and Grid Design." Paper presented at Tripartite Chemical Engineering Conference, Montreal, September 25, 1968.

PART IV

Application

SYSTEM THEORY IN THE SPOUTED BED TECHNICS

Dr. TIBOR BLICKLE and Dr. JENÖ NEMETH

The spouted bed technics is a variant of fluidization suitable for drying, cooling, mixing, etc. of solid materials which are too coarse for good fluidization and uniform in size. The general characteristics of a spouted bed are described and the conditions necessary for stable spouting are défined in the literature by K.B. MATHUR, N. EPSTEIN, P.G. ROMANKOV and co-workers /1, 2 /, but the hydrodynamic features, namely gas and particle velocities in the spout and in the annulus, as well as the concentration distribution in a continously operating spouted bed dryer by means of the so-called system theory has so far not been analyzed.

The spouted bed equipment together with the gas and sòlid streams has been divided longitudinaly into three sections to simplify its analitical handling. In reality, a certain mixing between the sections is unavoidable, because the central spout is surrounded by an annulus formed from the particles sliding downwards, nevertheless the character of the main conditions of movement is correct. According to the input w_b, c_b raw material and outlet w_k,

c_k product side, the annulus can be divided
into two sections. The solid flow between the
sections are w_2, w_1, w_3. The gas flow is
w_b, $= w_1$, $+ w_2$, $+ w_3$, $= w_1$, $+ 2 w_2$, and the
concentration changes in the various sections
is $\underline{M}$ /M_1, M_2, M_3/.

The most general information for the
realization of the various modifications of
the spouted bed technique and equipments on
the basis of system theory is gained from a
so-called structure study /3/. To this end
let us now introduce the following concepts
and symbols:

Both the loosening phase /L_1/ and the
loosened solid phase /L_2/ are fed continuosly
/t_1/ into the internal section where they
from a freely moving layer /l_4/ and move in
the same direction /π_1/. An inlay tube /ε_2/
can be used in the inner section to separate
the latter from the outer annulus section or
packing /ε_3/. A separating wall /ε_4/ will
facilitate heat exchange. There seems to be
no sence in the use of a separator /ε_5/ and
there is no information on this level the body
/ε_1/ of the apparatus and about the mode of
input and outlet /ε_6/. Hence, if the unit
element is $\underline{e}$

$$t_1 \wedge \pi_1 \wedge \varepsilon_2 \wedge (e \vee \varepsilon_3) \wedge (e \vee \varepsilon_4).$$ /1/

In this way we have surveyed the equivalency
class $\mathcal{E}$.

a/ Let us now consider in connection with
$\mathcal{E}_2$ the equivalency classes λ , $\mathcal{H}$, μ , ν :
the inlay, as a rule, does not move λ_1
its position is vertical λ_2
the transfer might be - conductin neither heat

nor the material μ_1

- heat conducting, but not

conducting the material μ_2

- woven μ_3

- perforated μ_5

- other,material conducting μ_6

- conducting the impulse,

but not the material μ_8

the shape of the inlay can be - plane ν_1

-curved surface ν_2

- cylinder ν_5

- cone,truncated

cone ν_6

- prism ν_9 .

Thus:

$$\mathcal{E}_2 \wedge \lambda_1 \wedge \mathcal{H}_1 \wedge \left(\bigvee_{i=1}^{3} \mu_i \vee \bigvee_{j=5}^{8} \mu_j \right) \wedge \left(\nu_1 \vee \nu_2 \vee \nu_5 \vee \nu_6 \vee \nu_9 \right) \quad /2/$$

b/ In connection with $\mathcal{E}_3$ let us consider
the information λ , μ , ν :
the packing can be: - non moving λ_1

- vibrated λ_2

- conducting neither

heat nor the

material μ_1

− heat conducting

but not conducting

the material μ_2

its shape can be :   − sphere ν_4

− cylinder ν_5

− other ν_{11}

Thus:

$$\mathcal{E}_3 \wedge \left(\lambda_1 \vee \lambda_2 \right) \wedge \left(\mu_1 \vee \mu_2 \right) \wedge \left(\nu_4 \vee \nu_5 \vee \nu_{11} \right). \quad /3/$$

c/ The information in relation with $\mathcal{E}_4$

are $\lambda, \varkappa, \mu, \nu$:

Separating wall: − non moving λ_1

− vertical $\varkappa_1$

− horizontal $\varkappa_2$

− slanting $\varkappa_3$

− transfers heat, but

not the material μ_2

its shape: − plane ν_1

− tube ν_5

− helix ν_7

Thus:

$$\mathcal{E}_4 \wedge \lambda_1 \wedge \left(\varkappa_1 \vee \varkappa_2 \vee \varkappa_3 \right) \wedge \mu_2 \wedge \left(\nu_1 \vee \nu_5 \vee \nu_7 \right). \quad /4/$$

From Eqs 1, 2, 3 and 4:

$$A_1 = t_1 \wedge \pi_1 \wedge \left[\mathcal{E}_2 \wedge \lambda_1 \wedge \mathcal{H}_1 \wedge \left(\bigvee_{i=1}^{3} \mu_1 \vee \bigvee_{j=5}^{8} \mu_j \right) \wedge \nu_1 \vee \nu_2 \vee \nu_5 \vee \nu_6 \vee \nu_9 \right] \wedge$$

$$\wedge \left\{ e \vee \left[\mathcal{E}_3 \wedge \left(\lambda_1 \vee \lambda_2 \right) \wedge \left(\mu_1 \vee \mu_2 \right) \wedge \left(\nu_4 \vee \nu_5 \vee \nu_{11} \right) \right] \right\} \wedge \qquad /5/$$

$$\wedge \left\{ e \vee \left[\mathcal{E}_4 \wedge \lambda_1 \wedge \left(\mathcal{H}_1 \vee \mathcal{H}_2 \vee \mathcal{H}_3 \right) \wedge \mu_2 \wedge \left(\nu_1 \vee \nu_5 \vee \nu_7 \right) \right] \right\}$$

The number of elements in the Bull polynomial
A_1 are:

$$35. \; 13. \; 10 = 4550$$

The applied phases can be:

$$b_1^* = \beta_1 \vee \beta_1 \longleftrightarrow \beta_1$$

$$b_2^* = \beta_1 \longrightarrow \beta_2$$

$$b_3^* = \beta_1 \longrightarrow \beta_3$$

$$b_4^* = \beta_2$$

$$b_5^* = \beta_2 \longrightarrow \beta_1$$

$$b_6^* = \beta_2 \longrightarrow \beta_2$$

$$b_7^* = \beta_2 \longrightarrow \beta_3$$

$$b_8^* = \beta_3$$

$$b_9^* = \beta_3 \longrightarrow \beta_1$$

$$b_{10}^* = \beta_3 \longrightarrow \beta_2$$

The loosening phase can be:

$$L_1 \wedge \bigvee_{i=4}^{10} b_i^{\ast} \quad . \tag{/6/}$$

The loosened phase can be :

$$L_2 \wedge \bigvee_{i=1}^{7} b_i^{\ast} \quad . \tag{/7/}$$

The forms of the phases are:

$$\Theta_1 \wedge \bigvee_{i=4}^{10} b_i^{\ast} \quad . \tag{/8/}$$

$$\Theta_2 \wedge \bigvee_{i=4}^{7} b_i^{\ast} \tag{/9/}$$

$$\Theta_3 \wedge \left(\bigvee_{i=1}^{3} b_i^{\ast} \vee \bigvee_{j=8}^{10} b_j^{\ast} \right) \tag{/10/}$$

and $$L_1 \wedge \Theta_1 \tag{/11/}$$

$$L_2 \wedge \left(\Theta_1 \vee \Theta_2 \vee \Theta_3 \right) . \tag{/12/}$$

From these correlations:

$$L_1 \wedge \Theta_1 \wedge \bigvee_{i=4}^{10} b_i^{\ast} \tag{/13/}$$

$$L_1 \wedge \Theta_3 \wedge \bigvee_{i=8}^{10} b_i^{\ast} \tag{/14/}$$

$$L_2 \wedge \Theta_1 \wedge \bigvee_{i=4}^{7} b_i^{\ast} \tag{/15/}$$

$$L_2 \wedge \Theta_2 \wedge \bigvee_{i=4}^{7} b_i^{\ast} \tag{/16/}$$

$$L_2 \wedge \Theta_3 \wedge \bigvee_{i=1}^{3} b_i^{\ast} \vee b_5^{\ast} \quad . \tag{/17/}$$

Introducing the following symbols:

$$B_1 = z_4 \wedge L_1 \wedge \mathcal{E}_2 \wedge \vartheta_5 \wedge \Theta_1 \qquad \text{/18/}$$

$$B_2 = z_4 \wedge L_1 \wedge \mathcal{E}_2 \wedge \vartheta_5 \wedge \Theta_3 \qquad \text{/19/}$$

$$c_1' = L_2 \wedge \mathcal{E}_2 \wedge \Theta_1 \qquad \text{/20/}$$

$$c_2' = L_2 \wedge \mathcal{E}_2 \wedge \Theta_2 \qquad \text{/21/}$$

$$c_3' = L_2 \wedge \mathcal{E}_2 \wedge \Theta_3 \qquad \text{/22/}$$

SUMMARIZING

$$A_1 \wedge /B_1 \vee B_2/ \wedge /c_1' \vee c_2' \vee c_3' /. \qquad \text{/23/}$$

B/ In the <u>outer section</u> the restrictions for the various phases are the same as for the inner section.

Introducing the following symbols:

$$B_1^{*} = L_1 \wedge \mathcal{E}_1 \wedge \Theta_1 \wedge (z_1 \vee z_2 \vee z_4) \qquad \text{/24/}$$

$$B_2^{*} = L_1 \wedge \mathcal{E}_1 \wedge \Theta_1 \wedge z_4 \qquad \text{/25/}$$

$$c_1^{*} = L_1 \wedge \mathcal{E}_1 \wedge \Theta_3 \qquad \text{/26/}$$

$$c_2^{*} = L_2 \wedge z_1 \wedge \Theta_1 \qquad \text{/27/}$$

where z_1 is a fixed bed and z_2 a loosened bed by stream;

the correlations

$$B_1^{*} \wedge c_1^{*} \qquad \text{/28/}$$

$$B_2^* \wedge c_2^* \tag{29}$$

$$B_1^* \wedge \bigvee_{i=4}^{10} b_i^* \tag{30}$$

$$B_2^* \wedge \bigvee_{i=8}^{10} b_i^* \tag{31}$$

$$c_1^* \wedge \bigvee_{i=1}^{3} b_i^* \vee b_5^* \tag{32}$$

$$c_2^* \wedge \bigvee_{i=4}^{7} b_i^* \tag{33}$$

are valid.

In the outer section beside the body of the apparatus / ε_1 / the separating wall / ε_4 / too might have to be considered.

a/ the body of the apparatus: —non moving λ_1

—vertical $\mathcal{H}_1$

—transfers neither heat nor the material μ_1

—transfers heat μ_2

its shape: —cylinder ν_5

—cone, truncated cone ν_6

— composed of the above $\nu_5 \wedge \nu_6$

—prism ν_9

—truncated pyramid ν_{10} composed of the above $\nu_9 \wedge \nu_{10}$

b/ the separating wall is like in the inner section,

thus

$$A_2 = \mathcal{E}_1 \wedge \lambda_1 \wedge \mathcal{H}_1 \wedge (\mu_1 \vee \mu_2) \wedge (\nu_5 \vee \nu_6 \vee \nu_5 \wedge /34/$$
$$\wedge \nu_6 \vee \nu_9 \vee \nu_{10} \vee \nu_9 \wedge \nu_{10}) \wedge$$
$$\wedge \left[e \vee \mathcal{E}_4 \wedge \lambda_1 \wedge (\mathcal{H}_1 \vee \mathcal{H}_2 \vee \mathcal{H}_3) \wedge \mu_2 \wedge (\nu_1 \vee \nu_5 \vee \nu_9) \right].$$

The number of elements in the Bull polynomial A_2 is:

$$12. \ 10 = 120$$

THE JOINT SYSTEM

Information relating to the equipment:

$$A_1 \ \wedge \ A_2 . \qquad\qquad /35/$$

Information relating to its operation:

the loosening phase is in parallel connection, added continuously with partial or total recirculation:

$$(L_1 \wedge \nu_1^*) \wedge \left[L_1 \wedge t_1 \wedge (\xi_0 \vee \xi_1 \vee \xi_2) \right]. \qquad /36/$$

The loosened phase might be added batch-wise os continuously, when partial or total recirculation might applied, but under all conditions with internal recirculation:

$$(L_2 \wedge \xi_3) \wedge \left[t_2 \vee t_1 \wedge (\xi_0 \vee \xi_1 \vee \xi_2) \right]. \qquad /37/$$

A unit equipment or one equipment can be used, in parallel or series, counter-current / 6_1/

direct-current / 6_2/ or cross-current / 6_3/ set-up,
that is:

$$v_1 \wedge \left[v_2 \vee v_2^* \vee (e \vee v^*) \wedge v_2^{\circ} \wedge (6_1 \vee 6_2 \vee 6_3) \right] \qquad /38/$$

From Eqs /35/ to /38/:

$$D = A_1 \wedge A_2 \wedge \left(L_1 \wedge v_1^* \right) \wedge \left[L_1 \wedge t_1 \wedge \left(\xi_0 \vee \xi_1 \vee \xi_2 \right) \right] \wedge$$
$$\wedge \left(L_2 \wedge \xi_3 \right) \wedge \left[t_2 \vee t_1 \wedge \left(\xi_0 \vee \xi_1 \vee \xi_2 \right) \right] \wedge v_1 \wedge \qquad /39/$$
$$\wedge \left[v_2 \vee v_2^* \vee (e \vee v_2^*) \wedge v_2^{\circ} \wedge \left(6_1 \vee 6_2 \vee 6_3 \right) \right].$$

The number of elements of the Bull polynomial
D is:

    4550 . 120 . 3 . 6. 8 = 78 624 000.

The set-up relations of the phase systems are:

$$\left(B_1 \vee B_2 \right) \wedge \left(B_1^* \vee B_2^* \right) \qquad /40/$$

$$c_3' \wedge c_1^* \qquad /41/$$

$$\left(c_1' \vee c_2' \vee c_3' \right) \wedge c_2^* . \qquad /42/$$

On the basis of this E_i can be formed, e.g.:

$$E_1 = \left(B_1 \wedge B_1^* \wedge \bigvee_{i=4}^{10} b_i^* \right) \wedge \left(c_3' \wedge c_1^* \wedge \bigvee_{i=1}^{3} b_i^* \right) \qquad /43/$$

with the number of elements 3.7.3. = 63.

$$E_2 = \left(B_2 \wedge B_1^* \wedge \bigvee_{i=8}^{10} b_i^* \right) \wedge \left(c_3' \wedge c_1^* \wedge \bigvee_{i=1}^{3} b_i^* \right) \qquad /44/$$

with the number of elements 3.3.3 = 27, and

$$E_{10} = \left[(B_1 \vee B_2) \wedge B_2^* \wedge \bigvee_{i=8}^{10} b_i^* \right] \wedge \left[(c_1' \vee c_2' \vee c_3') \wedge c_2^* \wedge b_5^* \right] \quad /45/$$

with the number of elements 3.3 = 9,

$$E_{11} = \left[(B_1 \vee B_2) \wedge B_2^* \wedge b_9^* \right] \wedge \left[(c_1' \vee c_2' \vee c_3') \wedge c_2^* \wedge \bigvee_{i=4}^{7} b_i^* \right] \quad /46/$$

with the number of elements 2.3.4 = 24.

In such phase systems the following alterations
can be realized:

in all systems transport, change in the form,
pressure, change in the phase, concentration
change, chemical change, whose heap is F_1 ;
in the system containing the solid component
change and biological change $\left(F_2 \right)$;
in the three-phase system change in the phase $\left(F_3 \right)$.
The relationship between the alterations and the
phase systems is:

$$F_1 \wedge \bigvee_{i=1}^{3} \quad E_i \qquad\qquad\qquad /47/$$

$$F_2 \wedge / E_1 \vee E_2 \vee E_{10} \vee E_{11} / \qquad\qquad /48/$$

$$F_3 \wedge \bigvee_{i=4}^{11} \quad E_i \ . \qquad\qquad\qquad /49/$$

Hence, summing up the structure of the spouted
bed operational units:

$$S_G = D \wedge \left[F_1 \wedge \bigvee_{i=1}^{3} E_i \vee F_2 \wedge \left(E_1 \vee E_2 \vee E_{10} \vee E_{11} \right) \vee F_3 \wedge \bigvee_{i=4}^{11} E_i \right] \ . \quad /50/$$

The number of elements in this structure is:

$$78\ 624\ 000\ \left[8\left(63+\ +\ 27+36\right)+2\left(63+27+9+\ +24\right)+\ 9+15+18+6+3+9+24\right]=$$

$$=105\ 906\ 528\ 000\ .$$

THE MATHEMATICAL MODEL

Next we shall narrow down the model and investigate only a drying equipment without a separating wall and packing, when neither the inlay nor the body of the apparatus transfers either heat or the material; the cross-section of flow remains unchanged in the direction of flow; the bottom of the apparatus is conical and operation steady-state.

Hence the number of the structure elements is:

$$36.\ 16.\ 126=64\ 512\ .$$

The inlay or tube should preferably be a pris; in case of drying or desorption the unchanged recirculation of the loosening medium is not expedient; theoretically no cross-flow arises. With these restrictions the number of the elements is:

$$2.\ 2.\ 1.\ 2.\ 8.\ 2.\ 126=16\ 128\ .$$

Let us now consider the possibilities of simplifying E_i . If

$$L_1 \longrightarrow \text{gas} ; \quad L_2 \longrightarrow \text{solid},$$

in this case the number of elements is:

$$128 \ / \ 12 + 12 \ / \ = 3 \ 072.$$

As indicated above, the simplified models
have still a great number of elements. This
leads to the conclusion that designing a
spouted bed dryer seems to be useful by means
of a numerical program. The steady-state work of
a spouted bed dryer can be oillustrated by the
so-called systemfunction. The system-function
expressing the concentration change /M/ in the
various sections involves the relationship
between the M-s.

Using the above mentioned model and symbols
the steady state operation can be characterized
by the following balance equations /the concentration of the gas is C_b' at the inlet and C' at
the outlet/:

$$W_b C_b + W_1 C_1 - W_2 C_2 = W_2' \left(C' - C_b' \right) \tag{/51/}$$

$$W_2 C_2 + W_3 C_3 - 2 W_1 C_1 = W_1' \left(C' - C_b' \right) \tag{/52/}$$

$$W_1 C_1 - W_3 C_3 - W_k C_k = W_2' \left(C' - C_b' \right). \tag{/53/}$$

Since $W_b = W_k ; \ W_2 = W_b + W_1 ; \ W_3 = W_1 - W_k$ and

$$C_3 = C_k$$

and introducing the following symbols:

$$\varphi_1 = w_1 / w_b \qquad\qquad \text{and} \quad p = \varphi_3 / (\varphi_2 + 2)$$
$$\varphi_2 = w_1' / w_2' \qquad\qquad\qquad s = \varphi_2 \varphi_3 / (\varphi_2 + 2)$$
$$\varphi_3 = w_b' / w_b \qquad\qquad\qquad r = C_b' / c_b$$

Eqs /51/ to /53/ can be written as

$$c_b + \varphi_1 c_1 - (1 + \varphi_1) c_2 = p(C' - C_b') \qquad\qquad /54/$$

$$(1 + \varphi_1) c_2 + (\varphi_1 - 1) c_k - 2\varphi_1 c_1 = s(C' - C_b') \qquad /55/$$

$$\varphi_1 (c_1 - c_k) = p(C' - C_b'). \qquad\qquad /56/$$

The known starting data of the operation
are w_b c_b and C_b' . From the fluid mechanical correlations of the spouted bed   1, 2
w_b' can be calculated, consequently $\underline{p}$ and $\underline{s}$
are also known. With the help of the balance
equations   /54/ to /56/ the three unknowns
c_1 , c_2 and C, can be determined for the given
parameter values of the concentration of the
product /c_k/ and internal   recirculation /φ_1/.
The changes in the concentration, $\underline{M}$ are as
follows:

$$M_1 = \left[c_b / (1 + \varphi_1) - c\,b/p - C_b' + \varphi_1 c_1 / (1 + \varphi_1) - \varphi_1 c_1 / p + \right. \qquad /57/$$
$$\left. + (1 + \varphi_1) c_2 / p \right] / (c_2 - C_b')$$

$$M_2 = \frac{c_1 + 2\varphi_1 c_1 / s - c_2 (1 + \varphi_1)/s - c_k (\varphi_1 - 1)/s - C_b'}{c_2 (\varphi + 1)/2\varphi_1 + c_k (\varphi_1 - 1)/2\varphi_1 - C_b'} \qquad /58/$$

$$M_3 = \left[c_1 - \varphi_1 c_1 / p + \varphi_1 c_k / p - C_b' \right] / (c_k - C_b') \qquad /59/$$

In general it can be assumed that $M_1 \cong M_3$.

Let us then produce the sought system-function in the form $M_2 = f/M_3/$. It can be proved that

$$\frac{c_k}{c_b} = \frac{\hat{B}_1 + r\,\hat{B}_2}{\hat{B}_1 + \hat{B}_2} \qquad /60/$$

$$\frac{\hat{B}_1}{\hat{B}_2} = \frac{c_b\,/1-r/}{c_b - c_k} = \frac{c_b - C_b'}{c_b - c_k} = \psi \quad , \qquad /61/$$

from which it is possible to express $M_2 = f/M_3/$:

$$M_2 = \frac{a\,M_3^2 - c\,M_3 + e}{d - b\,M_3} \qquad /62/$$

where $a = \psi p^2 (2\varphi_1 + s)$

$b = \psi ps (\varphi_1 - p + p/2\varphi_1)$

$c = \psi p \left[2\varphi_1 (p + s + \varphi_1 + 1) - p + s\right]$

$d = \psi s (1 + \varphi_1)(\varphi_1 - p) + \left[\varphi_1 - p + (\varphi_1 - p)^2 + \psi p^2\right] s /2\varphi_1$

$e = \psi \varphi_1 (1 + \varphi_1)(s + 2p) - \psi p^2 + \varphi_1 + p - (\varphi_1 - p)^2$.

For the numerical expansion of the sought equation /62/ the experimental data obtained during the drying of wheat were processed. The characteristic values calculated from the experimental data are shown in the following Table:

$c_{b\%}$	$c_{n\%}$	$c_{1\%}$	$c_{2\%}$	c_1/c_k	$C_b'\%$	$C'\%$
23,5	14,8	14,99	16,07	1,01	1	2,74

φ_1	ψ	p	s	M_1	M_2	M_3
5,83	1,58	0,5	4	0,83	0,85	0,83

The $\underline{M}$ values of the characteristic concentration change are smaller that one. The assumption that $M_1 \cong M_3$ is with good approximation true under the given experimental conditions.

REFERENCES

[1] Mathur, K.B., Epstein, N.: Spouted Beds,
         Academic Press, N. Y. 1974.

[2] Davidson, J. F.,Harrison,D.: Fluidization,
         Academic Press, N. Y. 1971.

[3] Seitz, K.,Blickle, T.: The structure of systems,
         K. Marx Univ. of Economics, Budapest,1974.

PARTICLE FORMATION AND AGGLOMERATION IN A SPRAY GRANULATOR

S. MORTENSEN and S. HOVMAND

ABSTRACT

Spray granulation is a simultaneous drying and particle forming process which is carried out in a fluidized bed by spraying a liquid feed into the fluidized layer of already dried particles. The technology of the continuous process is described.

The most characteristic and essential part of the spray granulation process is the formation of new particles and their growth in the fluidized layer. The process parameters and their influence thereupon are discussed.

A stable production of granules and agglomerates (0.5 - 2 mm) for a wide spectrum of materials has been achieved through control of the process by grinding a certain product fraction and recycling this to the fluid bed.

The spray granulation of sodium sulphate, ferrous sulphate, clay, and potassium sorbate is described. Microphotos of the granular, dried products are shown.

INTRODUCTION

The application of a fluid bed as a drier and a granulator at the same time (i.e. a liquid feed is sprayed into a fluidized layer) is often called Spray Granulation.

The principle of the continuous operation is shown in
Fig. 1. The liquid feed (this can be either a solution,
suspension, or melt) is atomized and sprayed into a fluidized layer of already dried or partially dried particles.
The fluidization medium is the drying air, and the fluid
bed is normally kept under vigorous fluidization. The
particles leaving the bed are classified. The granules of
desired particle size are discharged from the unit while
the oversize fraction is milled and recycled together with
undersize fraction and fines recovered in the cyclone for
exhaust air. Thus spray granulation includes much of the
technology from spray and fluidized bed drying.

The process is being used for a wide range of applications,
particularly as a batch granulation process in the pharmaceutical industry, see e.g. [1], [2]. The continuous spray
granulation process has only been applied on a large
industrial scale in relatively few cases, e.g. drying of
azo dye [3] , combustion of spent liquor from a sulphite
pulp mill [4] , [5], drying of manganese sulphate [6] , and
the fluid coking process [7] , [8] , in which a fluid bed
reactor, converting heavy petroleum residua into gasoline,
gas oil and coke is working after the spray granulator
principle. There is also the production of granulated clay
particles for porphyrized tiles [9] , [10] , see description below.

Other applications, however on a small scale, include de-
hydration and calcination of uranyl nitrate solutions [11]
and conversion of liquid radioactive wastes from the processing of nuclear fuels to solid form [12] .
Considering also the attention paid to this process in
patents, [13] - [16] , this technology will undoubtedly
become increasingly important.

When a liquid feed is either spray dried or spray cooled,
the shape and size of particles in the finished product

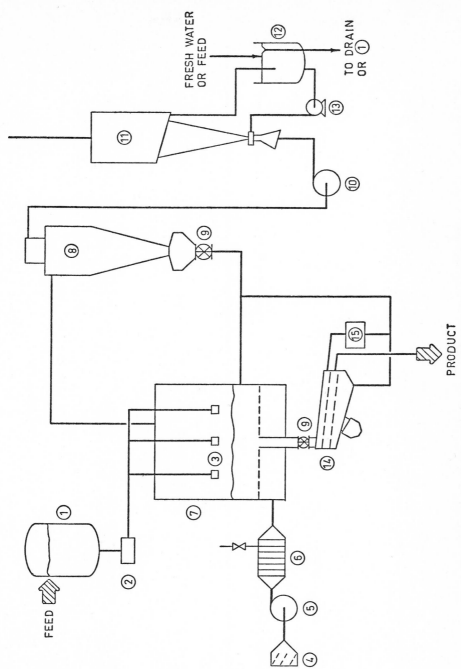

Fig. 1 Flow diagram of a typical continuous spray granulation process. (1) Feed tank. (2) Feed pump. (3) Nozzle Atomizers. (4) Inlet air filter. (5) Supply fan. (6) Air heater. (7) Fluidized bet unit. (8) Cyclone separator. (9) Rotary valve. (10) Exhaust fan. (11) Wet scrubber. (12) Scrubber tank. (13) Scrubber pump. (14) Sieve. (15) Grinder.

normally relate directly to the droplets created during
the atomization stage. As the particles must generally
be solid and non-sticky before reaching the chamber wall
of the spray drier, the average particle size obtained in
a spray drier ranges from 50 microns to 500 microns, see
Masters [17]. Thus, in certain cases the spray dried
product can be dusty, non-freeflowing, caking or difficult
to re-disperse in liquid. An improvement of the powder
characteristics might be achieved by various methods of
agglomeration, particularly developed for food products,
see Jensen [18] , [19].

In the spray granulator, the directly obtainable particle
size is generally one order of magnitude larger than for
spray dried products, i.e. 0.5 - 5 mm, and free-flowing,
dense particles can normally be produced. Here the particle formation is a result of the rather complex interaction between the atomized feed, the binding forces of
the solid(s) and various process parameters. When spray
granulating heat sensitive materials, the inlet drying
air temperature should be kept lower than by spray drying
in order to avoid heat damage of the product.

PARTICLE FORMATION AND GROWTH

The most characteristic and essential part of the spray
granulation process is the formation of new particles
and their growth in the fluidized layer. These are the
main parameters in determining the size distribution and
bulk density of the product.
New particles can either be formed inside or outside
the fluid bed.

Within the fluid bed, see e.g. [20] , [21]:

1) When the droplets from the spray dry sufficiently
   prior to contact with particles in the fluidized layer

or when they are conveyed directly out of the fluidized chamber.

2) Through attrition between the fluidized granules, small particles can be created by wear effects and larger particles by fracture. Special designs for using this mechanism in order to control particle size have been described, see e.g. [6] , [7] , [11] , [22].

3) When the liquid feed penetrates into the interior of a particle and the particle subsequently is moved to a zone with relatively high temperature, evaporation might then take place to rapidly that a kind of explosion ruptures the particle.

4) When the granules are exposed to varying temperatures, the stresses in the particles may cause fracture of the granules. This has been described as a control of particle size when uranyl nitrate was denitrated in a fluidized bed [3].

Outside the fluid bed:

1) From disintegration of part of the recycled granules.

2) From feeding solid particles to the fluid bed.

How large particles, defined as sizes from 0.3 - 2 mm, are built up during the process depends upon a number of parameters.

After wetting a particle with liquid in the fluidized bed, two mechanisms are possible:

1) A collision with another particle creating an agglomerate which, after drying, is kept together by a bridge of solid material.

2) Drying of particles before collision, thus particle growth by layering or granulation.

The first mechanism often leads to porous irregularly
shaped particles with relatively low bulk density and
strength. On the other hand, the second mechanism often
results in compact nearly spherical shaped particles
with relatively high bulk density and strength.

Whether loose agglomerates with relatively low bulk density or dense granules will be formed depends to a great
extent upon the characteristics of the material in question, such as rate of solubility of particles, stickiness
of wet particles, and liquid feed concentration. Also,
the condition of the fluidized layer has an influence
upon the product. For example, the tendency of the particles to agglomerate will normally increase significantly when the moisture content of the fluidized particles
is increased.

This complex of interactions governs the growth rate and
formation of the particles in the fluidized bed and the
particle structure of the product, and thus it also influences the necessary adjustments and controls that have
to be made for making a certain spray granulation process
into a reliable industrial process.

Mathematical models for particle growth rates under a
given set of conditions have been proposed by some authors,
see e.g. [7] , [21] , [23]. However, it has not been
possible to predict this complex behavious in general terms.

It has been found that even quite similar materials behave
very differently in the spray granulator, for example
sodium sulphate tends to form granular particles, Fig. [2],
whereas zinc sulphate and ferrous sulphate, Fig. [4],
have a great tendency to form agglomerates. Thus, the
design of an industrial spray granulator must be accompanied by experimental work.

CONTINUOUS SPRAY GRANULATOR DESIGN

The flow sheet of a typical continuous spray granulation process is shown in Fig. 1.

The fluidizing gas enters the fluid bed unit in the lower part normally in one or more separate chambers. The gas is distributed over the cross section by either a perforated plate (design of which, see [24]) or by tubes with nozzles [6] or as a spouting bed [25]. It is of great importance that the design of the underpart and the distributor plate results in a uniform distribution of the fluidizing gas. Areas of the fluidized layer with poor fluidization and simultaneously exposed to the feed spray might cause formation of lumps, which can eventually lead to complete breakdown of the operation.

The chamber above the distributor can consist of one or more compartments. The sectioning can have various purposes:

1) Creation of different conditions for formation and growth of particles. This is used in the NIRO ATOMIZER spray granulator producing clay granules for porphyrized tile production, [10], see more detailed description below. Sectioning can also be used to create a narrow product residence time distribution and therefore a narrow particle size distribution. Shakhora et al. [26] describe a nine stage spray granulator for urea using this principle.

2) Separate regions for pre or after treatments. For example, afterdrying may be necessary if granulation conditions result in the formation of a product of undesirably high volatile content or if the dew point of the outlet drying gas is substantially higher than normal ambient conditions. Examples of other kinds

of after treatments include calcining, sintering,
cooling, crystallisation, and classifying.

In many applications, the fluid bed chamber is conical.
This means that the velocity of the fluidizing gas is
considerably greater in the bottom of the fluidized bed
compared to that in the top part of the fluidized layer,
where the feed is sprayed into the bed and where granulation takes place. Thus, large granules will have a
tendency to segregate out within the lower region of the
granulating zone, however, they will be kept in fluidization, so overheating is avoided. These granules leave
the bed with the normal product stream, see e.g. [27].

The velocity of the fluidizing gas should be selected
to give a vigorous surface movement in order to avoid
formation of lumps in the contact zones between the spray
and the fluidized layer. Typical gas velocities are
1-2 m/s for granule sizes of 1-2 mm.

The vigorous fluidization often causes much elutriation
of smaller particles from the bed. However, this can be
counteracted by making the upper part of the fluid bed
with a de-entrainment section, e.g. [5].

The product withdrawal is generally by underflow, see
e.g. [5] , [8] , [15]. This has the advantage that buildup of larger particles and lumps (e.g. deposits from spray
nozzles) is avoided at the distributor plate.

The solids recycle enters through a single opening and
the solids are distributed throughout the fluidized bed
by the vigorous fluidization.

The feed is atomized either by a centrifugal pressure
nozzle or a pneumatic two-fluid nozzle. Whenever satisfactory atomization can be achieved with a pressure

nozzle this type is preferred because it is easier to install and to change during operation. Furthermore, it does not require auxiliary compressed air. However, the capacity and degree of atomization of a two-fluid nozzle is more flexible, and the chances for blockage in such a nozzle are less than for a pressure nozzle.

The nozzle position can either be above or below the surface of the fluidized layer. It is important that the position is variable, because it highly influences the particle formation and growth.

Fines carried out of the fluid bed by the gas are recovered either in a cyclone or a bag filter and the exhaust gas is cleaned in a wet scrubber. The scrubbing liquid can either be fresh water or the liquid feed to the plant. In the latter case, a preconcentration and thereby improvement of the heat economy is achieved and formation of an effluent stream is avoided.

The particle size distribution of the material in the

fluid bed can be wider than desired and often a recirculation of fine particles is required in order to control the particle size, see Process Applications below and e.g. [7]. Thus, the solids withdrawn from the fluid bed are classified, for example on a vibratory sieve, yielding a fines-, product-, and oversize fraction (in a few cases, air classifiers have been suggested, e.g. [15]). The oversize fraction is ground and recycled to the fluid bed together with the fines from the sieve and the cyclone.

PROCESS PARAMETERS FOR SPRAY GRANULATION PLANT

Parameter values for a spray granulation process have to be predicted from pilot plant work in order to ensure that a certain desired particle structure and size to-

gether with production capacity can be achieved. Some
parameters of the process must be controlled so the
fluidized bed in the industrial plant will operate under
reliable steady-state conditions. Some of the relevant
parameters and their influences upon the spray granulation process are discussed below:

1) Feed
    A) The concentration of solids in the feed is normally
       given and kept as high as possible in order to
       minimize the evaporative load.

       The influence of varying the feed concentration
       upon particle growth is not clear, as both form-
       ation of new particles and agglomeration might be
       promoted.

    B) The increase of feed temperature is reported to re-
       sult in decreasing particle size, [1], [20] in
       systems where interparticle binding is caused by
       evaporation of solution between particles. On the
       other hand, if cooling is required (melts) increased
       feed temperature is expected to result in increased
       particle size.

2) Atomization
    A) The degree of atomization has a direct influence on
       particle size: Coarser droplets give larger part-
       icles and a tendency to agglomeration [1]. The in-
       fluence is diminished if the size of the granules
       is relatively large compared with the size of the
       spray droplets.

    B) When two-fluid nozzles are applied, the atomizing
       air will have a drying effect upon the atomized
       feed. Thus the temperature and rate of the atom-
       izing air will have a similar influence upon the
       particle formation in the bed, as the temperature
       of the feed.

C) The fraction of the surface of the fluidized
   layer wetted by the spray is directly dependant upon
   the geometry of the spray in relation to the
   total fluid bed surface, i.e. number of nozzles,
   spray angle, spray pattern (hollow or solid cone
   spray) and vertical position of the nozzle.
   Generally, if the wetted proportion of the surface
   of the fluidized bed is decreased, the particle
   size will increase, as illustrated by sodium sulphate granulation below.

3) Velocity of the Fluidizing Gas

A very high degree of mixing in the fluidized layer
is required due to the very intense and localised
supply of feed on to the fluidized layer. It is
often apparent that when the gas velocity is decreased, the rate of agglomeration increases due to
a less rapid exchange of particles within the wetted
zone of the fluidized bed. However, excessive reduction in gas velocity will cause formation of lumps
and eventual termination of the fluidization. An
increase in gas velocity will cause less agglomeration
to take place due to the expansion of the fluidized
layer and the higher degree of impact occurring between particles causing attrition.

4) Particle Moisture Content

The moisture content of the particles in the fluid bed
depends upon the material itself (hygroscopicity,
water of crystallization) and the outlet drying air
temperature and dew point. Increased free moisture
content normally increases the tendency to agglomeration, see clay granulation below.

5) Outlet Drying Potential

Drying in a spray granulator does not only take place
in the fluidized layer but also above the surface un-

less the nozzles are located in the layer. The degree of drying above the bed depends among other
things upon the difference between the bed temperature and the dew point of the outlet gas. When
this difference is increased the predrying of the
spray droplets is increased, thus more new particles
are formed and there is a decreased tendency to particle growth, see potassium sorbate granulation below.

6) Residence Time Distribution for the Particles

The particle size distribution is dependent upon the
residence time distribution of particles in the fluidized layer. Shakhova et al. [26] made a product with
a narrower particle size distribution by changing the
solids flow pattern in the fluidized layer from back-
mix to plug flow conditions. However, the particle
size distribution of the product taken out of the
fluid bed must always reflect the need for a sufficient
amount of new particles to function as seeds in the
spray granulator.

7) Grinding

The extent to which the recycling solids are ground
has a strong influence on granular particle size.
Greater degrees of grinding give more particles for
the feed to adhere upon and/or agglomerate and consequently a smaller particle size, see e.g. [7].

Normally, the particle size distribution in the
granulator is not constant and control is necessary.
Among the process parameters discussed above some
are unsuitable for control purposes, for example the
feed conditions, the atomization, the fluidizing gas
velocity, and the residence time.

We have found that the most convenient control of part-

icle size is obtained by varying the amount of recycled ground particles. Hence, by grinding the oversize fraction and a controlled amount of fines from
the sieve and recycling, a stable operation is
achieved for a wide spectrum of materials (see examples below).

A signal suitable for control can be obtained in
various ways:

a) The most direct is to make a normal particle size
   analysis at regular time intervals. Nearly the
   same information can be obtained by installing
   weighing belts in the solid streams from the sieve.

b) In our normal design of the spray granulator, the
   product is withdrawn from the bottom through, for
   example, a rotary valve. The height of the fluidized layer is maintained constant by controlling
   the rotation of the valve to give a constant
   pressure drop across the layer. For a given
   product rate and particle size range of the product,
   the rotational speed of the valve is directly related to the granule particle size. Thus, increasing particle size in the fluidized layer is
   reflected in slower rotation of the valve which,
   in turn, can be applied for controlling the grinding stage.

c) It is well known that when a material, having a
   wide particle size distribution, is fluidized the
   coarser particles tend to segregate out close to
   the distributor plate. The vertical mixing in this
   layer is poorer than in the rest of the bed. Therefore, at constant inlet and outlet temperatures, the
   temperature in this layer is related to the amount
   of coarse particles and therefore also related to the
   particle size distribution. This could also be used

as a signal for control of the grinding stage. See
discussion of the clay spray granulator below.

PROCESS APPLICATIONS

A few examples of industrial process applications are
described below in order to illustrate the flexibility
of the process.

Inorganic Salts

Spray drying of inorganic salt solutions often yield
very dusty, low bulk density product which, for many
applications, is unacceptable. The spray granulation
process can eliminate these drawbacks and the production of anhydrous sodium sulphate and ferrous sulphate
monohydrate is outlined below as examples.

The main factor governing the performance of a spray
granulator is the product itself and the resulting
mechanism of particle formation and growth. The types
of particle structures obtained are shown on the microphotos, Figures 2 and 4, and the particle size distributions of the materials from the fluidized bed are given
in Figures 3 and 5. The sodium sulphate particles are
built up by layering of the spray droplets, whereas the
ferrous sulphate particles are formed partly by agglomeration and partly by granulation.

These mechanisms influence significantly the process
conditions, especially solids recycling rate, under which
spray granulation becomes stable and gives a constant
sized product.

Ferrous sulphate monohydrate: Under process conditions
where the nozzle position was 100 mm above the fluidized
layer, the amount of solids recycling from the sieve was

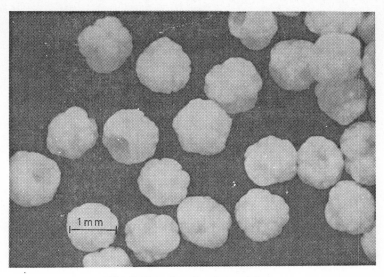

Fig. 2 Granules of sodium sulphate.

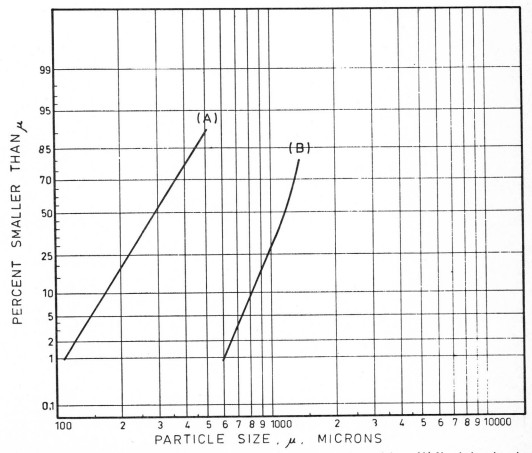

Fig. 3 Typical particle size distributions of spray granulated anhydrous sodium sulphate. (A) Nozzle location above the fluidized layer, (B) Nozzle submerged in the fluidized layer.

Fig. 4 Granules of ferrous sulphate monohydrate.

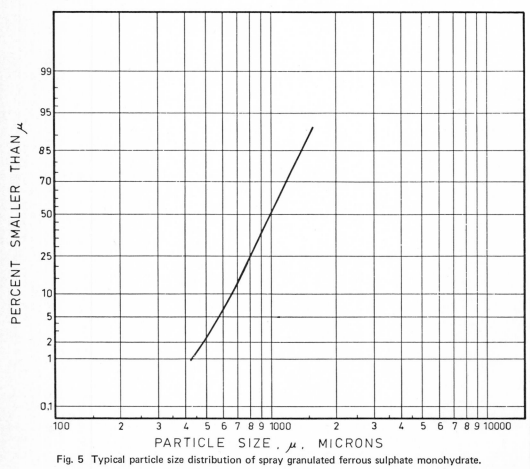

Fig. 5 Typical particle size distribution of spray granulated ferrous sulphate monohydrate.

two to three times the production rate. On average, an
amount of 50% of the production had to be ground in order
to maintain constant particle size. Typical particle
size distributions are shown in Fig. 5.

Sodium sulphate: Granulation was performed without solids
recycling and the obtained particle size distribution of
the product was sufficiently narrow to be suitable without sieving. The average particle size depends strongly
upon the nozzle position, see Fig. 3. Thus, with the
nozzle location above the fluidized layer, an average
particle diameter around 300 microns is obtainable, whereas the nozzle submerged in the layer yields, for example,
1000 micron particles.

Clay

The objective of the clay spray granulation process is
to achieve a clay particle size distribution and moisture
content suitable for pressing into two-coloured ceramic
tiles, porphyrized tiles, [9], [10].

Two stages are involved in the process, as shown on flow
sheet Fig. 6. The primary function of the first stage
is the formation of new small particles which, in the
second stage, are agglomerated into larger particles.

In order to achieve this, the moisture content is kept
relatively low (3-5%) in stage I and high (6-9%) in
stage II. Also,in stage II the nozzles should be located
close (approximately 100 mm) to the surface of the fluidized layer.

In this manner is obtained an agglomerated particle structure which is very suitable for pressing, see Fig. 7.
Solids recycling and grinding is thus substantial and
apart from cyclone fines, the recycle to product ratio
is approximately   2 to 3:1.

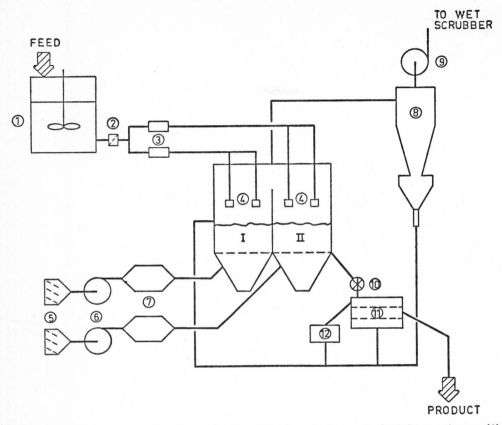

Fig. 6 Flow diagram of spray granulator for production of clay granules for porphyrized tile manufacture. (1) Storage tank. (2) Slip filter. (3) Slip pumps. (4) Nozzle atomizers. (5) Inlet air filters. (6) Supply fans. (7) Air heaters. (8) Cyclone separator. (9) Exhaust fan. (10) Rotary valve. (11) Sieve. (12) Grinder.

The particle size is controlled by varying the degree of grinding, which in its turn can be controlled by a temperature in the fluid bed layer close to the distributor plate. (This is because the plate temperature is sensitive to the particle size distribution in the bed). A typical inlet air temperature is 360°C. At the desired particle size distribution, the plate temperature is 210°C and even slight deviations in the particle size distribution will result in a change of approx. 10°C in plate temperature.

Fig. 7 Agglomerates of clay.

Other applications of clay granules, e.g. industrial
electro-porcelain, require a more granular product, see
Fig. 8. This particle structure can also be produced
in the plant layout described above, however, with
nozzles retracted to a position further above the fluidized layer.

Organic Products

All the products discussed above are characterized by
being rather non heat sensitive. When granulating
organic, mostly heat sensitive materials, the inlet drying air temperature is limited due to risk of heat degradation of product on prolonged contact with the
distributor plate.

Although relatively higher inlet air temperature can be
applied by spray drying, the same heat economy can often
be achieved by spray granulation because a certain (low)

Fig. 8. Granules of clay.

product moisture content is normally obtained at a lower
outlet temperature than by spray drying.

Potassium sorbate (a food preservative) is taken as an
example of a heat sensitive organic product.

This material can be spray dried to 0.1% residual moisture content at inlet/outlet temperatures of approximately 200/85°C and spray granulated at 120/60°C. The difference in heat economy (fuel consumption per amount of
final product) is within 10% in spite of the low inlet
temperature used in the spray granulator. The main advantage of the spray granulation process is improved
powder characteristics: non-dusty and free-flowing.

A typical particle size distribution of a sample from
the spray granulator is shown in Fig. 9. The particles are mainly built up by agglomeration, see Fig. 10,
requiring a solids recycle to product ratio of less than
1:1. Considering the particle structure, this ratio
appears rather low, but can probably be explained by two
factors:

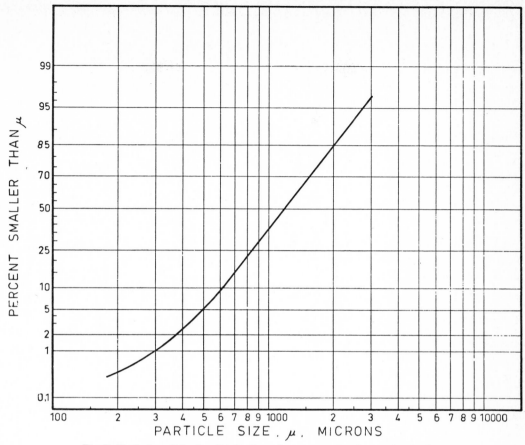

Fig. 9 Typical particle size distribution of spray granulated potassium sorbate.

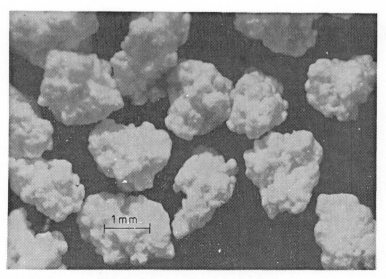

Fig. 10 Granules of potassium sorbate.

1) The strength of the particles is low compared with,
   for example, ferrous sulphate particles, as shown
   in Fig. 5. Thus particle collisions might well be
   a significant source of new particles.

2) The nozzle position is approximately 400 mm above
   the fluidized layer, causing about 25% of the drying
   to take place above the layer.

The scope of applications of the continuous spray granulation process is not limited to the examples already mentioned. The basic spray granulation process can also be
applied, with slight modifications, in other operations,
such as agglomeration of powdery materials and mixtures
and operations involving coating or encapsulation and
wetting of particles.

REFERENCE LIST

1   Thurn, U.   "Mischen, Granulieren und Trocknen
                           pharmazeutischer Grundstoffe in
                           heterogenen Wirbelschichten"

                           Diss Nr. 4511, Technische Hochschule,
                           Zürich, 1970

2   Davies, W.L.,
        Gloor, W.T.   "Batch Production of Pharmaceutical
                           Granulations in a Fluidized Bed",

                           J. of Pharm. Sci.

                           I: Effects of Process Variables on
                           Physical Properties of Final Granul-
                           ation.

                           Vol 60, No. 12, 1971, pp.1869-74

                           II: Effects of Various Binders and
                           Their Concentrations on Granulations
                           and Compressed Tablets.

                           Vol 61, No. 4, 1972,   pp.618-622

3   Romankov, P.G.   "Drying" in "Fluidization"

                          Edited by J.F. Davidson and D.Harrison,
                          Academic Press Inc., New York, 1971
                          p. 586

4   Priestley, R.J.   "High Temperature Reactions in a
                          Fluidized Bed",

                          In "Proceedings of the Symposium
                          on Fluidization",

                          Edited by Drinkenburg, A.A.H.,
                          Netherlands University Press,
                          Amsterdam, 1967, p. 704

5   Wall, C.J.,
       Graves, J.T.
          Roberts, E.J. "How to burn salty Sludges",

                          Chem. Eng., April 14, 1975, p. 77-82

6   Pictor, J.W.D.   "Solids from Solutions in one Step",

                          Process Eng., June 1974, pp. 66-67

7   Dunlop, D.D.,
       Griffin, Jr.,L.I.,
          Moser,Jr., J.F. "Particle Size control in fluid coking"

                          Chem. Eng. Prog., Vol. 54, No. 8, 1958,
                          pp. 39-48

8   Annon   "Coking process offers wide flexibility"
                          Chem. and Eng., Dec. 2, 1974, pp.17-18

9   NIRO ATOMIZER
       Copenhagen   Bulletin No. 14
                          "Start up of 1 t/h Clay Spray Granulator'

10   NIRO ATOMIZER   "Verfahren zur Herstellung eines grob-
                          körnigen Erzeugnisses mit relativ
                          hohem Feuchtigkeitsgehalt"

                          West German Patent Application
                          No. 2260723, July 1973

11   Björklund, W.J.,
       Offutt, G.F.   "Fluidized Bed Denitration of Uranyl
                           Nitrate"
                           Chem. Eng. Progr. Symp. Ser., 69, 1973,
                           pp. 123-129

12   Legler, B.M.   "Fluidized-Bed Processing in the
                           Nuclear Fuel Cycle"
                           Chem. Eng. Progr. Symp. Ser., 66, 1970,
                           p. 173

13   Kaspar, J.,
       Rosch, M.   "Eine neue Entwicklung auf dem Gebiet
                           der Wirbelschichtgranulation",
                           Chem. Ing. Techn. No. 10a, 1973,
                           pp. 736-739

14   Allied Chemical
       Corporation   "Production of Granular Hydrous
                           Sodium Silicate"
                           British Patent No. 1370578, Oct. 1974

15   CIBA-Geigy AG
     Basel   "Verfahren und Vorrichtung zur Her-
                           stellung eines Granulats"
                           West German Patent No. 2231445, Jan.73

16   BASF
     Ludwigshafen   "Verfahren zur Herstellung von nicht-
                           staubenden oder praktisch nicht-
                           staubenden Farbstoffkörner"
                           West German Patent No. 2263968,
                           July 1974

17   Masters, K.   "Spray Drying"
                           Leonard Hill Books, London, 1972

18   Jensen, J.D.   "Spray Drying and Rewet Methods for
                           Agglomerating and Instantizing",
                           Lecture given at the Technical Seminar
                           INSTANTISIEREN held at Zentralfach-
                           schule der Deutschen Süsswarenwirt-
                           schaft, Solingen, Germany, 23-25.
                           September 1974.

19 Jensen, J.D.   "Some Recent Advances in Agglomerating,
                            Instantizing, and Spray Drying",
                            To be published in Food Technology 29
                                                        (1975)

20 Markvart, M.,
       Vaněček, V.,
          Drbohlav, R.   "The Drying of Low Melting Point
                            Substances and Evaporation of
                            Solutions in Fluidized Beds",
                            Brit. Chem. Eng., Vol. 7, No. 7,
                            p. 503-507, 1962

21 Todes, O.M.   "Kinetik der Massenkristallisation (II)
                            Dehydration und Granulierung von
                            Lösungen in der Wirbelschicht",
                            Kristall und Technik 7, 1972,
                            pp. 729-753

22 Barsukov, E.Y.,
       Soskind, D.M.   "Jet Pulverization of Solid Particles in
                            a Fluidized Bed Apparatus",
                            Int. Chem. Eng., Vol. 13, No. 1, 1973
                            pp. 84-86

23 Grimmet, E.S.   "Kinetics of Particle Growth in the
                            Fluidized Bed Calcination Process",
                            A.I.Ch.E. Journal, Vol 10, No. 5, 1964
                            pp. 717-722

24 McAllister, R.A.,
       McGinnis,Jr., P.H.
          Plank, C.A.   "Perforated-Plate Performance",
                            Chem. Eng. Sci. Vol. 9, 1958, pp.25-35

25 Sul, Miter,
       Rashkovskaya,Romankov "Some Problems in the Modelling and
                            Design of Vortex-Bed Equipment for
                            Drying Process",
                            Int. Chem. Eng., Vol 14, No. 4, 1974,
                            pp. 700-703

26   Shakhova, N.A.,
        Yevdokimov, B.G.
          Ragozina, N.M.   "An Investigation of a Multi-Compart-
                             ment Fluid-Bed Granulator",

                             Process Tech. Int., Vol. 17, No. 12,
                             1972, pp 946-47.

27   Davidson, J.F.,
        Harrison, D.   "Fluidization",

                             Academic Press, London, 1971, p. 586

GRANULATION AND COATING IN MULTICELL FLUIDIZED BED

ZOLTAN ORMOS, BELA CSUKAS, and KAROLY PATAKI

In the present lecture the study of the continuous fluidization combined with atomization is dealt with in general and the process of build-up granulation and coating of granulates in a multi-cell fluidized bed in particular.

The main feature of this fluidization process is that a liquid (solution, suspension or melt) containing the appropriate solid components is atomized into the bed of the basic materials fluidized by gas (see Fig. 1). If the task of the process is granulation, the atomized liquid generally contains some binder hence the fluidized particles form granulates in a gradual build-up process [1, 2, 3]. If the task is coating of granulates, the atomized liquid contains such solid components that do not cause agglomeration but coat the particles [4, 5].

The fluidization process combined with atomization was first developed for the production of tablet-granulates in the pharmaceutical industry [2]. However, because of its many advantages its fields of application can be much wider: the method is capable of the granulation and coating of pesticides and complex fertilizers, of the granulation of foodstuffs and raw material mixtures of the silicate industries. Accordingly the continuous realization of the process is of growing importance. Also the requirement often arises for the realization of both the build-up granulation and the coating of the granulates within a single apparatus.

In the traditional technique of the continuous fluidized bed granulation [6, 7] the solid basic materials are fed into the upper part of the fluidized bed and the granulates or coated products are withdrawn from the bottom of the bed (see Fig. 2). Because of the intensive mixing of the bed there is a given probability of contamination of the product by ungranulated, uncoated or wet particles. This cannot often be eliminated even by employing great bed height/bed diameter ratios.

More problems emerge in granulation owing to the fact that the part-processes of the granulation (preheating, mixing, drying) are brought about in a batch process partly one after the other. On the other hand, in the batch process the gas velocity has to be increased with the growth of the particle size, so that the appropriate movement of the particles be maintained.

In the continuous granulation in a single bed the contradictory conditions of wetting and drying are to be ensured simultaneously: that is granulation has to be realized in a fluidized bed too "dry" with respect to the formation of granulates. The selection of the gas velocity is difficult as well, since the appropriate movement of the particles in the bed must be maintained and at the same time the elutration of the smallest particles of the raw material must be avoided.

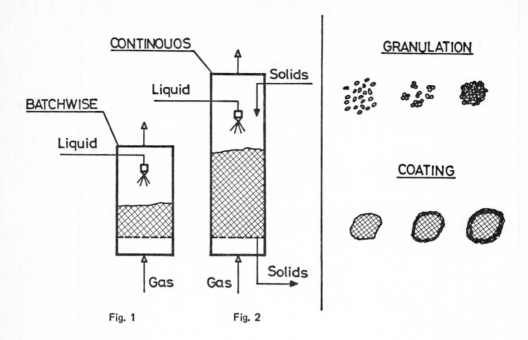

Fig. 1   Fig. 2

Similar problems emerge in the continuous coating too, since it requires a given moisture content in the bed and the fluidization velocity of the coated and uncoated particles is different.

We have developed the multi-cell fluidization process for the granulation and coating having all these in mind. The main feature of this process is that the part-processes of the granulation are separated in space by dividing the fluidized bed (that is, the apparatus) into a given number of sub-spaces (cells) using partitions providing for the required particle flow. The optimum circumstances can be ensured separately for the various part-processes in these cells of different functions.

The individual cells may have different gas velocities and temperatures so the moisture content of the bed can be adjusted according to the requirements of the individual part-processes and the elutration of the smallest particles can be avoided.

In Fig. 3, a schematic drawing of the granulation and coating in multi-cell fluidized bed is shown. Cell I serves for the mixing and preheating of the solid basic materials A and B. From this cell these materials get into Cell II, where the granulating liquid C atomized in the nozzle N_1 wets the particles and where the agglomeration takes place. The wet granulates are dried in Cell III, while in Cell IV the solution or suspension of the active ingredient D is atomized into the bed by nozzle N_2. In Cell V the solution or suspension of the coating material E is introduced by nozzle N_3. Cell VI serves for the final drying of the coated granulates P. The quantity and temperature of gas G introduced into the individual cells may be different, moreover its composition can be different.

The opening in the partition between Cells I and II is at the medium height of the bed to check the backmixing of the wet granulated particles and to avoid that pure components A and B get into Cell II. The openings in the partitions between Cells II-III, IV-V and V-VI are placed at the bottom of the bed, so they ensure that greater granulates gathered there can move further. The opening in the partition between Cells III and IV is placed somewhat above the underplate, therefore the overgranulated particles gathered at the bottom must be disintegrated before getting in Cell IV. This disintegration can be fostered by mechanical agitation if needed.

In Fig. 4, a laboratory five-cell fluidized bed apparatus is shown. In house (1) five cylindrical cells of a diameter of 0,12 m and a height of 0,4 m are formed. The size and location of the openings (3) in the partitions (2) can be varied according to the processes to be brought about in the adjacent cells and according to the physical properties of the solid material to be fluidized. The solid basic materials are fed into Cell I by the worm feeder (4). The liquids containing binder, active ingredient or coating material are atomized onto the surface of the fluidized bed by the two-fluid atomizer (5). The granulates or the coated products are withdrawn from the discharge cell (6). The particles under the prescribed size limit are recycled to one of the process cells by the air stream forced through the discharge cell. The fluidized bed is mechanically agitated by the mixing elements (7) if the materials processed are hard to be fluidized. Mechanical agitation improves the particle size distribution of the granulates and it helps in avoiding agglomeration of the particles in coating.

The main fluid mechanical parameters of the multi-cell fluidization process can be divided into three categories (see Fig. 5) namely:

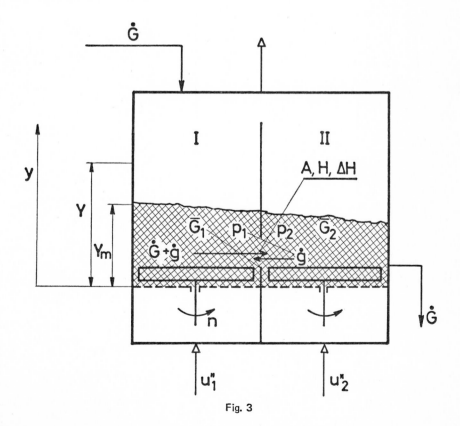

Fig. 3

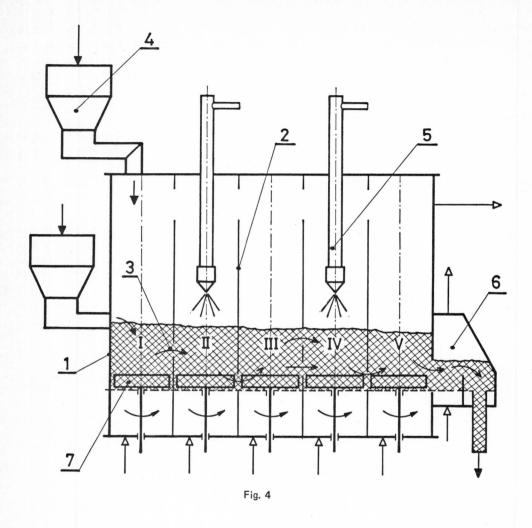

Fig. 4

— the characteristics of the fluidized bed: the extent of bed expansion (Y/Y_m), the equilibrium mass of beds, ($\bar{G}_i$), the pressure drop through the beds [Δp_i (y)], the mass flow ($\dot{G}$);

— the characteristics of the openings in the partition wall between, the cells: the shape, surface area (A), and height of location (H);

— the parameters of mechanical agitation used as an auxiliary process: the type of the stirrer, the speed of rotation (ω) the relative position of the openings and agitators (ΔH).

The mass flow characteristic of the back-mixing through the opening between the cells ($\dot{g}$) is vital with respect to the practical realization of various series of processes. The residence time spectra of multi-cell fluidization apparatus was studied by a radioisotopic tracing technique and an elution method for the determination of the back-mixing mass flow. It was found that for the evaluation of the residence time spectrum the backmixing cascade model [8] can be applied since the individual cells can be regarded as completely mixed cells. Hence

$$\dot{g} = \psi \dot{G} \tag{1}$$

where

$$\psi = \frac{2N - \Phi}{2\Phi} \tag{2}$$

where Φ is the solution of the implicit equation

$$\frac{\sigma^2}{\bar{\tau}^2} = \frac{2}{\Phi^2} \left\{ \Phi - \left\{ \left(1 - \frac{\Phi^2}{4N^2}\right) \left[1 - \left(\frac{2N - \Phi}{2N + \Phi}\right)^N\right] \right\} \right\} \tag{3}$$

In the equations

σ^2 = is the variance of the residence time spectrum
N = the number of the cells
$\bar{\tau}$ = the average residence time

The determination of the fluctuation in the pressure on the two sides of the opening in the partition between the cells and that of the average pressure showed that backmixing is due to the pressure fluctuation, and at the same time the dirving force of the mass flow $\dot{G}$ is the difference between the average hydrostatic pressures on both sides of the opening:

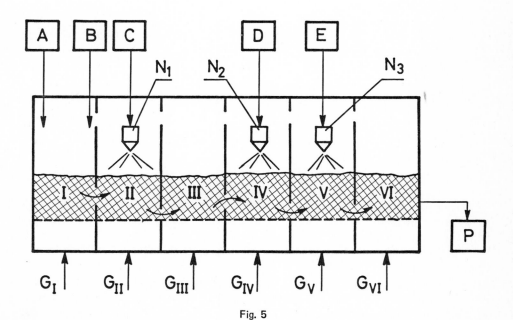

Fig. 5

$$G = kA \ [\overline{p_1(H)} - \overline{p_2(H)}] \tag{4}$$

Constant k is dependent mainly upon the parameters of the fluidization (extent of the bed expansion) and on the parameters of the mechanical agitation (the height of location, type and speed of rotation of the stirrer).

The performance of the granulating cell (or cells) can be characterized by the balance equations based on the over-all number of particles:

$$\dot{N}_1 = \dot{N}_0 - q_N \tag{5}$$

and on the particle surface area:

$$\dot{S}_1 = \dot{S}_0 - q_S \tag{6}$$

Switching over to the specific number and area based on mass and assuming complete mixing within the cell, the balance equations can be written as:

$$n_1 = n_0 - \frac{\overline{G}}{G} \ q_n \ (n_1, s_1) \tag{7}$$

and

$$s_1 = s_0 - \frac{\overline{G}}{G} \ q_s \ (n_1, s_1) \tag{8}$$

The source terms, q_n and q_s change in function of the specific surface area, the specific number and the process parameters of the granulation. The specific surface area and the specific number determine the particle size distribution of the granulates (which can be well approximated by a log-normal distribution function [9].

The determination of the source terms, q_n and q_s (the detailed derivation goes beyond the scope of this lecture) is done of the following physical model:

- In non-steady-state granulation the particle-size distribution approaches a limit distribution, that is a dynamic equilibrium develops owing to the growth of the particle size and to the disintegration (abrasion) of the particles.

- Greater particles can be formed from the primary particles and from the smaller agglomerates formed of the primary particles directly. Above a certain particle size, however, there is no effective collision between two greater particles (granulates of the shape of a dumb-bell consisting of two greater particles do not get stabilized).

- The size increase of greater particles is brought about by the "adhesion" of the primary particles or smaller agglomerates formed of the primary particles on the surface of the greater particles.

    — Disintegration is characterized by the coming off of primary particles or smaller agglomerates formed by primary particles from the surface of the greater particles. Disintegration is more pronounced with greater particles. Abrasion is characteristic of disintegration.

The performance of the coating cell or cells can be easier described on the basis of the mass balance, assuming an uniform bark-like surface growth.

In the multi-cell fluidization the granulation and coating feasible in the traditional (single-cell) fluidized bed can be realized under more advantageous process conditions and the physical properties of the products can be well controlled (e.g., particle size distribution, porosity, strength, etc.).

Our experiences showed that this method extends considerably the fields of application of the fluidization combined with atomizing, since the problems mentioned for the traditional process are eliminated by the separation of the part-processes in space.

The most important field of application for the multi-cell fluidization process is the realization of complex technological problems (series of processes). In the following the practical possibilities of the practical applications will be illustrated by a few examples.

In Fig. 6, the schematic diagram of the production of coated granulates containing a pesticide ingredient. The first cell of the six-cell apparatus serves for the mixing and preheating of the basic materials of the solid carrier. Binder containing liquid is atomized into Cell II and the wet granulates formed here are dried in Cell III. In the fourth cell the solution or suspension of the active ingredient is sprayed onto the carrier and in Cell V the granulates already containing the needed ingredient are coated. Cell VI serves for the drying of the coated granulates. Coating lessens the

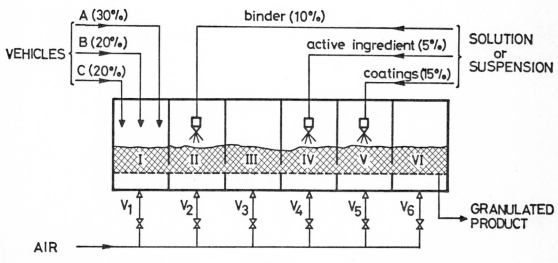

Fig. 6

dermic toxicity of the granulates of pesticides and the environmental pollution and loss caused by crumbling during use.

In Fig. 7, the schematic diagram is shown of the production of pharmaceutical tablet-granulates. The first cell of the five-cell apparatus serves for the drying of component "A" fed in directly from the centrifuge, while Cell II for the mixing and preheating of components "A" and "B". The solution of the active ingredient of small quantity is atomized into Cell III. In the meanwhile the product is pregranulated here by the ethyl-cellulose content of the solution. The final particle size is developed in Cell IV where aqueous gelatine solution is sprayed in, and the dry granulates are withdrawn from Cell V. Because of the heat sensitivity of ingredient "C" air of lower temperature should be introduced into Cells III and V. By increasing the number of the cells the coating of the granulates (the production of micro-dragées) is also possible.

In Fig. 8, the production process of the granulates of an NPK fertilizer containing small amounts of active ingredients is shown. Cell I serves for the mixing and preheating of the triphosphate urea and potassium salt. There are two atomizers in Cell II: one for spraying the solution of the trace elements and urea used for pregranulation, the other for spraying the solution of various pesticides. The additional urea needed for the grain formation is atomized into Cell III. Cell IV serves for coating or colouring of the granulates. The granulates of the complex fertilizer are withdrawn from Cell V. The advantage of the method is that in a given apparatus products of different composition satisfying the actual requirements can be produced by appropriate changes in the process parameters.

Pharmaceutical granulation

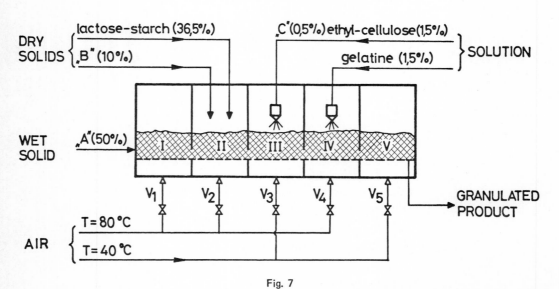

Fig. 7

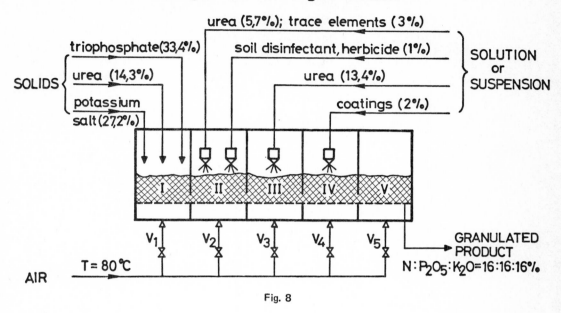

Fig. 8

LIST OF SYMBOLS

A	surface area of openings
$\dot{g}$	backmixing mass flow
$\dot{G}$	mass flow
$\overline{G}_i$	equilibrium mass of i-th bed
H	height of location of opening
k	mass transfer coefficient
n	specific number of particles
N	number of cells (in Eqs. 2, 3)
$\dot{N}$	stream of particles by number
q	source terms in balance equations by number or by surface area
s	specific surface area of particles
$\dot{S}$	stream of particles by surface area
u_i''	linear gas velocity in i-th cell
y	height co-ordinate in bed
Y	bed height
Y_m	minimum bed height
ΔH	relative position of the opening and the stirrer
$\Delta p_i(y)$	pressure drop function in i-th cell
σ	variance of the residence time spectrum
$\overline{\tau}$	average residence time
Φ	parameter of Eq. 3.
ψ	backmixing ratio coefficient
ω	rotational speed of the stirrer

LITERATURE

1. Wurster, D. E.: J. Pharm. Sci. *49*, 82 (1960).
2. Davies, W. L., Gloor, Jr. W. T.: J. Pharm. Sci. *60*, 1869 (1971).
3. Ormós, Z., Pataki, K., Csukás, B.: Hung. J. Ind. Chem. *1*, 307 (1973).
4. Wurster, D. E.: J. Pharm. Sci. *48*, 451 (1959).
5. Schultz, G.: Chem. Ing. Techn. *33*, 1382 (1968).
6. Scott, M. W. et alii: J. Pharm. Sci. *53*, 314 (1964).
7. Rankell, A. S. et alii: J. Pharm. Sci. *53*, 320 (1964).
8. Dechwer W., Popovic, M.: Chem. Ing. Techn. *45*, 984 (1973).
9. Beke, B.: Theory of grinding, Akadémiai Kiadó, Budapest, 1963.

QUENCHING METHANE FLAMES IN FLUIDIZED BEDS

G. DONSI, L. MASSIMILLA, and G. RUSSO

SCOPE OF THE WORK In acetylene manufacturing by partial combustion of methane, water is sprayed in oxygen-methane premixed flames at the point where acetylene content in the reactant mixture has reached a maximum. Quenching must be fast enough to prevent degradative reactions of acetylene.

Acetylene consumption is involved in the reaction with water to form carbon monoxide and hydrogen. Other reactions lead to the formation of soot, which gives rise to troublesome disposal problems. Another draw back of ordinary water quenching is the low temperature at which heat is recovered.

Dry, fluid bed quenching both improves energy balance and meets pollution problems. Bed particles act as collectors and as heat sink as well, and there are no limitations on bed temperatures other than those related to quenching fastness. Combustion heat of deposits can also be easily recovered.

Fluid bed quenching has already been applied to acetylene production for low speed, laminar oxygen-methane flames (1,2). Present work is directed to extend operation towards conditions of industrial interest with burners operating at gas speed larger than the laminar propagation velocity of flames in the fuel rich oxygen-methane mixture.

RESULTS Experiment scale has been related to the maximum available feed rate of reacting mixture. This was of 4 Ncu.m./h, with O_2/CH_4 ratios from about 0.6 to 0.85. The fluid bed quencher, operating at about 200°C, was

150 mm I.D., with a static bed height above the burner of about 30 cm. A
spent fluid cracking catalyst, at incipient fluidization, was the quenching material. Two burners have been used: one, 30 mm I.D., discharging
in the bed downwards; the other, 12 mm I.D., discharging horizontally.
With the latter, flame speed at the burner outlet was about 100 m/sec.

The optimal O_2/CH_4 ratio was between 0.65 and 0.70, at the maximum
available flow rate. At such a ratio, concentration of acetylene in exit
gas and selectivity to acetylene were, respectively, 5.5 and 26%, which
favourably compare with data from larger scale units considering they re
fer to an operation with an unpreheated reacting mixture. Besides that,
acetylene concentration and selectivity to acetylene were the same as
those determined with microprobes directly at the burner mouth, this
showing that acetylene degradation in the jet and in the bed was negligi
ble.

It has also been shown that fluidized bed is very effective in reta
ining sooty particles. More than 90% of soot produced in the flame was
taken up from outlet gas and found as deposits on the particle surfaces.
Re-entrainment of collected soot in the gas stream was negligible. A
steady increase of carbonaceous deposits was found even after many hours
of operation, up to soot contents in the bed of the order of 10%.

Interaction between combustion products and fluidized bed controls
both gas quenching and cleaning, but the bed is not equally involved in
performing the two objectives. Due to high rates of heat exchange at the
burner mouth and to high values of activation energies of degradative
reactions of acetylene, quenching is confined in the jet zone. Cleaning,
on the contrary, is a much slower process and develops throughout the
bed.

Investigation on the extent of acetylene conversion to be expected
in the jet zone has been based on a tubular reactor model of the space
occupied by the jet in the bed. Accordingly, equations of conservation of
acetylene and heat have been integrated along the jet, using in the lat-

ter heat transfer coefficients at the jet boundary as given in previous
works based on the same model (3,4). For given inlet jet temperature
(1300°-1500°C) and acetylene concentration, integration has been extended to jet lengths as obtained from available relationships on the jet
penetration in fluidized beds (5,6). In agreement with experimental results, calculations indicated a negligible degree of acetylene conversion in the jet space (less than 0.25%). Heat exchange with the bed was
so fast as to bring gas temperature below the minimum for further conver
sion (900°C with the reacting mixture in consideration) at about jet penetration length or even before that, respectively, with vertical or horizontal jets.

Theoretical estimates of bed efficiency as a filter have been made
by considering: 1) soot exchange between bubbles produced by flame jet
and the particulate phase and 2) soot collection by bed particles in con
tact with gas.

Regarding the first step, information from similar fluidization con
ditions indicate that the bed and flowing gas can be considered as perfe
ctly mixed. Then, fractional removal of soot has been expressed in terms
of the ratio of bed height to particle diameter and collection efficiency for an individual bed particle. Due to the small ratio of soot to bed
particle diameter (<0.01), such efficiency has been evaluated only in
terms of diffusion contribution to removal(7) disregarding both inertia
and interception effects. Calculations gave percentage of collected soot
as 90-95%, in agreement with experimental data.

COMMENTS

Results above are not conclusive because of scale limitations.

Concerning collection of carbonaceous deposits on bed particles, an
increase of gas flow rate per unit bed cross section should not change
present findings. The effect of an increase of fluidizing velocity on
soot exchange between bubble and particulate phase can be contained by

proper design of bed and internals to extend performance of other fluid bed units to the fluid bed filter.

According to calculations, an increase of jet diameter by ten would increase acetylene degree of conversion by about the same factor if, on a larger scale, heat transfer coefficients at the jet boundary were as those expected from experiments with jet diameters of 10-30 mm. Then, the size of a single burner orifice would be determined, once the loss in acetylene which can be afforded is fixed.

This approach, however, leaves doubts about scale up effects on quenching fastness. The problem is how far assumptions underlying the tubular reactor model might be extended to larger operation. Thus, another model is being studied, based on the theory of submerged jets. This accounts for the existence of a jet potential core at the burner outlet, which is uniformly at the inlet temperature of hot gas. Preliminary evaluation of degree of conversion of acetylene gives, with these assumptions, values 2-3 times larger than those obtained from tubular reactor model of the jet space.

Results so far obtained on quenching methane flames in fluidized beds are reported in detail elsewhere (8,9).

REFERENCES

1) F. Fetting, E. Wicke: Chemie Ing.Techn., 28, 88 (1956)

2) G. Russo, L. Massimilla: II Int. Symposium on Chemical Reaction Engineering, Amsterdam, 1972

3) M. Baerns, F. Fetting: Chem.Eng.Sci., 20, 273 (1964)

4) L.A. Behie, M. Bergougnou, C.G.J. Baker: The Canadian Journal of Chemical Engineering, in press

5) J.M.D. Merry: Trans.Instn.Chem.Engrs., 49, 189 (1971)

6) L. Massimilla, G. Russo: Symposium on Fluidization and its Applications, Toulouse, 1973

7) S.K. Friedlander: Jour. of Coll. and Interf. Sci., 23, 157 (1967)

8) G. Donsi, G. Russo, L. Massimilla: La Rivista dei Combustibili
   29, 5/6 (1975)

9) G. Donsi, G. Russo, L. Massimilla: "Proceedings of the Fluidized
   Combustion Conference", The Institute of Fuel, London,September, 1975

PYROLYSIS OF POLYSTYRENE CHIPS
IN FLUIDIZED BEDS

SHIGEO MITSUI, HIROKI NISHIZAKI and KUNIO YOSHIDA

SCOPE

How to treat waste materials is a world-wide problem to
be solved as promplty as possible. Even though every com-
munity wants to be rid of wastes no matter what origin they
come from, relatively small number of investigations have
been reported on re-utilization of these waste materials.

Advantages of fluidized bed reactor have been gradually
recognized these days, but its practical application to
processing of waste materials has been still restricted to
fluidized bed incinerators. (Reviews by Priestley, 1967;
Reh, 1968 and Hanway, Jr., 1971)

Heat of oxydation of plastic material is much higher
than that of decomposition. Therefore, only a part of the
plastic material is enough to be burnt by the air to supply
necessary heat for decomposition of the rest. This is the
basis of feasibility for partial oxidation processes of
plastic wastes by the air, which can be operated without
heat supply.

In the present study, pyrolysis of polystyrene chips
was carried out in several kinds of fluidized bed reactors,
which inner diameters being 3.5, 15, 30, 50 and 120cm respectively. Conditions for stable operation as well as necessary data for scale-up procedure are discussed.

CONCLUSIONS AND SIGNIFICANCE

Present investigation was carried out to solve a environmental problem arising from plastic industries. In Japan,

total production of plastic materials was about 6,000,000 ton
in 1973.   37% of the above amount is estimated to become
scrap within a year and 71% may be re-utilized within three
years, circulated and consumeds. If the recovery and recycling utilization of the plastic materials is not performed,
the plastic industry would probably become the most troublesome prodigal of raw materials.

Pyrolysis of polystyrene chips from spent bottles of
yoghurt was carried out in the fluidized beds, in which
necessary heat was produced by partial oxidation. Temperatures of these fluidized beds were maintained easily in the
range from 400 to 510°C, and almost all of the polystyrene
chips were decomposed to styrene monomer and dimer. On the
otherhand, when the bed was heated up above 550°C the polystyrene chips were completely gasified. Products obtained by
the pyrolysis were found suitable for re-circulation as a
good raw material of polymerization after destillation or for
fuel oil of a good quality.

This technique proposes some of more promizing application of fluidized bed reactor to re-utilization of various
wastes such as car tire and garbage.

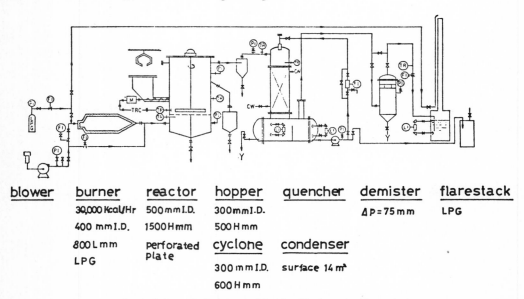

blower	burner	reactor	hopper	quencher	demister	flarestack
	30,000 Kcal/Hr	500mmI.D.	300mmI.D.		ΔP=75mm	LPG
	400 mmI.D.	1500Hmm	500Hmm			
	800Lmm	Perforated	cyclone	condenser		
	LPG	Plate	300mmI.D.	surface 14m²		
			600Hmm			

Fig. 1 Schematic flowsheet of fluidized bed reactor 50 cm. i. d.

Table 1 Experimental Conditions and Results

Reactor D_t cm	Sand L_c cm	Air U_o cm/s	LPG kg/h	pstyr. feed kg/h	Bed temp °C	O_{2in} %	O_{2out} %	Oil produced kg/h	n_c -	α -	β -	γ -	δ -	ω -
15	19.0	2.0	-	1.4	460	21.	11.6	0.89	0.77	0.83	-	.045	-	-
15	19.0	4.0	-	3.1	480	21.	12.9	2.08	0.80	0.85	0	.039	-	-
15	19.0	2.0	-	2.3	480	21.	9.8	1.62	0.81	0.86	0	.035	-	-
15	19.0	2.0	-	1.9	460	21.	13.3	1.53	0.89	0.92	0	.035	-	-
15	19.0	2.0	-	2.1	430	21.	15.6	1.19	0.70	0.78	0	.029	-	-
15	19.0	2.0	-	4.2	400	21.	16.3	3.54	0.92	0.94	0	.026	-	-
15	16.2	4.0	-	1.6	460	21.	14.3	0.68	0.61	0.71	0	.047	-	-
30	30.0	7.0	-	25.9	500	21.	1.1	14.9	0.69	0.77	0	.039	.033	0
30	30.0	10.0	-	30.7	500	21.	1.1	17.1	0.69	0.77	0	.042	.057	0
30	30.0	8.0	-	23.8	470	21.	1.1	15.6	0.75	0.81	0	.042	.042	0
50	35.2	7.3	-	40.9	512	21.	5.0	20.3	0.66	0.75	0	.050	.140	0
50	34.5	7.2	2.30	33.3	490	8.4	7.6	18.7	0.56	0.68	.049	.073	.088	0
50	35.2	4.6	-	8.0	490	21.	5.6	5.43	0.92	0.94	0	.096	.113	0
50	35.2	7.0	1.90	33.3	450	9.7	5.7	28.9	0.90	0.93	.044	.060	.025	0
50	35.2	7.0	2.73	33.3	435	5.5	4.5	26.6	0.83	0.87	.054	.054	.065	0
50	35.2	7.0	2.12	27.2	442	9.0	6.7	21.7	0.87	0.91	.055	.053	.084	0
50	35.9	7.0	1.92	31.1	430	10.1	6.0	25.8	0.88	0.91	.047	.060	.059	0
50	28.1	6.8	2.02	27.0	465	9.1	2.5	20.0	0.81	0.86	.053	.069	.084	0
50	28.1	5.7	-	42.7	470	21.	1.4	35.0	0.79	0.84	0	.048	.078	0
120	90.0	4.2	-	190.	480	21.	0.5	143.	0.77	0.83	0	.048	.012	0
120	120.	5.5	-	214.	480	21.	0.5	152.	0.82	0.87	0	.052	.012	0

Experimental

Fluididized bed reactors used: 3.5(batch), 15, 30, 50 and 120 cm i.d. respectively

In Table 1, values of $\alpha, \beta, \gamma, \delta$ and ω were calculated from material and heat balances, which will be given in detail in presentation. These values are defined respectively as

α =(combustion heat of carbon) (carbon efficiency)/Q_g

β =amount of preheat/Q_g

γ =heat amount of exit gas/Q_g

δ =heat loss/Q_g

ω =heat input/Q_g

where Q_g=calorific value of polystyrene.

Tables 2-4 show the typical examples of the compositions and properties of liquid and gas obtained.

In addition, properties of polystyrene chips and fluidizing sands, operation technique and conditions, and then some analysis useful for scale-up of reactor shall be described in presentation.

ACKNOWLEDGEMENTS

The authors express their thanks to Mr. T.Ise, Nippon Gasoline Co., for assisting with the experimental works, and to Prof. K. Endo of Hokkaido University for his adequate suggestions.

Table 2 Representative Composition of Oil Produced

C_6H_6	0.023 wt%
$C_6H_5-CH_3$	0.810
$C_6H_5-CH_2-CH_3$	0.402
$C_6H_5-CH=CH_2$ (monomer)	62.525
Dimer	9.43
Trimer	20.44
residue	6.37

Table 3 Representative Composition of Exit Gas

N_2	68.4 wt%
O_2	1.0
CH_4	2.0
CO_2	22.9
H_2O	0.6
$C_6H_5-CH=CH_2$	5.1

Table 4 Representative Properties of Products as Fuel Oil

density	0.948 g/cm^3,25°C
flash point	37.0 °C
pour point	-40.0 °C
viscosity	1.33 p. at20°C
calorific value	9510 kcal/kg
C remained	5.20 wt%
S content	0.02 wt%
ash	0.06 wt%
water	0.50 wt%

APPLICATION OF FLUIDIZED BED REACTOR
TO CHLORINATION AT HIGH TEMPERATURES

AURÉL UJHIDY, JÁNOS SZÉPVÖLGYI, and OSZKÁR BORLAI

Many metal chlorides of industrial importance e.g. $AlCl_3$, $TiCl_4$, $FeCl_3$ etc. can be produced economically only from oxide of the said metal or mixture of metal oxides by reductive chlorination at elevated temperature. Chlorination of pure metal oxides usually can be carried out relatively easily however chlorination of silicate materials forming bulk of raw materials means problem. In order to ensure appropriate conversion during chlorination presence of gaseous and/or solid reducing agent is needed in the case of silicate materials. This fact—additionally taking into consideration the chlorine gas—involves necessity of intensive solid-solid and solid-gas contacts.

Chlorination takes place from 400°C to 1200°C depending on metal oxides to be chlorinated.

In order to ensure favourable heat and mass transfer conditions and uniform temperature distribution, too our experiments were carried out in a batch fluidized bed reactor of 2 inch in diameter and 25 inch in height made of quartz glass.

The scheme of apparatus is shown on Fig. 1. The reactor (6) was heated by a heating wire wound on the wall. Temperature was regulated by a temperature regulator connected with a thermocouple butted in the fluidized bed.

Sample of 100 grams and prepared suitably was fed at the top of cold reactor then it was heated up to the temperature of reaction in a N_2 stream. After reaching this temperature N_2 was replaced by CO-Cl_2 gas stream. Amount of gases was measured by rotameters (1).

Temperature of fluidized bed was increased of about 15-20°C following inlet of CO-Cl_2 gas mixture. This effect was compensated by inlet of N_2 gas.

The gas stream leaving the system through the head (8) and containing products of reaction and small amount of flying dust, too was allowed to flow through cyclone (8) in order to separate flying dust.

Products of chlorination partly were separated in the condenser (12) cooled by air, partly in cooled collectors (14) and (15).

Georgia clay was chosen as model and we wanted to gain its Al_2O_3 content as $AlCl_3$. The composition of raw material was the following: 39.1% Al_2O_3, 43.8% SiO_2, 0.9% Fe_2O_3, 2.6% other metal oxides and 13.6% loss on ignition. Application of this clay as model material is verified by

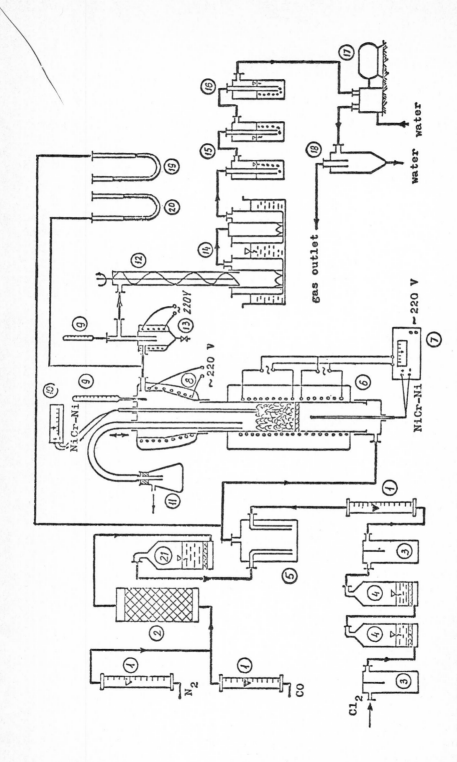

Fig. 1 Batch fluidization apparatus for chlorination. 1. Rotameters. 2. Dryer with silica gel. 3. Spray catcher. 4. Dryer with concentrated H_2SO_4. 5. Bottle for mixing of gases. 6. Reactor. 7. Temperature regulation. 8. Head for partial separation of flying dust. 9. Thermometers. 10. Reading device. 11. Sampler. 12. Condenser. 13. Cyclone. 14. Cooled product separators. 15. Scrubbers with water. 16. Scrubbers with NaOH solution. 17. Vacuum pump. 18. Gas-liquid separator. 19. Differential manometer. 20. Differential manometer for measuring of pressure in the reactor. 21. Gas scrubber with CCl_4.

the fact that with decreasing amount of bauxite applicable in the Bayer-Hall-Heroult process arose demand of elaboration of processes using low grade, mainly silicate materials as raw materials of aluminum production. One of the possible processes opening of new vistas is production of aluminum through aluminum chloride.

The clay contains physically and chemically bounded water being chlorine consumer and increasing the corrosion so the raw material has to calcined at 800-850°C before chlorination in fluidized bed reactor.

Numerous methods are known for chlorination of aluminous materials. The most important processes coming up as industrially realizable ones were reproduced by us in laboratory scale and the main problems in connection with these processes mean the long reaction time and relatively low conversion (Al_2O_3 conversion of 40-80% through 4-8 hours) and the almost equal chlorination rate of Al_2O_3 and SiO_2, respectively.

Our activity aimed at decrease of reaction time, and increase of conversion and selectivity that is decrease of silica conversion.

Preceeding chlorination experiments in fluidized bed we had to determine fluidization characteristics of material to be chlorinated. Fluidization parameters of clay and brown coal coke applied as solid reductant were measured in the chlorination reactor itself in $CO-Cl_2$ stream. Clay and coke having mean particle size below 50 microns can't be fluidized properly so mean particle size of grain fractions used was always above this value. Data of fluidization measurements are summarized in Table 1.

Studying effect of particle size on reaction time and conversion there was stated that by change of particle size within the given limits conversion doesn't change at a given temperature and reaction time when the raw material was chlorinated in absence of solid reducing agent with $CO-Cl_2$ gas mixture. At temperature above 800°C Al_2O_3 was converted to $AlCl_3$ in 25% through 2 hours. Conversion wasn't increased by change of reaction parameters (temperature, quality and quantity of alkali and alkali earth metal chlorides applied as catalysts, etc.) within rational limits.

Conversion wasn't increased considerably even if chlorination was carried out in presence of solid reducing agent only, so necessity of simultaneous application of gaseous and solid reductant became evident. When clay and solid reducing agent were fed without previous mixing into the reactor—particle size were chosen of course so that the solid reductant was dispersed uniformly in the fluidized bed—even application of $CO-Cl_2$ gas mixture didn't increase conversion above 40%. It means that presence of solid reducing agent isn't enough; it has to be contacted with the clay intimately.

For this reason granules of appropriate composition and particle size were prepared from clay and coal milled below 63 microns. Chlorinating these granules with $CO-Cl_2$ gas mixture at 850°C and 930°C, the conversion of Al_2O_3 was 80,2 and 79,9% respectively through 1 hour. Diminution of grain means the main problem during chlorination in fluidized bed; it is partly consequence of mechanical effects but mostly results from loss of weight in the chemical reaction. Amount of unreacted but crushed particles leaving the system can be reduced by continuous regulation of gas velocity and by appropriate reactor construction. In order to decrease conversion of

Table 1   Fluidization Parameters of Materials Used in Chlorination Experiments

Particle size /μ/	Weight /g/	V^x /m/s/	Δp /water head/	Flying dust /%/	Height of layer/cm/
		Calcinated clay			
63–250	100	0,021	50	3,15	7,5
250–800	100	0,040	57	1,82	6,0
		Brown coal coke			
63–250	100	0,015	52	1,26	7,2
250–800	100	0,045	55	0,83	6,3
	Granule of 20 % coke and 80 % calcinated clay				
63–250	100	0,027	51	3,62	7,3
250–800	100	0,045	58	2,04	6,2

where V^x linear gas velocity

      Δp pressure drop within the layer

silica mixture containing 5% sodium chloride was prepared. Application of this catalyst decreased conversion of silica with 50 relative per cent.

If amount unreacted dust leaving the system is minimized process can be transformed to a continuously operating method. Products of chemical reaction can be removed at the top of the reactor while the crushed residue of chlorination is separated from the gas by cyclones.

The authors wish to express their thanks to Toth Aluminum Company, New Orleans, La. and Tatabánya Coal Mines, Tatabánya, Hungary for financial assistance of researches and the useful consultations.

FLUIDIZED BED STEAM GENERATOR CONCEPTS FOR LMFBR

D. L. KEAIRNS, R. A. NEWBY, and D. H. ARCHER

ABSTRACT

A fluidized bed steam generator, FBSG, system is being developed for the liquid metal fast breeder reactor, LMFBR. The FBSG eliminates the need for an intermediate sodium loop and heat exchanger utilized in the current design and isolates the sodium and water thus reducing the possibility of a sodium-water reaction. The FBSG system offers the potential to reduce plant cost, increase plant availability and improve plant safety.

Alternative fluidized bed steam generator concepts are presented and design and operating parameters reviewed. An experimental program is underway to develop design data and demonstrate concept feasibility. The experimental program objectives are reviewed.

INTRODUCTION

The liquid metal fast breeder reactor, LMFBR, utilizes an intermediate sodium heat transfer loop and heat exchanger in addition to a steam generator. The intermediate system receives heat from the primary sodium, which removes heat from the nuclear reactor, and transfers it in turn to the steam which drives the turbine generator. Figure 1 illustrates this heat transfer system.

---

* The Ⓦ Advanced Reactor Division is directing the development of alternative steam generator concepts for the LMFBR. Work on the FBSG concept reported in this paper was funded by Westinghouse. Since September 1974, work on this program has been carried out under contract AT(11-1)-3045, Task 30, with the Energy Research and Development Administration and with Ⓦ support.

Proceedings International Fluidization Conference, Asilomar Conf. Grounds, Calif., June 1975. (Engineering Foundation Conference).

D. L. KEAIRNS, R. A. NEWBY, and D. H. ARCHER

Dwg. 6352A44

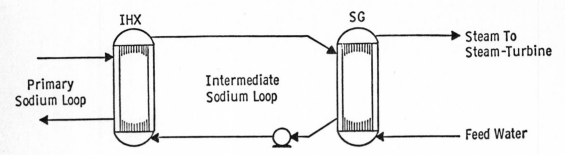

Fig. 1 Conventional LMFBR heat transport system concept.

The intermediate sodium heat transfer loop helps to assure
plant operability and safety. It prevents direct interaction of radioactive sodium with steam or water in the case of a single leak or
breech of the heat transfer surface. This loop increases plant cost and
does not avoid a sodium-steam/water reaction in the steam generator in
the case of a single leak.

The functional and safety requirements for the LMFBR reactor
cooling and steam generator systems were reviewed to explore alternative
arrangements to the intermediate sodium heat transfer loop. The fluidized
bed steam generator is one system which was conceived to replace this
loop.

Fluidized Bed Steam Generator System Concept

The FBSG concept permits the transfer of heat directly from
the radioactive sodium to the non-radioactive steam system. Four general
design alternatives are illustrated in Figure 2. All of the concepts
utilize bed materials which are inert to both sodium and steam and an
inert fluidization gas, such as helium, to transport heat between primary-
sodium and steam. In the first concept, conventional pneumatic transport
methods are used to circulate the bed material between two fluidized beds.
One unit functions analogously to the current IHX, the other as a SG.
Maximum separation between sodium and steam is realized by this concept.
The other three concepts place the primary sodium tube-bundle and the

steam tube-bundle into a single vessel, each with a different degree of
separation between the primary sodium and steam. The second concept

arranges the tube-bundles side-by-side with a physical barrier between them. The fluidized bed is circulated by "natural bed circulation" (i.e., by operating with a higher fluidizing velocity on one side of the physical barrier than on the other, which causes bed circulation by the

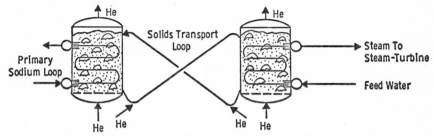

Concept 1—Fluidized Bed SG Concept with Pneumatic Transport of Solids

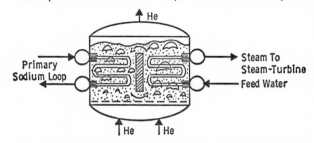

Concept 2—Fluidized Bed SG Concept with Natural Bed Circulation

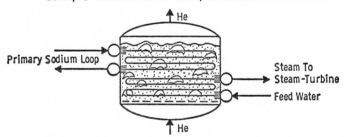

Concept 3—Fluidized Bed SG Concept with Stacked Tube-Bundles

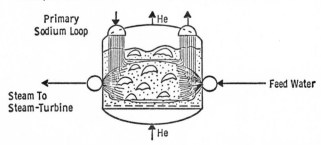

Concept 4—Fluidized Bed SG Concept with Uniformly Distributed Tubes

Fig. 2 Alternative fluidized bed steam generator concepts for the LMFBR.

induced non-uniformity in bed density). The third concept arranges the
tube-bundles with one directly above the other. Heat is transported
between the sodium and steam by normal fluidized bed mixing. The last
concept illustrated, which provides the smallest degree of separation
between sodium and steam, distributes the sodium- and steam-tubes uniformly
in the bed. Concepts which combine aspects of concepts 3 and 4 and
distribute separate sodium and steam tube-bundles uniformly in the bed
are also being considered.

      The overall fluidized bed steam generator system includes the
steam generator, a closed fluidizing gas loop to provide the fluidizing
gas for the steam generator, a materials handling system for the bed
material, sodium and steam leak detection equipment, an instrumentation
and control system, and provision for maintenance. The circulating gas
system is the primary auxiliary system. Figure 3 illustrates the composite
fluidized bed steam generator and circulating gas system.

Design Parameters

      The design of a fluidized bed steam generator requires
specification of a number of critical parameters.[1] The tube bundle
configuration and orientation must be selected. The conceptual diagrams
in Figure 2 illustrate some of the options. Tube diameter and spacing
are then selected to be compatible with the configuration, fabrication
and inspection and cleaning requirements. Horizontal tube configurations
have been selected in the initial designs to avoid excessive bed depths.
Selection of the tube bundle configuration will determine the overall bed
volume. There is a choice of bed depth, width and number of modules.
The fluidizing gas pressure is constrained to be lower than the sodium
pressure in the tube bundles in order to prevent gas bubbles from
entering the reactor core. This pressure is in the order of 150 psi.

      Helium was selected for the fluidizing gas. Helium is inert
and permits high heat transfer coefficients. Nickel and fused magnesium
oxide powders have been selected as candidates for the FBSG based on the
consideration of the compatibility of the powder with sodium, steam and
fluidizing gas impurities (O_2, N_2, water vapor). Powder wetting,
solubility, chemical reaction, hardness, strength, and physical phase
changes were compared in selecting nickel and magnesium oxide and small-

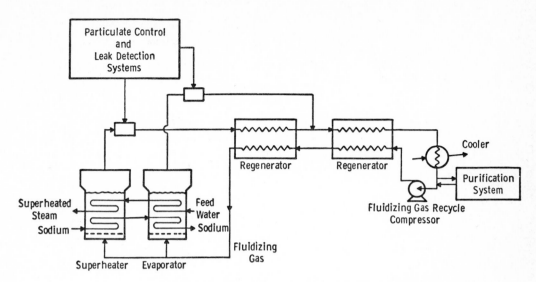

Fig. 3 Fluidized bed steam generator and circulating gas system.

scale screening tests were carried out. Fluidization behavior (minimum
fluidization velocity, quality of fluidization, heat transfer rates) and
power requirements have been examined in selecting powder average diameters
of about 40 to 80 microns for the nickel and magnesium oxide. Particle
attrition and tube erosion due to bed mixing and potential agglomeration
characteristics are also important factors in final selection of the powder
material and operating conditions. The powder cost and auxiliary power
requirements for the circulating-helium loop are of practical importance
in the selection. Nickel powder cost is presently about $2/lb and fused
magnesium oxide is about $0.25 to $0.50/lb. Nickel powder, being about
3 times as dense as MgO powder, requires a higher helium pressure drop
across the bed and, thus, higher auxiliary power requirements. These
factors must be balanced against the higher heat transfer coefficients
achieved with nickel. Other factors are still being investigated; e.g.,
particle attrition.

The heat transfer coefficients achieved in the steam generator
are dependent on the fluidizing gas, particle characteristics, fluidizing
gas velocity, tube diameters, tube packing configuration and the operating
temperature. Experimental tests have been carried out to obtain bed-to-tube
heat transfer coefficients for the projected design. Bed to tube heat
transfer coefficients of about 400 Btu/hr-ft^2-°F are projected for fused

magnesium oxide powder and about 500 Btu/hr-ft^2-°F for nickel powder at
commercial bed temperatures of 600–1000°F and fluidizing velocities of
0.3 to 0.5 ft/sec.[1]

Development Program

An experimental program is continuing to investigate candidate
powder fluidization properties, fluidized bed heat transfer to submerged
tubes, natural fluidized bed circulation and bed mixing, particle attrition,
tube erosion, the effects of sodium leaks and steam leaks on fluidized
bed behavior, and the effect of jet impingement on a tube as the result
of a leak in an adjacent tube.

Each of the four fluidized bed steam generator concepts shown
in Figure 2 has apparent advantages and potential disadvantages when
compared. At this point in the development program it is not possible
to quantify the true value of the apparent advantages or the potential
disadvantages.

Concept 1 may not be economically attractive because of the
high capital costs and power requirements associated with
the pneumatic transport system. The greatest mechanical
separation between sodium and steam is possible with this
concept.

Concept 2 may be less economical than Concepts 3 and 4 due
to the high helium rates required for sufficient bed circulation
or the excessively large equipment size resulting from limited
circulation rates. Non-uniformity of bed circulation and the
scale-up of bed circulation data are potential problems.
Mechanical separation between sodium and steam is realized
with this concept.

Concept 3 may be the easiest to commercially fabricate. The
rate of bed mixing, usually assumed to provide isothermal bed
conditions in most chemical process. may reduce the economic
potential of this concept as compared to concept 4. Good
separation between sodium and steam is possible.
Concept 4 is not limited by the bed mixing rate, but the
erosion effects of a steam-jet issuing from a tube leak and

impinging on a neighboring sodium tube may limit the concept feasibility. Though this may be the most compact concept, fabrication may be more difficult than concepts 1, 2 or 3.

Other concepts have been concluded which are combinations of these four concepts. For instance, multiple tube bundles may be arranged in the bed to provide a compromise between concepts 3 and 4.

Design studies are underway to produce preliminary designs for the FBSG and the auxiliary systems. These designs are based on the results from the experimental program. Safety, maintenance, control, containment design, decay heat removal and plant cost are being assessed to assure the development of an integrated system. The design study shows that

The proposed fluidized bed SG concepts can potentially operate with sodium or steam leaks without damage to the SG.

Leak detection capability is excellent in the system by monitoring sodium vapor or water vapor in the helium loop.

In the unlikely event of simultaneous sodium and steam leaks, the reaction will be less violent than in the conventional SG due to the dilution effect and the thermal capacity of the bed.

The fluidized bed SG is not as compact as the conventional SG, but the IHX, intermediate piping system, and other components in the plant are eliminated.

CONCLUSION

Development and application of a fluidized bed heat transport system offers the prospect for improved LMFBR power plant safety, reliability and economics. The preferred concept will be selected based on analyses of the power plant design and operating constraints and the analyses of results from the present experimental test program, projected large scale tests and the preliminary design studies.

APPLICATION OF FLUIDIZED BEDS
TO COATING PROCESSES

C. GUTFINGER and N. ABUAF

ABSTRACT

The fluidized bed coating process is reviewed and the main areas of
application are outlined. The two cases considered are coating of objects
immersed into a fluidized bed of plastic powder and coating of granular particles during fluidization in a bed.

The highlights of the fluidized bed powder coating theory are presented.
Comparison of the theory with some experimental results shows good agreement.

NOMENCLATURE

A_w - surface area of object

c_c - specific heat of coating material

c_w - specific heat of object

h - convective heat-transfer coefficient

k_c - thermal conductivity of coating material

L - half thickness of object

m_w - mass of object

t - time

T - temperature

T_m - softening temperature of coating material

T_w - object temperature

T_{wo} - initial object parameter

x   - position coordinate measured from wall

Z   - dimensionless parameter

Greek Symbols

δ - coating thickness

Δ.   - dimensionless coating thickness

Δ_f - dimensionless final coating thickness

θ - dimensionless temperature

τ - dimensionless time

ρ_c - density of coating material

ρ_w - density of object

INTRODUCTION

When a heated object is dipped into a fluidized bed of plastic particles, the particles will stick to it forming a coating subject to the body
being heated above the softening temperature of the plastic powder.

Fluidized bed coating is one of the modern solventless coating processes which include also electrostatic powder spraying, flame powder spraying and electrostatic fluidized bed coating. The rapid development of these
processes is due to their inherent advantages. One of the main advantages
is the absence of the solvent, common to conventional coating techniques.
In this age when pollution control and abatement has become a true necessity,
if not for guarding our health than at least for staying in business, a solventless coating process is quite welcome. The absence of solvents is also
advantageous because there is no need for solvent recovery systems. The
fire hazards associated with solvent handling are also dramatically reduced.

Fluidized-bed-applied coatings are thick, robust and durable. Only
one coat is needed to achieve the final coating, and due to its increased

thickness it possesses good abrasion and corrosion resistance and improved electrical insulating properties. Moreover, polymers insoluble in solvents may be applied to a surface by a powder coating technique. The main disadvantage of the fluidized bed coating process is in the necessity of heating up the coated object. Another disadvantage is the poor coatings obtained on very thin objects having a low heat content. Uniform coatings below 100μm in thickness are also difficult to achieve by this method.

Figure 1 presents a typical fluidized bed coating system. It consists of a powder container equipped with a false porous bottom through which fluidizing air is being introduced from a low pressure air supply. The part to be coated is first preheated in an oven and then dipped into the fluidized bed for a period of 5-15sec during which time a coating is formed on its surface. The coating ability of the fluidized bed is affected by its two

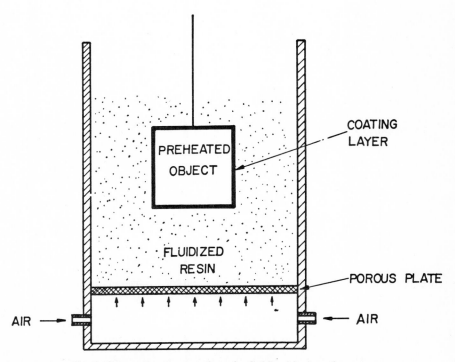

Fig. 1 Schematic representation of a fluidized bed coating set-up.

main features, i.e., its fluidity and its high contacting ability which results in high heat transfer coefficients. The fluidity is beneficial as well
as essential for obtaining a coating, because without fluidization the object could not have been immersed into the powder. On the other hand, the
high heat transfer coefficient typical of a fluidized bed is a definite drawback for the coating process, as it results in increased heat losses from the
object to the bed and thinner coating thicknesses. Therefore, the coating
technologist is interested in working under conditions of minimum heat transfer losses while still retaining complete fluidity. Thus, it is advantageous
to operate the fluidized bed close to its minimum fluidization velocity. An
additional advantage of operating in this region is in minimizing bubble formation.

We will proceed now and list some of the applications of fluidized bed
coating. Here we will distinguish between plastic powder coatings applied
for the purpose of improving the wear, corrosion, electrical and decorative
properties of the surface and other coatings applied for specialized purposes. Coating of fluidized particles will also be reviewed. We will then
describe the theoretical work on fluidized bed coating and compare it with
actual coating experiments.

APPLICATION OF FLUIDIZED BEDS TO PLASTIC
POWDER COATING

One of the main uses of the fluidized bed coating process is in applying a polymeric protective layer on a metallic object. The process was in-
vented in Germany by E. Gemmer in 1953 and a patent was issued to his company
Knapsack Griesheim in 1955 [1]. Early development of the process continued
in Germany and in the U.S. and was subject to a fairly large number of patents [2,3,4,5,6]. Initial reports in the literature describing the fluidized

bed process were presented by Gemmer [7], Gaynor [8,9] and others [10,11,12, 13].

An excellent comprehensive review of the literature on fluidized bed coating covering the work performed through 1963 was prepared by Landrock [14] who listed in his review 139 references. More recent general reviews describing the areas of application and the range of products being coated with plastic powders in fluidized beds are those of Yakovlev and Okhrimenko [15], Landrock [16] and Feigenbaum and Lefort des Ylouses [17].

The main process parameters affecting the quality, uniformity, and thickness of the coating obtained in a fluidized bed are the object temperature, the immersion time, the size, shape and size distribution of the plastic particles, the fluidizing gas velocity and the physical properties of object, powder and carrier gas. In addition to these, the object surface properties and its treatment will affect the adherence and general quality of the coating but not its thickness.

Metals are the most popular substrate materials for fluidized bed coating. These include various grades of steel and iron, aluminum, magnesium, copper, brass, tin, etc. Although the commercial literature recommends the fluidized bed coating process to such materials as wood, glass, concrete, and leather, the present authors after gaining some practical coating experience would be reluctant to recommend this method over some other techniques more suitable for these materials.

Metals are ideally suited for fluidized bed coating due to their high thermal conductivity. The success of the fluidized bed coating technique depends largely on being able to maintain a high surface temperature of the coated object. This condition is fulfilled quite well with metals. Procedures for surface preparation which include sand blasting, cleaning and degreasing are well described in the commercial literature supplied by coating powder manufacturers.

Both thermoplastic and thermosetting resins may be used for fluidized bed coating. The thermoplastic types include: polyamides [18] such as nylon 6/6, nylon 11 and nylon 12; PVC [19,20]; polyolefins such as polyethylene [21], and polypropylene [22]; cellulosics [23] such as cellulose acetate butyrate (CAB) [24] and cellulose acetate propionate (CAP), cellulose nitrate and ethylcellulose; fluorocarbons [25] such as teflon; chlorinated poly-ethers; acrylics; polyphenylene sulfide [26] and thermoplastic polyesters [27]. The thermosetting coating materials include the epoxies [28] and thermo-setting polyesters [29]. Conte [30] presented an up-to-date guide for selec-ting the right powder for a given coating application. Research and develop-ment is continuing in the area of powder formulation with the purpose of im-proving its relevant properties such as strength of adhesion to substrate, protective qualities, case of fluidization, etc. Some of the modern work is directed in developing multicomponent powder systems [31] or powders of co-polymers [32,33], other compositions include inorganic pigments, stabilizers, lubricants, etc. [34].

There are many different applications of the fluidized bed coating pro-cess, the range of objects coated grows steadily and it seems like it is only limited by the imagination of the coating technologist.

The fluidized bed process is used for applying decorative and protective coatings on household and commercial products. Some examples of these are door and cabinet handles, dishwasher racks, refrigerator shelves and drawers, handtools, lawn furniture, fencing, etc. In many cases where corrosion and wear resistance is important the fluidized bed process is chosen. Typical applications include external coating of pipes [35] with PVC, epoxy coating of coil springs [36], fan blades, valves, machinery housings, etc.

Koury et al. [37] review the aircraft applications of powder coating techniques. Nylon 11 coated spline assemblies are listed. Elastomeric coatings such as neoprene, hypalon and elastomeric :polyurethane are evaluated experimentally for use in leading edge abrasion protection. Improved resistance to metal fatigue is reported for coated specimens. Coatings on aluminum and magnesium previously reported [38,39] are re-evaluated.

The fluidized bed coating technique has found vast applications in the electrical and electronics industry for applying electrical insulation and protective coatings [40]. Items such as transformer windings, housings, etc., motor windings, motor stators, and other components [41] aircraft battery cases, electrical resistors [42] and capacitors, were all successfully coated. However, the fluidized bed coating process has not been found reliable for coating bare transistors and other active semi-conductors devices, mainly because of the high temperatures required for fusing the coating material. Licari in his book on plastic coatings for electronics reviews also the fluidized bed applications to this field [43].

Plastic coatings can also be applied by the electrostatic fluidized bed process. This process combines some of the elements of the fluidized bed with the electrostatic spray process. Here the object to be coated is not preheated. Instead, the plastic particles in the bed are given a negative charge by applying a high voltage in the range of 40-90kv. The grounded object is passed over the bed attracting the charged particles that form a coating on its surface. These are then fused in a curing oven [44,45]. The process has several advantages such as the fact that the substrate need not be preheated, or immersed in the bath. Thus, the fluidized bed does not have to be much larger than the coated object. Moreover, the process can be used to coat thin, low mass parts which cannot be coated by the ordinary fluidized bed process due to their poor heat retention. Several experimental investigations of this process were recently reported [46,47]. In these, the thickness and rate of deposition of the coating were evaluated as a function of electrode voltage, gas flow rate, particle diameter and other parameters.

Vibrating the coating vessel was found to affect favorably the coating performance and several commercial vibrating beds are offered on the market. The effect of vibrations on coating was evaluated both for the ordinary [48] and the electrostatic fluidized bed process [47].

OTHER APPLICATIONS

Until now we have described primarily the coating of metallic objects by plastic powders. However, the fluidized bed coating process has been used for many other applications. Continuous filament tow used for tobacco smoke filters was uniformly coated with fine particulate matter by passing it through a fluidized bed of the additive [49]. Other uses of the fluidized bed coating process include application of prefluxed solder powder to a metallic substrate [50], deposition of a carbon layer on iron [51] and coating metals by means of other metals [52,53,54].

In all the applications described above the particles in the fluidized bed served as the coating material. However, other possibilities may be conceived. The fluidized bed can also be used as a medium in which the coated object is suspended. In this case the coating material is usually the gas used for fluidization or is admixed to it. This gas may be either adsorbed on the surface of the substrate or chemically decomposed and reacted with the substrate. The bed is usually operated at high temperatures. Using this principle ceramic components were suspended in a fluidized bed of inert molybdenum granules and coated with an electrically resistant layer of carbon, which was obtained from in-bed pyrolysis of methane with 3% hydrogen at 1000°C [55]. Another application describes the coating of cutting tools with titanium carbide (TiC), the tool being suspended in a bed of tungsten granules, the fluidizing agent being argon at $1000^{\circ}C$ to which $TiC\ell_4$, CH_4, and H_2 were injected [56].

COATING OF FLUIDIZED PARTICLES

When the objects to be coated are of small size and large quantity, one may consider to use them as the fluidized particles in a bed, the coating being performed during fluidization. The coating agent may either be a gas that reacts with the fluidized granules or a liquid solution either sprayed into the bed or entrained by the fluidizing gas. Some of the desirable features of the fluidized bed system make this coating technique quite attractive. Of these, one may note the fluid-like behavior of the bed that allows continuous introduction of uncoated and removal of coated particles [57]. One may note that close to minimum fluidization velocity the position of the particles inside the fluidized bed changes very little relatively to the other particles [58]. Thus, there is no intermixing between the particles in the bed in this flow regime. If one, therefore, continuously introduces uncoated particles at the top of the bed and removes coated particles at its bottom, he may achieve operating conditions approaching those of an ideal plug flow reactor. These conditions are very favorable for obtaining a uniform coating thickness on the particles.

An important area of application of particle coating during fluidization is the preparation of nuclear fuel particles coated with pyrolytic carbon. The carbon for the coating is obtained from pyrolytic decomposition of gaseous hydrocarbons at high temperatures. Several publications and patents describe different methods for pyrolytic carbon coating [59,60,61,62,63,64,65]. Equations were derived for the rate of carbon deposition both for turbulent and

laminar flow regimes [66]. These were based on a series of experiments performed under controlled conditions.

Fluidization was also applied in chemical vapor deposition of refractory metal claddings on microspheres [67] and metal coating of solid particles [68]. Edible granular products/pharmaceutical pills were also coated
during fluidization [69,70].

Experiments were performed by Harada et al. in order to evaluate the
maximum feed rate and coating thickness distribution of spherical particles[71],
and the rate of coating growth and coating efficiency [72]. Equations were
proposed for correlating these quantities [73].

Fluidized beds may be used for removal of mist or dust from industrial
flue gases [74,75]. The dust particles or mist droplets are removed from the
gas and deposited on the fluidized filter elements upon passage of the dusty
gas through the bed. Although the purpose of this operation is entirely
different than that of fluidized particle coating, the mechanism of dust deposition in a fluidized bed filter [76,77] is very similar to the latter.

THEORY OF PLASTIC POWDER COATING

From the mathematical stand point the fluidized bed coating process is
very involved and one has to apply many simplifying assumptions and approximations in order to make the problem tractable. In general, the practicing
coating technologist will be primarily interested in a time versus coating
thickness relationship and its dependence on the various coating parameters.
On the other hand, he may not be concerned   with details of the coating
process, such as the temperature distribution in the film at any instant, the
hydrodynamics of the fluidized bed or its temperature variation.

The problem of fluidized bed coating may be   viewed as a variation on
the well known Stephan problem. Here a hot object immersed in a fluidized

bed loses part of its internal energy as heat to the bed, while another part is being used to melt the plastic powder forming the coating layer. This heat transfer process can be described by the simultaneous solution of the heat conduction equation in the coated object and the conduction equation in the growing coating layer with the proper matching boundary conditions.

If the coated object possesses a very large heat content, its temperature could be taken as constant during the coating process. The case of constant-wall-temperature fluidized bed coating was solved by Gutfinger and Chen [78,79]. The more general case is that of the variable object temperature. Typically one coats metallic objects of high thermal conductivity with a plastic coating of relatively high thermal resistivity. Here the object temperature, although dropping with time, will remain fairly constant throughout the body at any time; thus the object may be represented as a lumped parameter system. The solution of the case of variable object temperature, which is applicable to fluidized bed coating of thin plates was obtained by Elmas [58,80] and Abuaf and Gutfinger [81]. All these solutions were developed for a regular fluidized bed under the assumption of a known heat transfer coefficient between the object, or coating, and the bed [82].

Consider a flat plate with an area A_w and halfwidth L, which is dipped vertically in a fluidized bed, (Figure 2). The object is initially at temperature, T_{wo}, which is higher than the softening temperature of the coating material, T_m. The plastic coating material in contact with the object surface will melt and begin to form a layer on the plate. The process now involves the transfer of heat from the plate to the continuously growing film, and then into the fluidized bed, held at constant temperature T_∞. The heat transfer within the coating is given by the one-dimensional heat conduction equation:

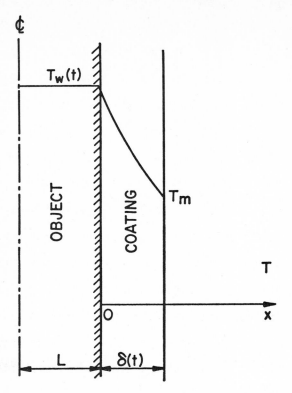

Fig. 2 Schematic representation of the mathematical model.

$$\rho_c c_c \frac{\partial T}{\partial t} = k_c \frac{\partial^2 T}{\partial x^2} \tag{1}$$

with the initial and boundary conditions:

$$T(0,0) = T_{wo} \tag{2}$$

$$T(0,t) = T_w(t) \tag{3}$$

$$T(\delta,t) = T_m \tag{4}$$

$$- k_c \left.\frac{\partial T}{\partial x}\right|_{x=\delta} = h(T_m-T_\infty) + [\rho_c c_c (T_m-T_\infty) + \lambda]\frac{d\delta}{dt} \tag{5}$$

$$\frac{m_w c_w}{A_w}[T_{wo}-T_w(t)] = ht(T_m-T_\infty) + \rho_c c_c \int_o^{\delta(t)} (T-T_m)dx +$$

$$+ [\rho_c c_c (T_m-T_\infty) + \lambda]\delta \tag{6}$$

and

$$\delta(0) = 0 \tag{7}$$

Equation (5) is the heat balance on the surface of the coating film. It equates the heat conduction to the surface with the heat convected into the fluidized bed plus the heat absorbed by the coating material which sticks to the plate and forms the film. Equation (6) expresses the fact that the heat loss by the body during the time interval $(0,t)$ is equal to the heat transferred to the fluidized bed, plus the heat consumed in bringing the coating film temperature from its initial value T_∞ to its final value T.

There exist several approaches to the solution of the system of Equations (1)-(7). All of them are approximate in nature due to the nonlinearity of this system. Moreover, it is usually assumed that the latent heat of the fusion, λ, may be neglected. This assumption is reasonable for non-crystalline polymers usually used in coating. In these materials one observes no heat of fusion and a softening range instead of a sharp melting point. In this range the specific heat is not constant and therefore one has to work with heat capacities averaged over the temperature range studied.

The problem defined by Equations (1)-(7) may be rewritten for the case $\lambda = 0$, in terms of dimensionless coordinate ξ, time τ, temperature θ, and coating thickness Δ, respectively:

$$\xi = \frac{x}{\delta} \quad ; \quad \tau = \frac{k_c t}{\rho_c c_c} \left[\frac{h}{k_c} \frac{(T_m - T_\infty)}{(T_{wo} - T_\infty)} \right]^2$$

$$\theta = \frac{T - T_\infty}{T_{wo} - T_\infty} \quad ; \quad \Delta = \frac{\delta h}{k_c} \left(\frac{T_m - T_\infty}{T_{wo} - T_\infty} \right)$$

and solved by a heat balance integral technique similar to the one used by Goodman [83]. The numerical solution results in plots of dimensionless coating thickness, Δ, as a function of dimensionless time τ, dimensionless melt-

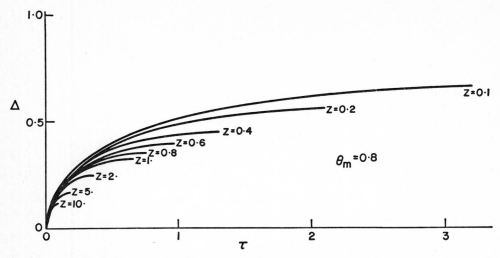

Fig. 3 Theoretical predictions for dimensionless coating thickness as a function of dimensionless time.

ing temperature θ_m, and a coating parameter, Z:

$$\Delta = \Delta(\tau : \theta_m, Z)$$

where

$$\theta_m = \frac{T_m - T_\infty}{T_{wo} - T_\infty} \qquad ; \qquad Z = \frac{\rho_c c_c k_c (T_{wo} - T_m)}{\rho_w c_w hL(T_m - T_\infty)}$$

Figure 3 provides a typical plot of dimensionless coating thickness, Δ, as a function of time, τ, for the special case of dimensionless melting temperature $\theta_m = 0.8$. Figure 4 provides a comparison between the theoretical solution and some experimental results obtained in our laboratory [48]. As seen, the agreement is fairly good.

Experimental investigations of fluidized bed coating with plastic powder were also performed by Pettigrew [84,85] and Richart [13]. They have investigated the effect of different parameters on the coating process. Their results agree quite well with the theoretical approach [78] derived along the lines described here.

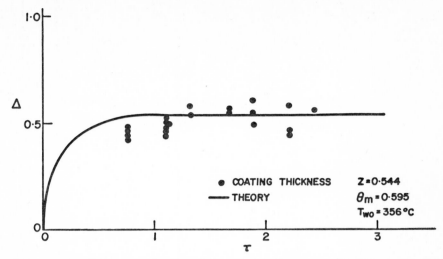

Fig. 4 Dimensionless coating thickness as a function of dimensionless time—comparison between theory and experimental data.

For any set of coating parameters, there is a certain maximum achievable thickness beyond which no coating can be deposited. This maximum or final thickness is represented by the horizontal parts of the curves in Figures 3 and 4.

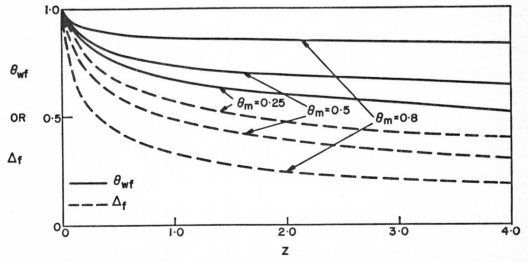

Fig. 5 Plot of final dimensionless coating thickness Δ_f and wall temperature θ_{wf} versus dimensionless parameter Z for various dimensionless melting temperatures θ_m.

Figure 5 provides a plot of the dimensionless final thickness and object temperature as a function of the coating parameter Z for different values of θ_m. This plot was obtained from the solution of Equations (1)-(7). The information given in this plot is the one that the practicing coating technologist will be mostly interested in, because it shows the upper limit of the coating thickness for a given set of coating parameters as well as the drop in wall temperature at the point where this thickness is achieved.

CONCLUDING REMARKS

In the present paper we have reviewed the applications of fluidized beds to coating processes. The literature cited attests to the widespread use of the fluidized bed in coating applications. The literature lists several ways that the fluidized bed can be applied to coating. The most extensive use is that of fluidizing the coating powder into which the hot object to be coated is dipped. This method allows solventless coating resulting in many advantages as listed above. In coating of large numbers of granular objects such as pills, catalyst pellets, etc., the bed may be used to fluidize the granules themselves during the coating process. Continuous addition and removal of the granules is possible. Another way the fluidized bed may be applied to coating is in serving as an inert support for the coated object. This method is particularly suitable for high temperature coatings.

The theoretical analysis of coating processes has developed mainly for the case of object coating with fluidized powder. The analytical approaches to the problem of coating of fluidized particles are of a semi-empirical nature and more basic work is still needed.

The fluidized bed coating processes were developed fairly recently and maybe are not as popular as other applications of fluidized beds; still, the many areas these processes are being applied at present are a good testimony for their future potential.

REFERENCES

1. Gemmer, E.: Process and apparatus for the preparation of protective coatings from pulverulent synthetic thermoplastic materials. German Patent 933019, September 1955.

2. Davis, W. J.: Methods and apparatus for applying protective coatings. U.S. Patent 3004861, October 1961.

3. Dettling, C. J.: Fluidized bed coating method. U.S. Patent 2974060, March 1961.

4. Dettling, C. J. and R. E. Hartline: Process and apparatus for producing continuous coatings. U.S. Patent 2987413, June 1961.

5. Gemmer, E.: Fluidized bed coating process. U.S. Patent 2974059, March 1961.

6. Gemmer, E.: Fluidized bed coating process for coating with thermosetting materials. U.S. Patent 3090696, May 1963.

7. Gemmer, E.: Vortex Sintering. *Kunstoffe*, 47, 510, (1957).

8. Gaynor, J.: Some fundamental aspects of fluidized bed coating. *SPE J.*, 15, (12), 1059, (1959).

9. Gaynor, J.: Fluidized bed coating. *Chem. Eng. Progr.*, 56, 75, (1960).

10. Van den Hoeven, A.: Het verwarmde wervels interbad. *Plastica*, 14, 1130, (1961).

11. Van den Hoeven, A.: Het verwarmde wervels interbad. *Plastica*, 15, 36, (1962).

12. Sharretts, R. R.: A report on the fluidized bed coating system. Part 1, *Plastics Design and Processing*, pp. 15-21, June (1962).

13. Richart, D. S.: A report on the fluidized bed coating system. Part 2, *Plastics Design and Processing*, pp. 26-34, July (1962).

14. Landrock, A. H.: Fluidized bed coating with plastics: Technology and potential for military applications. Plastec Report No. 13, Plastics Technical Evaluation Center, Picatinny Arsenal, Dover, New Jersey, (1964).

15. Yakovlev, A. D. and I. S. Okhrimenko: Powdered polymer materials and coatings deposited as aerodispersions. *Lakokrasoch Mater. Ikh. Primen.*, 5, pp. 13-18, (1966).

16. Landrock, A. H.: Fluidized bed coating with plastics. *Chem. Eng. Progr.*, 63(2), pp. 67-74, (1967).

17. Feigenbaum, L. and D. Lefort des Ylouses: Fluidized bed coating. *Labo. Pharma-Probl. Tech.*, 19(198), pp. 68-74, 79-80, (1971).

18. Pascoe, W. R.: Nylon fusion coating. *Society of Manufacturing Engineers*, Paper No. FC72-940, March 1972.

19. Renkis, A. I.: Vinyl powder versatility: products, application methods and use. *Society of Manufacturing Engineers*, Paper No. FC72-936, March 1972.

20. Polymer Corp., Coating composition for a fluidized bed, Brit. Amend. Pat. 899,394 , 9th April, 1970.

21. U.S.I. Chemicals Data Sheet on Microthene FN500, available from U.S. Industrial Chemicals Company, 99 Park Avenue, New York.

22. Everson, R. and Rothschild, M. V.: Polypropylene. *Society of Manufacturing Engineers*, Paper No. FC72-939, March 1972.

23. Harrington, R. C., Jr. and Hood, J. D.: Plastic powders and coatings of Tenite Butyrate and Propionate. *Society of Manufacturing Engineers*, Paper No. FC72-938, March 1972.

24. Harrington R. C. and Hood, J. D.: Plastic grade butyrate. *Paint Varn. Prod.*, <u>58</u>(7), 39-43, (1968).

25. Anonymous: Fluorocarbon plastics. *Materials in Design Engineering*, Vol. 59, No. 2, pp. 94-98, 1964.

26. Anonymous: PPS - a plastic with unique properties. *The Sulfur Institute Journal*, <u>7</u>, No. 3, Fall 1971.

27. Christensen, J. K.: Polyester powder coatings. *Society of Manufacturing Engineers*, Paper No. FC72-941, March 1972.

28. Bullivant, J. G.: What epoxy powder coatings can and can't do. *Products Finishing*, <u>36</u>, No. 1, pp. 48-58, (1971).

29. Taft, D. D. and R. Hong: Thermosetting polyester and acrylic powder coatings. *Society of Manufacturing Engineers*, Paper No. FC72-942, March 1972.

30. Conte, A. A. Jr.: Choosing a powder for coating applications. *ASME*, publication no. 73-DE-14, (1973).

31. Nishikage, T. and Hattori, K.: Fluidized bed coating compositions. *Japan Kokai*, <u>73</u>, 95.437, 7th December, 1973, 6pp.

32. Sato, K., Saito, F., and Aso, T.: Vinylidine chloride copolymer fluidized bed coating materials. *Japan Kokai*, <u>73</u>, <u>101,441</u>, 20th December, 1973, 4pp.

33. Nagel, F. J. and Gruber, H. M.: Powdered poly (vinyl chloride) mixture for whirl sintering. German Patent 1,223,085 to Knapsack A.G., August 18, 1966.

34. Volken, K.: Coating objects with a plastic layer. German Patent 1,571,048, 20th September 1973.

35. Gemmer, E.: External coating of pipes with PVC by a fluidized bed dip coating method. *Kunststoffe*, <u>57</u>(1), pp. 21-24, (1967).

36. Rockwell, J. B.: Epoxy coating applied by fluidized bed technique. *Industrial Finishing Magazine*, October 1962.

37. Koury, A. J., Conte,A. A. Jr., and Devine, M. J.: Aircraft applications of powder coating technology. *Proceedings of the Second North American Conference on Powder Coating*, Toronto, Canada, March 13-15, (1972), pp. 91-110.

38. Blessin, F.: Friction and wear properties of plastic coatings applied by the fluidized bed process. Technical Report, 61-3143, Rock Island Arsenal, Rock Island, Illinois, August 1961.

39. Landrock, A. H.: The coating of aluminum with plastics by the fluidized bed and electrostatic powder techniques. Plastec Note 18, Plastics Technical Evaluation Center, Picatinny Arsenal, Dover, New Jersey, (1968).

40. Guilbault, L. J.: Pulverulent coating composition and fluidized bed coating. U.S. Patent 3,485,789, 23 December, 1969, 5pp.

41. Sharetts, R. R.: Plastic coatings: the fluidized bed process. *Mechanical Engineering*, __(3), pp. 57-61, (1966).

42. Modowski, J., Troszynski, A., and Tuzin, T.: Protective coating of resistors. Polish Patent 55,024, 17th April, 1968.

43. Licari, J. J.: *Plastic Coatings for Electronics*. McGraw Hill Book Co., (1970).

44. Toth, W. L.: Coating products with plastics. *Automation*, (9), 98-104, (1970).

45. Downey, B. L.: Custom coating with powder fusion epoxies. *Industrial Finishing*, April 1966.

46. Garin, V. N.: Size of powdered polymer particles during application by an electrostatic fluidization method (in Russian). *Lakokrasoch. Mater. Ikh. Primen.*, pp. 21-22, May 1973.

47. Belyi, V. A., Klimovich, A. F., and Pleskachevskii, Yu. M.: Method for studying deposition of powdered polymers in a high voltage electrostatic

field (in Russian). *Lakokrasoch. Mater. Ikh. Primen.*, pp. 53-55, May 1973.

48. Abuaf, N. and Gutfinger, C.: Experimental coating and heat transfer studies in a vibrating fluidized bed. *Int. J. Multiphase Flow*, 1, pp. 683-695, (1974).

49. Terrell, R. L. and Stevens, J. B.: Application of particulate additives to continuous filament tow for tobacco-smoke filters. U.S. Patent 3,354,013, 21 November, 1967.

50. Maierson, T., Roe, H. L., and Williams, J. E.: Prefluxed solder powder. U.S. Patent 3,703,254, 21st November, 1972, 6pp.

51. Segura, M. A.: Depositing carbon on iron. U.S. Patent 3,619,262, 9th November, 1971, 4pp.

52. Reinke, F., Stengel, E., and Emde, F.: Coating of metal parts with metals. German Offen 2,148,779, 5 April, 1972, 11pp.

53. Jacobson, H. W.: Coating metal articles in a fluidized bed. U.S. Patent 3,405,000, 8 October, 1968, 3pp.

54. Sokhor, M. I., Syrkin, V. G., Uelskii, A. A., Drui, M. S., and Genvarskaya, B.A.: Phase compsoition of tungsten coatings obtained by a carbonyl method in a fluidized bed (in Russian). *Zh. Fiz. Khim.*, 48, 155-7, (1974).

54. Phillips, N. V.: Applying thin layers to substrates by chemical conversion of gaseous compounds. *Neth. Appl.*, 70 10, 060, 11 January, 1972, 12pp.

56. Benesovsky, F., Schintlmeister, W.: Wear resistant coatings (in German). German Offen, 2,306,402, 13th September, 1973, 9pp.

57. Souillart, C.: Fluidized bed chamber for continuous operation. German Offen 2,030,833, 17 December, 1970.

58. Elmas, M.: Fluidized bed powder coating. Powder Advisory Centre, 10 St. John's Road, London, 1973.

59. Beatty, R. L.: Pyrolytic carbon deposited from propane in a fluidized bed. U.S. Atomic E. Comm., ORNL-TM-1649, 124pp, (1967).

60. Huschka, H. and Popp, W.: Coating of feed and breed particles in a fluidized bed under reduced pressure. *J. Nucl. Mater.*, 23(1), 109-10, (1967).

61. U.S. Atomic Energy Commission: Improved fluidized bed contactor. British Patent 1,120,003, 17 July, 1968.

62. Wallroth, C. F., Gyarmati, E., and Nickel, H.: Model studies on conical fluidized beds for coating nuclear fuel particles. *Chem. Ing. Tech.*, 43(24), 1298-1304, (1971).

63. Wallroth, C. F.: Model investigations on fluidized beds. Coating of nuclear particles. *Ber. Kernforschungsanlage Juelich*, 1973, JUEL 669-RW, 135pp (in English).

64. Haange, R., Gyarmati, E., and Nickel, H.: Fluidized bed coating of nuclear fuel kernels. *Ber. Kernforschungsanlage Juelich*, 1973, No. 946-RW, 104pp, (in German).

65. McCreary, W. J., Baxman, H. R., Bard, R. J., Bertino, J. P., O'Rourke, J. A., and Hayter, S. W.: Development of particles coated with pyrolytic carbon. Report LA-4771, 26pp, (1972). Los Alamos Sci. Lab., Los Alamos, New Mexico.

66. Klevtzur, S. A., Sedelnikov, T. Kh., Samoteikin, V. V., and Kuleshov, I. M.: Gas phase deposition of a protective coating on a spherical particle in a fluidized bed. *Zasch. Pokritiya Metal*, 1970, (3), 27-42, (in Russian).

67. Sump, K. R. and Howard, B. D.: Chemical vapor deposition of refractory metal claddings on microspheres. *Chem. Vapor Deposition, Int. Conf. 2nd*

1971, 521-35, ed. by Blocker, J. M. Jr., Electrochem. Soc., New York, N.Y.

68. West DeWitt, H., Simpson, A. B., and Simms, R. L.: Metal coating/solid particles. British Patent 1,138,864, 1 January, 1969, 7pp.

69. Nack, H.: Fluidized bed coating of fragile bodies. U.S. Patent, 3,382, 093, 7 May, 1968, 8pp.

70. Garcia, L. A. and Gulden, E. E.: Fluidized bed particle coating (in Spanish). *SAFYBI*, 8(30), 252-70, (1968).

71. Harada, K. and Fujita, J.: Particle coating in the fluidized bed. Maximum feed rate and coating ratio of particles. *Kagaku Kogaku*, 32, 349-53, (1968).

72. Harada, K. and Fujita, J.: Particle coating in the fluidized bed. Particle growth rate and coating efficiency. *Kagaku Kogaku*, 31, 790-4, (1967).

73. Harada, K.: Particle growth during simultaneous coating and agglonuration in fluidized beds. Size distribution and steady state growth of particles. *Kagaku Kogaku*, 36, 1237-43, (1972).

74. Meissner, H. P. and Mickley, H. S.: Removal of mists and dusts from air by beds of fluidized solids. *I&EC*, 1238-1242, (1949).

75. Black, C. H. and Boubel, K. W.: Effectiveness of a fluidized bed in removing submicron particulate from an air stream. *I&EC, Proc. Des. Dev.*, 8, 573-578, (1969).

76. Tardos, G., Gutfinger, C., and Abuaf, N.: Deposition of dust particles in a fluidized bed filter. *Israel J. Tech.*, 12, 184-190, (1974).

77. Tardos, G., Abuaf, N., and Gutfinger, C.: Diffusional filtration of dust in a fluidized bed. TME 248, Technion, Haifa, Israel, (1975).

78. Gutfinger, C., and Chen, W. H.: Heat transfer with a moving boundary. Application to fluidized bed coating. *Int. J. Heat Mass Transfer*, 12, 1097, (1969).

79. Gutfinger, C. and Chen, W. H.: An approximate theory of fluidized bed coating. Presented at the AIChE Annual Meeting, December 1968, Los Angeles, California, and published in *Fundamental Processes in Fluidized Beds, CEP Symposium Series*, No. 101, <u>66</u>, 91-100, (1970).

80. Elmas, M.: On heat transfer with moving boundary. *Int. J. Heat Mass Transfer*, <u>13</u>, 1625-1627, (1970).

81. Abuaf, N., and Gutfinger, C.: Heat transfer with a moving boundary - application to fluidized bed coating of thin plates. *Int. J. Heat Mass Transfer*, <u>16</u>, 213-216, (1973).

82. Gutfinger, C. and Abuaf, N.: Heat transfer in fluidized beds. *Advances in Heat Transfer*, <u>10</u>, J. P. Hartnett and T. F. Irvine, editors, Academic Press, New York, 1974.

83. Goodman, T. R.: The heat balance integral and its application to problems involving a change of phase. *Trans. Am. Soc. Mech. Engrs.*, <u>80</u>, 335-342, (1958).

84. Pettigrew, C. K.: Fluidized bed coating. Part 1, *Modern Plastics*, <u>44</u>, 111-116, (1966).

85. Pettigrew, C. K.: Fluidized bed coating. Part 2, *Modern Plastics*, <u>44</u>, 150-156, (1966).

ATTENDEES

1975 ENGINEERING FOUNDATION CONFERENCE

INTERNATIONAL CONFERENCE ON FLUIDIZATION

Asilomar Conference Grounds

Pacific Grove, California

June 15–20, 1975

S. B. ALPERT   Electric Power Research Institute, 3412 Hillview Avenue, Palo Alto, California

ROGER E. ANDERSON   Aerojet Liquid Rocket Company, P.O. Box 13222, Sacramento, California 95813

H. ANGELINO   Institute Genie Chimique, Chemin de La Loge, Toulouse 31400, France

DAVID H. ARCHER   Westinghouse Research Laboratories, Beulah Road, Pittsburgh, Pennsylvania 15235

M. M. AVEDESIAN   Noranda Research Centre, 240 Hymus Boulevard, Pointe Claire, Quebec, Canada

SURESH P. BABU   Institute of Gas Technology, 3424 South State Street, Chicago, Illinois 60559

C. G. J. BAKER   The University of Western Ontario, London, Ontario N6A 5B9, Canada

JOHN A. BAZAN   Foster Wheeler Energy Corporation, John Blizard Research Center, 12 Peach Tree Hill Road, Livingston, New Jersey 07039

S. R. BECK   Atlantic Richfield Company, P.O. Box 2819, Dallas, Texas 75221

JOHN M. BEGOVICH   Oak Ridge National Laboratory, P.O. Box X, Oak Ridge, Tennessee 37830

L. A. BEHIE   Domtar Ltd., Research Centre, Senneville, Quebec H9X 3L7, Canada

M. A. BERGOUGNOU   University of Western Ontario, London, Ontario N6A SB9, Canada

OSZKAR BORLAI   MTA Muszaki Kemiai Kutato, Intezet, Veszprem, Hungary, Torokvesz ut 12/a, Budapest 10022, Hungary

J. S. M. BOTTERHILL   University of Birmingham, Department of Chemical Engineering, Birmingham B15 2TT, England

COLIN I. BRADLEY   British Gas Corporation, Midlands Research Station, Wharf Lane, Solihull, Warwickshire, England

T. E. BROADHURST   Imperial Oil Enterprises Ltd., Research Department, Sarnia, Ontario, Canada

GARY L. BROWN   Union Carbide Corporation, P.O. Box 8361, South Charleston, West Virginia 25303

RICHARD W. BRYERS   Foster Wheeler Energy Corporation, John Blizard Research Center, 12 Peach Tree Hill Road, Livingston, New Jersey 07039

P. H. CALDERBANK   University of Edinburgh, Chemical Engineering Department, Edinburgh, Scotland

G. S. CANADA   General Electric Corporation R & D, P.O. Box 43, Schenectady, New York 12301

ERWIN L. CARLS   Argonne National Laboratory, 9700 South Cass Avenue, Argonne, Illinois 60439

K. E. CARMICHAEL   Union Carbide Corporation, P.O. Box 8361, Technical Center, South Charleston, West Virginia 25303

J. CARVALHO   University of Cambridge, Department of Chemical Engineering, Pembroke Street, Cambridge CB2 3RA, England

JOHN CHEN   Lehigh University, Mechanical Engineering, Building 19, Bethlehem, Pennsylvania 18015

HERMAN C. T. CHENG   DuPont Experimental Station, Wilmington, Delaware 19810

CHARLES K. CHOI   Garrett Research & Development, 1855 Carrion Road, La Verne, California 91750

DONALD K. CLARKE   Stearns-Roger Incorporated, P.O. Box 5888, Denver, Colorado 80217

SANDFORD S. COLE   Director, Engineering Foundation, 345 East 47th Street, New York, New York 10017

R. C. DARTON   ICI Research Fellow, University of Cambridge, Pembroke Street, Cambridge CB2 3RA, England

J. F. DAVIDSON   University of Cambridge, Pembroke Street, Cambridge CB2 3RA, England

LAWRENCE T. DENK   M. W. Kellogg Company, 1300 Three Greenway Plaza East, Houston, Texas 77046

SHELTON EHRLICH   Pope, Evans, and Robbins, 320 King Street, Suite 503, Alexandria, Virginia 22314

NORMAN EPSTEIN   Department of Chemical Engineering, University of British Columbia, Vancouver 8, British Columbia, Canada

ANDREW ERDMAN, JR.   Dorr-Oliver, 66 Jack London Square, Oakland, California 94607

L. T. FAN   Department of Chemical Engineering, Kansas State University, Manhattan, Kansas 66506

THOMAS FITZGERALD   Program Director, Chemical Processes, National Science Foundation, 1800 G Street, N.W., Washington, D.C. 20550

STEPHEN FREEDMAN   U.S. ERDA, 2100 M Street, N.W., Room 126, Washington, D.C. 20037

MASAHISA FUJIKAWA   Hokkaido University, Kita 13, Nishi 8, Sapporo, Japan

ROBERT J. GARTSIDE   Stone & Webster Engineering Corporation, P.O. Box 2325, Boston, Massachusetts 02107

DEREK GELDART   Bradford University, Great Horton Road, Bradford Yorks, BD7 1DP, England

M. J. GLUCKMAN   City University of New York, Steinman Hall, 140 Street & Convent Avenue, New York, New York 10031

NELLO DEL GOBBO   National Science Foundation, 3943 Wilcoxson Drive, Fairfax, Virginia 22030

W. R. A. GOOSSENS   S.C.K./C.E.N., Boeretang 200, MOL, Belgium 2400

JACK GORDON   MITRE Corporation, Westgate Research Park, McLean, Virginia 22101

JOHN R. GRACE   McGill University, P.O. Box 6070, Montreal H3C 3G1, Canada

ROBERT A. GRAFF   City College of New York, Department of Chemical Engineering, New York, New York 10031

EARL H. GRAY   Phillips Petroleum Company, Building 94G, Phillips Research Center, Bartlesville, Oklahoma 74004

JOHN L. GUILLORY   Envirotech Corporation, One Davis Drive, Belmont, California 94002

CHAIM GUTFINGER   Associate Professor of Mechanical Engineering, Technion, I.I.T., Haifa, Israel

JOSEPH GWOZDZ   Copeland Systems Incorporated, 2000 Spring Road, Suite 300, Oak Brook, Illinois 60521

JOHN S. HALOW   Exxon Research & Engineering Company, P.O. Box 101, Florham Park, New Jersey 07932

DAVID HARRISON   University of Cambridge, Department of Chemical Engineering, Pembroke Street, Cambridge CB2 3RA, England

JOHN HART   PPG Industries, P.O. Box 1000, Lake Charles, Louisiana 70601

T. D. HEATH   Dorr-Oliver, Incorporated, 77 Havermeyer Lane, Stamford, Connecticut 06904

CLAIR E. HILDEBRAND   Allied Chemical Company, P.O. Box 2105 R, Morristown, New Jersey 07960

W. C. A. HOLTKAMP   SASOL, P.O. Box 1, Sasolburg, South Africa

MASAYUKI HORIO   West Virginia University, Department of Chemical Engineering, Morgantown, West Virginia 26506

SVEND HOVMAND   NIRO Atomizer, 305 Gladsaxevej, Soborg 2860, Denmark

C. L. JOHNES   Esso Petroleum Company, Victoria Street, London, England

DAVID H. JONES   The Badger Company, Incorporated, One Broadway, Cambridge, Massachusetts

ALBERT A. JONKE   Argonne National Laboratory, 9700 South Cass Avenue, Argonne, Illinois 60439

M. R. JUDD   University of Natal, Department of Chemical Engineering, King George V Avenue, Durban 4001, South Africa

DAVID JUNGE   Research Associate, Department of Mechanical Engineering, Oregon State University, Corvallis, Oregon 97330

KUNIO KATO   Department of Chemical Engineering, Gunma University, Tenjin-cho, Kiryu-shi, Gunma 376, Japan

DALE L. KEAIRNS   Westinghouse Research Laboratory, Beulah Road, Pittsburgh, Pennsylvania 15235

MELISSA KEENBERG   Engineering Foundation, 345 East 47th Street, New York, New York 10017

PATRICK G. KELLEY   Tennessee Valley Authority, 503 Power Building, Chattanooga, Tennessee 37401

GEORGE H. KESLER   Engineering Consultant, 2758 South Olympia Circle, Evergreen, Colorado 80439

MASAO KITO   Department of Chemical Engineering, Gunma University, Tenjin-cho, Kiruyshi, Gunma 376, Japan

TED M. KNOWLTON   Institute of Gas Technology, 4201 West 36 Street, Chicago, Illinois

EIICHI KOJIMA   Department of Chemical Engineering, Kansas State University, Manhattan, Kansas 66506

C. R. KRISHNA   Brookhaven National Laboratory, Building 526, Upton, New York 11973

KIROSHI KUBOTA   Tokyo Institute of Technology, 1-26-3 Aobadai Midri-ku, Yokohama 227, Japan

DAIZO KUNII   University of Tokyo, 7-3-1. Hongo, Bunkyo-ku, Tokyo 113, Japan

CHARLES E. LAPPLE   Stanford Research Laboratory, 333 Ravenswood Avenue, Menlo Park, California 94025

BERNARD S. LEE   Institute of Gas Technology, 3424 South State Street, Chicago, Illinois 60616

DAVID LEE   PPG Industries, P.O. Box 1000, Lake Charles, Louisiana 70601

RICHARD H. C. LEE   The Aerospace Corporation, P.O. Box 92957, Los Angeles, California 90009

PIERRE LE GOFF   Centre de Cinetique du CNRS, Route de Vadoeuvre, Villers le Nancy 54600, France

L. S. LEUNG   Queensland University, Chemical Engineering Department, Brisbane 4067, Australia

OCTAVE LEVENSPIEL   Oregon State University, Chemical Engineering Department, Corvallis, Oregon 97331

EDWARD K. LEVY   Lehigh University, Packard Laboratory, Building No. 19, Bethlehem, Pennsylvania 18015

HOWARD LITTMAN   RPI, 123 Ricketts Building, Troy, New York 12181

DWIGHT N. LOCKWOOD   Graduate Student, Oregon State University, 2365 South West Pickford Street, Corvallis, Oregon 97330

MIKE MAAGHOUL   Electric Power Research Institute, 3412 Hillview Avenue, Palo Alto, California 94304

YVES MARTINI   Graduate Student, University of Western Ontario, London, Ontario N6A 5B9, Canada

LEOPOLDO MASSIMILLA   University of Naples, Facolta Inqegneria, Piazzale Tecchio, Naples, Italy

KISHAN B. MATHUR   University of British Columbia, Department of Chemical Engineering, Vancouver, B.C. V6T 1W5, Canada

JOHN M. MATSEN   Exxon Research & Engineering Company, P.O. Box 101, Florham Park, New Jersey 07932

CORNELIUS L. McNALLY   M. W. Kellogg Company (R & D), P.O. Box 79513, Houston, Texas 77079

OTTO MOLERUS   University Erlangen-Nurnberg, 852 Erlangen, Martensstr. 9, West Germany

SHIGEKATSU MORI   Nagoya Institute of Technology, Gokiso-cho, Showa-ku, Nagoya, Japan

SHIGEHARU MOROOKA   University of Kentucky, 3300 Montavesta Road No. A51, Lexington, Kentucky 40502

G. MOSS   Esso Petroleum Company, 78 Harpes Road, Oxford, England

HERMAN NACK   Battelle-Columbus Laboratories, 505 King Avenue, Columbus, Ohio 43201

KOZO NAKAMURA   Department of Agriculture, University of Tokyo, Yayoi 1-1-1-, Bunkyo-ku, Tokyo 113, Japan

ALLEN S. NEULS   Allied Chemical Corporation, 550 2nd Street, Idaho Falls, Idaho 83401

RALPH H. NIELSEN   Teledyne Wah Chang Albany, P.O. Box 460, Albany, Oregon 97321

RICHARD C. NORTON   Stone & Webster Engineering Corporation, 225 Franklin Street, Boston, Massachusetts 02107

M. S. NUTKIS   Exxon Research & Engineering Company, Box 8, Linden, New Jersey 07036

KATSUYA OHKI   Tokyo Institute of Technology, 2-12-1 Ookayama, Meguro-ku, Tokyo 152, Japan

KNUD OSTERGAARD   Technical University of Denmark, Department of Chemical Engineering, Building 229, DTH, 2800 Lyngby, Denmark

W. BRENT PALMER   Allied Chemical Corporation, 550 2nd Street, Idaho Falls, Idaho 83401

ROBERT PFEFFER   City College of New York, Department of Chemical Engineering, 140 Street & Convent Avenue, New York, New York 10031

BERT PHILLIPS   NASA Lewis Research Center, 21000 Brookpark, Cleveland, Ohio 44135

OWEN POTTER   Monash University, 835 Riversdale Road, Camberwell 3124, Victoria, Australia

B. B. PRUDEN   Department of Chemical Engineering, University of Ottawa, Ontario K1N 6N5, Canada

DHARAM V. PUNWANI   Institute of Gas Technology, 3424 South State Street, Chicago, Illinois 60616

D. L. PYLE   Imperial College, Department of Chemical Engineering, Prince Consort Road, London SW7 2BY, England

WILLIAM E. REASER   Assistant Director of Conferences, Engineering Foundation, 345 East 47 Street, New York, New York 10017

LOTHAR REH   Lurgi Chemie and Huetten, Technik GMBH, Gervinusstr 17-19, 6 Frankfurt/Main, West Germany

LOUIS F. RICE   C. F. Braun & Company, 1000 South Fremont Avenue, Alhambra, California 91801

K. RIETEMA   Department of Physical Technology, Technical University, F.T.-hal, P.O. Box 513, Eindhoven, Netherlands

GILLES ROCHE   Graduate Student, Ecole Polytechnique de Montreal, Park Avenue 6028 No. 11, Montreal (PQ), Canada

P. N. ROWE   Ramsay Memorial Professor, University College, Torrington Place, London WC E 7JE, England

LAWRENCE A. RUTH   Exxon Research & Engineering Company, P.O. Box 8, Linden, New Jersey 07036

S. C. SAXENA   University of Illinois, Box 4348, Chicago, Illinois 60680

B. SCARLETT   Loughborough University, Loughborough, Leicestershire, England

A. R. SCHAEDAL   Field Services Engineer, Copeland Systems, Incorporated, 2000 Spring Road, Oakbrook, Illinois 60521

KLAUS W. SCHATZ   Mobil Research & Development Corporation, Research Department, Paulsboro, New Jersey 08066

T. SHINGLES   SASOL, P.O. Box 1, Sasolburg, South Africa

TAKASHI SHIRAI   Tokyo Institute of Technology, O-okayama, Meguro-ku, Tokyo 152, Japan

A. E. SKRZEC   Stauffer Chemical Company, East Research Center, Dobbs Ferry, New York 10522

LARRY M. SOUTHWICK   C. F. Braun & Company, Alhambra, California 91802

ARTHUR M. SQUIRES   City College of New York, 245 West 104 Street, New York, New York 10025

FRED W. STAUB   General Electric Company, I River Road, Schenectady, New York 12345

CHARLES V. STERNLING   Shell Development Company, 3737 Bellaire Boulevard, Houston, Texas 77001

J. D. A. STONES   SASOL, 16 Goeie Hoop, Retief Street, Sasolburg, South Africa

KAZUO SUGIMOTO   Kurita Water Industries, C. Ito & Company, Incorporated, 270 Park Avenue, New York, New York 10017

STEPHEN SZEPE   University of Illinois at Chicago Circle, Box 4348, Chicago, Illinois 60680

MICHEL TASSART   Graduate Student, Ecole Polytechnique de Montreal, Park Avenue 6028 No. 1, Montreal (P.Q.), Canada

A. I. THOMPSON   Imperial Chemical Industries, Mond Division, P.O. Box 14, The Heath, Runcorn, Cheshire WA6 91Z, England

HOSHENG TU   Standard Oil Company (Ohio), 4440 Warrensville Center Road, Cleveland, Ohio 44128

F. VERGNES   Centre de Cinetique du CNRS, Route de Vandoeuvre, Villers les Nancy, 54500, France

DIDIER VIEL   S.N.I. Aerospatiale, Route de Verneuil, Les Mureaux 78130, France

STERLING N. VINES   Pre-doctoral student, University of Virginia, 25 University Circle, Charlottesville, Virginia 22903

GORDON L. WADE   Combustion Power Company, 1346 Willow Road, Menlo Park, California 94025

PARVEZ H. WADIA   Union Carbide Corporation, P.O. Box 8361, Building 740-5320, South Charleston, West Virginia 25303

JACK S. WATSON   Oak Ridge National Laboratory, P.O. Box X, Oak Ridge, Tennessee 37830

DAVID F. WELLS   E.I. duPont de Nemours Incorporated, Engineering Department, Wilmington, Delaware 19898

C. WEN   West Virginia University, Evansdale Campus, Morgantown, West Virginia 26506

JOACHIM WERTHER   University of Erlangen, Martensstrasse 9, D 8520 Erlangen, West Germany

BASIL WHALLEY   Department of Energy Mines & Resources, 562 Booth Street, Ottawa, Ontario K1A OG1, Canada

MARVIN E. WHATLEY   Oak Ridge National Laboratory, P.O. Box Y, Oak Ridge, Tennessee 37830

WEN-CHING YANG   Westinghouse Electric Corporation, Research & Development Center, Pittsburgh, Pennsylvania 15235

JOSEPH YERUSHALMI   Chemical Engineering Department, City College of New York, 140 Street & Convent Avenue, New York, New York 10031

F. A. ZENZ   Particulate Solid Research, Incorporated, P.O. Box 205, Garrison, New York 10524

INDEX

Numbers in *italics* refer to pages in Volume I; numbers in **bold type** refer to Volume II.